Angiotensin-(1-7)

Robson Augusto Souza Santos
Editor

Angiotensin-(1-7)

A Comprehensive Review

 Springer

Editor
Robson Augusto Souza Santos
Universidade Federal de Minas Gerais
Belo Horizonte, MG
Brazil

ISBN 978-3-030-22698-5   ISBN 978-3-030-22696-1   (eBook)
https://doi.org/10.1007/978-3-030-22696-1

This Springer imprint is published by the registered company Springer Nature Switzerland AG
The registered company address is: Gewerbestrasse 11, 6330 Cham, Switzerland

Acknowledgment

I would like to acknowledge Prof. Marco Antonio Peliky Fontes for all his assistance in communicating with the authors of this book. His help was instrumental for accomplishing this task. This book was written in part during my senior fellowship at the Max Delbrück Center under the support of the Alexander von Humboldt Foundation (Georg Forster Research Award). The assistance of Stephanie Viana is also acknowledged. Special thanks to the Foundation for Research Support of the State of Minas Gerais (FAPEMIG), National Council for Scientific and Technological Development (CNPq), and Coordination for the Improvement of Higher Education Personnel (CAPES).

Introduction

Angiotensin-(1-7) is a heptapeptide formed from angiotensin I or angiotensin II by the action of endo- or carboxypeptidases, especially ACE2, neutral-endopeptidase (NEP), prolylendopeptidase (PEP), and thimet oligopeptidase, reviewed in Chappell et al. [1] and Santos et al. [2]. This book addresses many aspects of angiotensin (1-7). Its history started against a background of a strong dogma: the concept that angiotensin II is the only biologically active peptide of the renin-angiotensin system. The research on angiotensin-(1-7) was a groundbreaker leading to a considerable advance in our understanding of the still enigmatic, renin-angiotensin system. In a recent review, we have addressed many aspects of the angiotensin-(1-7) history, which is still going on soundly. Table 1 taken from the review summarizes some of the landmarks of the evolution of angiotensin-(1-7), from inactive to an important biologically active component of the RAS. We tried to address in the book the actions of this heptapeptide in many organs and systems with the collaboration of very well-established investigators in the field. As quoted by Kastin and Pan, "Schopenhauer realized that discovery frequently undergoes three stages: ridicule, opposition, and acceptance as self-evident," with JBS Haldane adding a fourth stage: "I always said so" [3]. After more than 25 years, the research on angiotensin-(1-7) is still bouncing between opposition and acceptance with predominance of the last. This book may contribute to force it to the last stage.

I hope you enjoy reading.

Table 1 Taken from the American Physiological Society: "The ACE2/Angiotensin-(1-7)/MAS Axis of the Renin-Angiotensin System: Focus on Angiotensin-(1-7)", Santos et al.

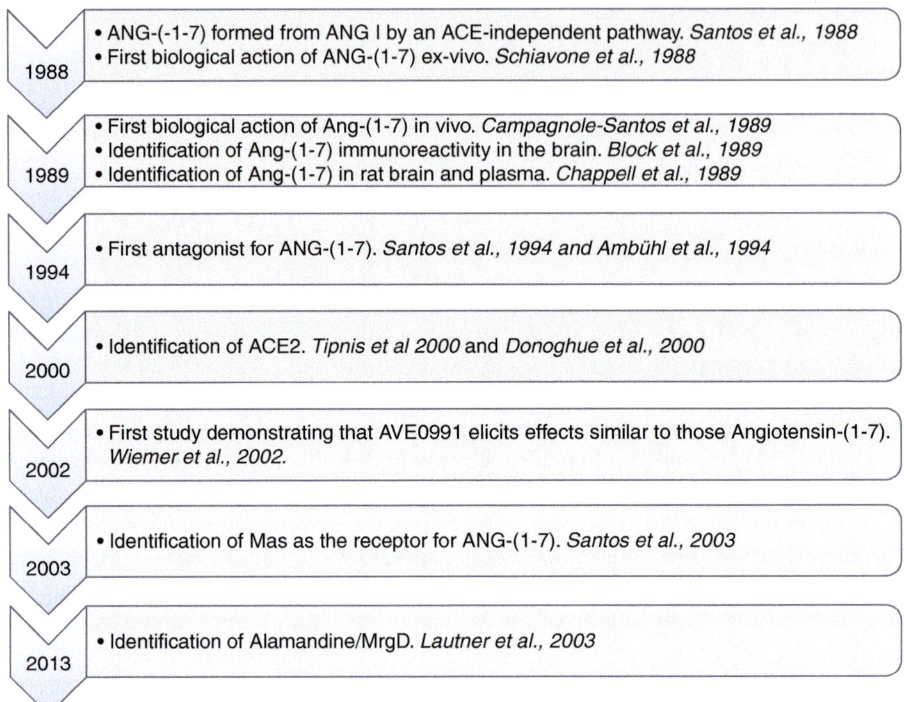

References

1. Chappell MC, Marshall AC, Alzayadneh EM, Shaltout HA, Diz DI. Update on the Angiotensin converting enzyme 2-Angiotensin (1-7)-Mas receptor axis: fetal programing, sex differences, and intracellular pathways. Front Endocrinol. 2014;4:201. https://doi.org/10.3389/fendo.2013.00201.
2. Santos R, Sampaio W, Alzamora AC, Motta-Santos D, Alenina N, Bader M, Campagnole-Santos M. The ACE2/Angiotensin-(1-7)/MAS axis of the renin-angiotensin system: focus on Angiotensin-(1-7). Physiol Rev. 2018;98:505–553. https://doi.org/10.1152/physrev.00023.2016.
3. Kastin AJ, Pan W. Concepts for biologically active peptides. Curr Pharm Design. 2010;16:3390–400. https://doi.org/10.2174/138161210793563491.

Contributors

María José Acuña Centro de Envejecimiento y Regeneración, CARE Chile UC y Departamento de Biología Celular y Molecular, Facultad de Ciencias Biológicas, Pontificia Universidad Católica de Chile, Santiago, Chile

Centro de Biología y Química Aplicada (CIBQA), Universidad Bernardo O Higgins, Santiago, Chile

Natalia Alenina Max-Delbrück-Center for Molecular Medicine (MDC), Berlin, Germany

DZHK (German Center for Cardiovascular Research), Partner Site Berlin, Berlin, Germany

Michael Bader, PhD Max-Delbrück-Center for Molecular Medicine (MDC), Berlin, Germany

DZHK (German Center for Cardiovascular Research), Partner Site Berlin, Berlin, Germany

Berlin Institute of Health (BIH), Berlin, Germany

Charité - University Medicine, Berlin, Germany

Institute for Biology, University of Lübeck, Lübeck, Germany

Enrique Brandan Centro de Envejecimiento y Regeneración, CARE Chile UC y Departamento de Biología Celular y Molecular, Facultad de Ciencias Biológicas, Pontificia Universidad Católica de Chile, Santiago, Chile

Maria Jose Campagnole-Santos, PhD Departamento de Fisiologia e Biofísica, INCT-Nanobiofar, Instituto de Ciências Biológicas, Universidade Federal de Minas Gerais, Belo Horizonte, MG, Brazil

Carlos Henrique Castro, PhD Department of Physiological Sciences, Biological Sciences Institute, Federal University of Goiás, Goiânia, GO, Brazil

Mark C. Chappell, PhD, FAHA Cardiovascular Sciences Center, Wake Forest University School of Medicine, Winston-Salem, NC, USA

Che Ping Cheng, MD, PhD Cardiovascular Center, Winston Salem, NC, USA

Carlos M. Ferrario, MD Department of Surgery, Winston Salem, NC, USA

Anderson José Ferreira, PhD Department of Morphology, Biological Sciences Institute, Federal University of Minas Gerais, Belo Horizonte, MG, Brazil

Marco Antônio Peliky Fontes, PhD Departamento de Fisiologia e Biofísica, INCT-Nanobiofar, Instituto de Ciências Biológicas, Universidade Federal de Minas Gerais, Belo Horizonte, MG, Brazil

Izabela Galvão Departamento de Bioquímica e Imunologia, Instituto de Ciências Biológicas, Universidade Federal de Minas Gerais, Belo Horizonte, MG, Brazil

Mariela M. Gironacci, PhD Departamento de Química Biológica, IQUIFIB (UBA-CONICET), Facultad de Farmacia y Bioquímica, Universidad de Buenos Aires, Buenos Aires, Argentina

Giselle S. Magalhães Department of Physiology and Biophysics, Federal University of Minas Gerais, Belo Horizonte, MG, Brazil

Post-Graduation Program in Health Sciences, Medical Sciences Faculty of Minas Gerais, Belo Horizonte, MG, Brazil

Aline Silva de Miranda, MD, PhD Interdisciplinary Laboratory of Medical Investigation, Faculty of Medicine, Federal University of Minas Gerais (UFMG), Belo Horizonte, MG, Brazil

Laboratory of Neurobiology, Department of Morphology, Institute of Biological Sciences, UFMG, Belo Horizonte, MG, Brazil

Isabella Zaidan Moreira Departamento de Análises Clínicas e Toxicológicas, Faculdade de Farmácia, Universidade Federal de Minas Gerais, Belo Horizonte, MG, Brazil

Daisy Motta-Santos National Institute of Science and Technology in Nanobiopharmaceutics, UFMG, Belo Horizonte, MG, Brazil

Sports Department, School of Physical Education, Physiotherapy, and Occupational Therapy (EEFFTO), UFMG, Belo Horizonte, MG, Brazil

Flavia Rago Departamento de Bioquímica e Imunologia, Instituto de Ciências Biológicas, Universidade Federal de Minas Gerais, Belo Horizonte, MG, Brazil

Maria da Glória Rodrigues-Machado Post-Graduation Program in Health Sciences, Medical Sciences Faculty of Minas Gerais, Belo Horizonte, MG, Brazil

Walyria O. Sampaio, PhD University of Itaúna, Biological Sciences Institute, Itaúna, MG, Brazil

Sérgio Henrique Sousa Santos, PhD Institute of Agricultural Sciences, Food Engineering College, Universidade Federal de Minas Gerais (UFMG), Montes Claros, MG, Brazil

Laboratory of Health Science, Postgraduate Program in Health Sciences, Universidade Estadual de Montes Claros (UNIMONTES), Montes Claros, MG, Brazil

Robson Augusto Souza Santos, MD, PhD National Institute of Science and Technology in Nano Biopharmaceutics – Department of Physiology and Biophysics, Biological Sciences Institute, Federal University of Minas Gerais, Belo Horizonte, MG, Brazil

Department of Physiology and Pharmacology, Biological Sciences Institute, Federal University of Minas Gerais, Belo Horizonte, MG, Brazil

Ana Cristina Simões e Silva, MD, PhD Department of Pediatrics, Interdisciplinary Laboratory of Medical Investigation, Faculty of Medicine, UFMG, Belo Horizonte, MG, Brazil

Interdisciplinary Laboratory of Medical Investigation, Faculty of Medicine, Federal University of Minas Gerais (UFMG), Belo Horizonte, MG, Brazil

Mauro Martins Teixeira Departamento de Bioquímica e Imunologia, Instituto de Ciências Biológicas, Universidade Federal de Minas Gerais, Belo Horizonte, MG, Brazil

Rhian M. Touyz, MBBCh, PhD Institute of Cardiovascular and Medical Sciences, BHF Glasgow Cardiovascular Research Centre, University of Glasgow, Glasgow, UK

Jasmina Varagic, MD, PhD Department of Surgery, Winston Salem, NC, USA

Wake Forest School of Medicine, Winston Salem, NC, USA

The Angiotensin-(1-7) Axis: Formation and Metabolism Pathways

Mark C. Chappell

Introduction

The renin-angiotensin system (RAS) was originally characterized as a circulating endocrine system initiated by the protease renin to hydrolyze angiotensinogen to the inactive peptide Angiotensin I (Ang I) and the subsequent conversion to Ang II by angiotensin-converting enzyme (ACE) (Fig. 1). Ang II recognizes the angiotensin type 1 receptor (AT_1R) to invoke both peripheral and central mechanisms in the regulation of blood pressure. Chronic activation of the ACE-Ang II-AT_1R pathway may also be associated with various pathological responses including fibrosis, inflammation, metabolic dysregulation, heart failure, cancer, aging, and diabetic injury [1–6]. Although the blockade of the Ang II axis through ACE inhibitors or AT_1R receptor antagonists are effective therapies for the treatment of cardiovascular disease, there is now abundant evidence for alternative pathways within the RAS that may contribute to the beneficial actions of conventional RAS blockade. Indeed, our original identification and quantification of the endogenous expression of Ang-(1-7) in the brain, circulation, and peripheral tissues provided a compelling case for a functional non-classical RAS pathway [7]. Subsequent studies in both experimental models and humans that ACE inhibition augments the circulating levels of Ang-(1-7) further supported the concept that Ang-(1-7) may oppose the actions of the Ang II-AT_1R pathway [8]. Blockade of the AT_1R also increases the formation of Ang-(1-7) through ACE2-dependent conversion of Ang II, as well as shunts Ang II to the AT_2R pathway that shares similar properties to the Ang-(1-7) system [9–11]. As the functional actions of the RAS now reflect a far more complex array of peptide ligands and distinct receptors than previously envisaged, we present a comprehensive review of the enzymatic pathways involved in the formation and

M. C. Chappell (✉)
Cardiovascular Sciences Center, Wake Forest University School of Medicine, Winston-Salem, NC, USA
e-mail: mchappel@wakehealth.edu

© Springer Nature Switzerland AG 2019
R. A. S. Santos (ed.), *Angiotensin-(1-7)*,
https://doi.org/10.1007/978-3-030-22696-1_1

1

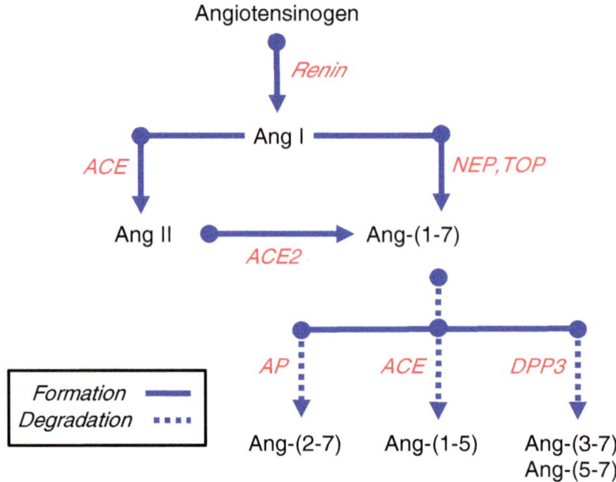

Fig. 1 Processing cascade for Angiotensin-(1-7). Renin cleaves angiotensinogen to angiotensin I (Ang I), which is further processed to the biologically active peptide Ang II by angiotensin-converting enzyme (ACE). Ang II undergoes further processing by the carboxypeptidase ACE2 to form Ang-(1-7). Ang-(1-7) is also formed through non-Ang II pathways by the direct processing of Ang I by the endopeptidases neprilysin (NEP) and thimet oligopeptidase (TOP). Ang-(1-7) is subsequently metabolized by ACE to Ang-(1-5), dipeptidyl peptidase 3 (DPP3) to Ang-(3-7) and Ang-(5-7), or by aminopeptidase A (APA) to Ang-(2-7). (Adapted from Chappell [89])

metabolism of the Ang-(1-7) axis in the circulation, central and peripheral tissues as well as cerebrospinal fluid (CSF) and urine. The review includes the roles of renin and ACE in regard to the generation of Ang I and Ang II as substrates for the subsequent processing to Ang-(1-7), as well as the influence of ACE and dipeptidyl peptidase I (DPP3) to metabolize Ang-(1-7). The pathways for the generation of the novel analog [Ala^1]-Ang-(1-7) (alamandine) whereby the aspartic acid is decarboxylated to alanine and the functional consequences of this substitution are discussed in Chap. 2.

Angiotensin-(1-7) Forming Pathways

Renin

Renin [EC 3.4.23.15, 35 kDa] belongs to the class of aspartyl-type acid proteases but exhibits a more neutral pH optima [6.5–8.0] than other proteases in this group. The only known substrate for renin is angiotensinogen and the enzyme hydrolyzes the Leu^{10}-Leu^{11}/Val^{11} bond of the precursor protein to form the inactive peptide Ang I. Renin-dependent formation of Ang I is considered the enzymatic initiator of the RAS cascade to ultimately generate Ang II (Fig. 1). The enzyme is synthesized predominantly in the juxtaglomerular (JG) cells of the kidney and is stored in both inactive (pro-renin) and active forms for the regulated release of the protease into the circulation. The collecting duct (CD) cells are another source of renin within the kidney that primarily secretes the active form of renin into the tubular fluid [12–14]. The tubular secretion of renin may contribute to the processing of angiotensinogen and subsequent Ang II and Ang-(1-7) formation to influence distal nephron function. The regulation of JG renin appears to be quite distinct from CD renin; Ang II negatively regulates JG renin release through inhibition of cAMP levels that constitutes the short negative feedback loop while CD renin release is stimulated by Ang II that may reflect increased protein kinase C (PKC) activity [14].

In regard to the generation of Ang-(1-7), renin is required for the formation of the peptide via the processing of Ang I or Ang II. Since renin is typically secreted from the kidney and other tissues, the formation of Ang-(1-7) is likely to occur in the extracellular compartments including the blood, interstitial fluid, renal tubular fluid, and CSF. However, there is evidence for alternative renin isoforms in the kidney, brain, heart, and adrenal gland [15–17]. The renin isoform lacks the pre-pro domain of the protein that includes the secretory signal, thus the isoform should reside within the cell. Peters and colleagues originally reported that truncated renin localized to the mitochondria and that the renin isoform was internalized by the mitochondria [18–21]. We recently reported the presence of active renin in a purified preparation of mitochondria from the sheep renal cortex, as well as evidence for mitochondrial Ang II and Ang-(1-7) [50–60 fmol/mg protein] [22]. Moreover, the endopeptidases neprilysin and thimet oligopeptidase processed Ang I directly to Ang-(1-7) in mitochondria suggesting an Ang I-dependent pathway for the intracellular formation of Ang-(1-7) (*see Neprilysin and Thimet Oligopeptidase sections*). In renal NRK-52 epithelial cells, immunocytochemical staining for renin with the Inagami antibody was evident in the nucleus of these cells and active renin was confirmed by aliskerin-sensitive conversion of angiotensinogen to Ang I. Isolated nuclei also contained Ang I, Ang II, and Ang-(1-7) [5–20 fmol/mg protein]; however, the processing pathways within the nuclear compartment remain to be defined [23]. Ishigami et al. reported a truncated renin transcript expressed in the proximal tubules of the mouse kidney [17]. Targeted expression of this renin isoform within the proximal tubules was associated with a sustained elevation in blood pressure, but no change in the circulating levels of renin [17]. Although the intracellular localization of the renin isoform or tubular content of angiotensins were not determined, overexpression of renin in tubules augmented blood pressure that suggests the primary intracellular generation of Ang II rather than Ang-(1-7). Indeed, Zhou and colleagues find that the intracellular expression of Ang II in the proximal tubules also resulted in a sustained increase in blood pressure [24] that is consistent with the demonstration of the intracellular AT_1R and a local RAS within the kidney [25–28].

Non-renin pathways for the processing of angiotensinogen to Ang I or Ang II include tonin, cathepsin G, and cathepsin A; however, the participation of these enzymes in the endogenous generation of angiotensins has yet to be firmly established [29]. Moreover, it is not known whether there is a direct pathway for the formation of Ang-(1-7) from angiotensinogen and whether this contributes to the intracellular content of the peptide. Non-renin pathways for the formation of

Ang-(1-7) (or Ang II) may also potentially occur through the processing of Ang-(1-12), a novel precursor peptide originally identified in rat urine by Nagata and colleagues [30]. In this regard, neprilysin converts Ang-(1-12) to Ang-(1-7) in renal cortical membranes and by recombinant forms of both human and mouse neprilysin by a two-step process that involves Ang I as an intermediate product; however, the role of Ang-(1-12) in the generation of endogenous Ang-(1-7) in the circulation and tissues is presently unknown [31].

Endopeptidases

Neprilysin

Neprilysin (EC 3.4.24.11; 95 kDa) is a metalloendopeptidase of the type II membrane-anchored family of enzymes [32]. The peptidase was initially characterized in brain homogenates or membrane preparations to hydrolyze the opiate pentapeptide enkephalin which explained its original characterization as an "enkephalinase." Inhibitors against neprilysin were originally developed to prolong the analgesic actions of the opiate peptides, although neprilysin was subsequently found to be expressed in a number of peripheral tissues [32]. Cardiovascular interest in neprilysin initially reflected its ability to metabolize the family of natriuretic peptides including ANP, BNP, and uroguanylin. Moreover, McKinnie et al. report that neprilysin inactivates apelin, suggesting that beneficial cardiovascular effects of neprilysin inhibition may also reflect the protection of endogenous apelin [33]. Neprilysin inhibitors were considered a potential therapeutic approach to prolong the natriuretic and vasorelaxant properties of these peptides; however, Ang II is also a substrate for neprilysin hydrolysis of the Tyr^4-Ile^5 bond to form Ang-(1-4) and combined inhibitors to neprilysin and ACE termed vasopeptidase inhibitors were developed to prevent the potential increase in circulating and renal Ang II by neprilysin inhibitors alone [34–36]. Although the vasopeptidase inhibitors were potent agents to lower blood pressure and reduce cardiac and renal damage, the first clinical agent omapatrilat was withdrawn due to a greater incidence of angioedema in patients. Subsequently, a new generation of agents have been developed that combine a neprilysin inhibitor and an AT_1R antagonist to obviate the inhibitory effects on ACE, yet maintain blockade of the Ang II-AT_1R axis. Indeed, the combined neprilysin/AT_1R blockade may be a promising therapeutic approach for the treatment of heart failure [37].

Neprilysin is located on the vascular surface of blood vessels and is responsible for the direct conversion of Ang I to Ang-(1-7) in the circulation, particularly under conditions of chronic ACE inhibition (Fig. 1) [32]. Neprilysin hydrolyzes the Pro^7-Phe^8 bond of Ang I to generate Ang-(1-7), as well as the Tyr^4-Ile^5 bond to form Ang-(1-4) that is consistent with the enzyme's preference for aromatic and hydrophobic residues [38]. The generation of circulating Ang-(1-7) from infused Ang I in ACE-blocked Wistar Kyoto (WKY) and spontaneously hypertensive rats (SHR) was abolished by the potent and selective neprilysin inhibitor SCH39370 [39]. We further demonstrated that administration of another selective neprilysin inhibitor CGS24592 partially attenuated the blood pressure-lowering effects of the ACE inhibitor lisinopril in the SHR [40]. The neprilysin inhibitor SCH39370 also reduced Ang-(1-7) levels in the rat hindlimb preparation perfused with Krebs buffer containing the ACE inhibitor lisinopril [41]. Finally, Campbell and colleagues found that a dual ACE/neprilysin inhibitor lowered circulating levels of Ang-(1-7) in the SHR using a combined HPLC-RIA approach to quantify an array of angiotensins using an N-terminally directed antibody [42].

Neprilysin is also highly expressed on the apical surface of proximal tubules within the kidney [32]. We demonstrated a neprilysin-dependent pathway in isolated proximal tubules from sheep kidney and rat cortical membranes for Ang I processing to Ang-(1-7) (Fig. 2) [44, 45]. In isolated tubules, we note that neprilysin blockade

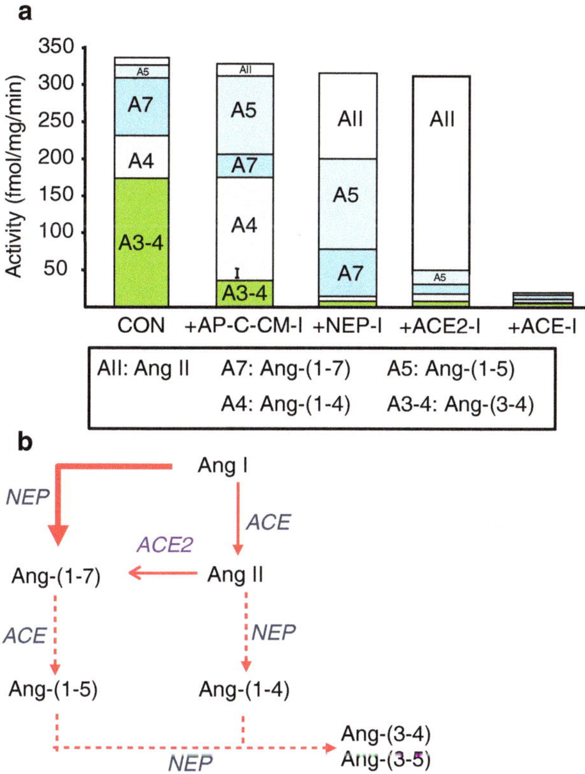

Fig. 2 Ang I processing to Ang-(1-7) and Ang II in sheep proximal tubules. ^{125}I-Ang I (AI) was incubated with 50 μg of proximal tubules membranes for 30 minutes at 37 °C and the metabolites identified by HPLC. Panel **a**: Quantification of the peptidase activities for ^{125}I-AII metabolism from the sheep proximal tubule membranes expressed as the rate of metabolism products formed (fmol/mg/min). Conditions: Control (no inhibitors); +AP,CYS,CHM-I (inhibitors for aminopeptidase, carboxypeptidase, chymase, cysteine proteases); +NEP-I (addition of neprilysin inhibitor); +ACE-I (addition of ACE inhibitor); +ACE2-I (addition of ACE2 inhibitor). Panel **b**: Metabolism pathway for Ang I to Ang-(1-7), Ang II and degradation products in sheep proximal tubules. Data shown are mean values; $n = 5$. (Adapted from Shaltout et al. [43])

did not abolish the formation of Ang-(1-7) as ACE-dependent generation of Ang II contributed to Ang-(1-7) via the ACE2 pathway (Fig. 2). Velez and colleagues applied HPLC-MS analysis of Ang I processing in rat glomeruli, isolated mouse podocytes, and human glomerular endothelial cells to reveal the predominant conversion of Ang I to Ang-(1-7) by neprilysin [46–48]. Poglitsch and colleagues recently show that the renal levels of Ang-(1-7) were significantly reduced in mice chronically treated with the neprilysin inhibitor LBQ657, but increased in ACE2-deficient mice utilizing UHPLC-MS to quantify angiotensins in the kidney [43]. Moreover, the neprilysin inhibitor markedly reduced Ang I to Ang-(1-7) conversion in both mouse and human kidney homogenates [43]. Indeed, these studies support our earlier findings that despite almost complete depletion of renal Ang II in ACE knockout mice, kidney levels of Ang-(1-7) were not diminished supporting an ACE (Ang II)-independent pathway for the generation of Ang-(1-7) in this tissue [49]. Finally, neprilysin activity in isolated mitochondria from the sheep kidney was partially responsible for conversion of Ang I to Ang-(1-7), and was the sole Ang-(1-7) forming activity in a 100,000 xg membrane fraction consistent with the membrane-anchored form of the peptidase [22].

Thimet Oligopeptidase

Thimet oligopeptidase (EC 3.4.24.15, 80 kDa) is a soluble metalloendopeptidase that resides within intracellular compartments of the cell. Similar to neprilysin, thimet oligopeptidase is an endopeptidase that prefers aromatic and hydrophobic residues and cleaves multiple peptide substrates [50]. Thimet oligopeptidase hydrolyzes Ang I exclusively at the Pro^7-Phe^8 bond to form Ang-(1-7) (Fig. 1). In contrast to neprilysin, thimet oligopeptidase does not process Ang I to Ang-(1-4), although the peptidase does cleave Ang II to Ang-(1-4) (Fig. 1). The current data suggest that thimet oligopeptidase may be responsible for the intracellular processing of Ang I to Ang-(1-7). Pereira et al. find that thimet oligopeptidase was the primary activity in tissue homogenates of brain medulla that processed Ang I to Ang-(1-7) [51]. Both neprilysin and thimet oligopeptidase contributed to the processing of Ang I to Ang-(1-7) within isolated mitochondria, and thimet oligopeptidase was the sole Ang-(1-7) forming activity in the 100,000 xg soluble fraction of renal mitochondria [22]. In human proximal tubule HK-2 cells, cytosolic thimet oligopeptidase was the primary activity responsible for the generation of Ang-(1-7) from exogenous Ang I [52]. Chronic treatment of the HK-2 cells with the cell-permeable metallopeptidase inhibitor JMV-390 reduced the intracellular levels of Ang-(1-7) which may reflect the inhibition of thimet oligopeptidase, although the JMV inhibitor may target other metallopeptidases or cell mechanisms that influence the intracellular expression of the peptide [53]. Moreover, thimet oligopeptidase activity in isolated nuclei of NRK-52 renal epithelial cells processed Ang I exclusively to Ang-(1-7) and this peptidase may contribute to the nuclear levels of Ang-(1-7) within the cell [23]. Suski et al. reported that Ang I was primarily converted to Ang-(1-7) in vascular smooth muscle cells (VSMC) as characterized by HPLC-MS analysis [54]; these data confirm our earlier study that thimet oligopeptidase directly processed Ang I to Ang-(1-7) in rat VSMC by HPLC characterization [55]. Although the RAS was originally characterized as a classic endocrine or circulating system, there is compelling evidence for the intracellular expression of angiotensins and angiotensin receptors that include Ang-(1-7) and the MasR [9, 56–61]. The mechanisms for intracellular expression of Ang-(1-7) are currently unknown; however, thimet oligopeptidase should be considered as one candidate processing enzyme in the cellular Ang-(1-7) axis.

Prolyl Oligopeptidase

Prolyl oligopeptidase (EC 3.4.24.16, 75 kDa), also known as prolyl endopeptidase (PEP), is a soluble intracellular serine peptidase that cleaves the Pro^7-Phe^8 bond of both Ang I and Ang II to form Ang-(1-7). Thus, prolyl oligopeptidase may function as both an endopeptidase and a monocarboxypeptidase in the processing of angiotensins, as well as the hydrolysis of other peptides that contain a Pro-XX motif including TRH, substance P, oxytocin, bradykinin and vasopressin [62]. Although prolyl oligopeptidase is considered a cytosolic peptidase, membrane and nuclear forms of the enzyme have been described in both central and peripheral cells suggesting a role for the enzyme in the intracellular expression of peptides [62]. Santos et al. report that prolyl oligopeptidase activity was involved in the extracellular conversion of Ang I to Ang-(1-7) in human aortic endothelial cells, although it remains unclear as to whether the peptidase was secreted into the media or that the peptidase resides on the plasma membrane of endothelial cells [63]. Domenig et al. found that although neprilysin was the predominant Ang-(1-7) forming activity in the kidney, prolyl oligopeptidase activity contributed approximately 20% and 10% to Ang-(1-7) in the mouse and human kidney, respectively, based on the inhibition by Z-prolyl prolinal (ZPP) [43]. Prolyl oligopeptidase would appear to be an ideal enzymatic candidate for the processing of both Ang I and Ang II to Ang-(1-7); however, treatment approaches with more selective inhibitors against the peptidase have generally revealed beneficial effects that contrast with the expected actions of Ang-(1-7), particularly regarding inflammation [62, 64, 65].

Carboxypeptidases

Angiotensin-Converting Enzyme 2

ACE2 is a membrane-bound monocarboxypeptidase (EC 3.4.17.23; 90-120 kDa) that converts Ang II directly to Ang-(1-7) (Fig. 1). ACE2 was initially characterized as an ACE homolog (~40% homology) that cleaved Ang I to the nonapeptide Ang-(1-9), but not directly to Ang II [66]. Subsequent studies found that Ang II exhibits far better kinetic values as a substrate for human ACE2 that would favor processing of Ang II over that of Ang I [67]. Among a number of peptide substrates (> 100) that were screened for human ACE2, Vickers et al. reported that only apelin 13 exhibited comparable kinetic values to that of Ang II [68]. In the murine heart, Ang II was primarily converted to Ang-(1-7) by ACE2 and that in the presence of the ACE2 inhibitor MLN-4760 or in ACE2 null mice there was little metabolism of Ang II [68]. In contrast, under identical kinetic conditions, Ang I was primarily converted to Ang-(1-9) by carboxypeptidase A and not ACE2 in both the wild-type and ACE2 knockout mice [68].

In comparison to ACE, the circulating levels of ACE2 are typically quite low and the extent this reflects a reduced degree of shedding or simply the lower vascular expression of the peptidase is not clear. Rice et al. reported that the molar concentration of ACE in human serum at 7 nM while ACE2 content was >200-fold lower at 33 pM and detectable in <10% of their patient population [69]. In comparison, circulating neprilysin content (290 pM) was also lower than ACE and evident in <30% of these patients [69]. Serum and urinary ACE2 activities are elevated in diabetes, heart failure, and hypertension [70–72]. Circulating ACE2 activity increased approximately three-fold in the diabetic hypertensive mRen2.Lewis rat; however, serum ACE activity also increased in the diabetic rats and may mitigate against the elevated levels of ACE2 (Fig. 3) [70]. As measured under identical kinetic conditions, serum ACE activity for Ang I was far higher than ACE2 for Ang II suggesting that the capacity to generate Ang II [or metabolize Ang-(1-7)] remains greater than the capability to form Ang-(1-7) from Ang II (Fig. 3) [70]. Moreover, the serum ACE:ACE2 ratio in both male and female normotensive Lewis and female mRen2. Leiws essentially reflected the ratio found in human plasma [69, 70]. Serum ACE2

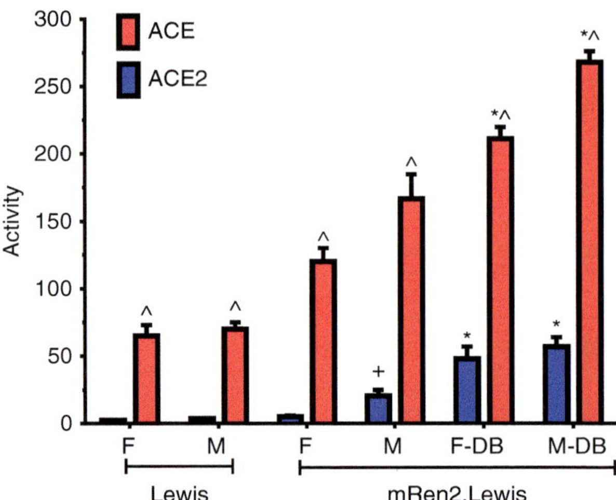

Fig. 3 Diabetes increases both circulating ACE and ACE2 activities in the hypertensive mRen2. Lewis rat. Comparison of circulating ACE2 and ACE activities in female and male normotensive Lewis, hypertensive mRen2.Lewis, and diabetic (DB) mRen2.Lewis rats. Data are mean ± SEM; 5-6 per group *P < 0.05 versus non-diabetic mRen2.Lewis; ^P < 0.05 versus ACE2; +P < 0.05 versus female mRen2.Lewis. Activities are expressed as fmol/minute/ml serum. (Adapted from Yamaleyeva et al. [71])

activity in the male hypertensive mRen2.Lewis was significantly higher than that in the females, despite higher blood pressure and greater circulating levels of Ang II in the males [70]. Whether the higher serum levels of ACE2 in the male mRen2.Lewis reflect a compensatory effect to reduce hypertension and vascular damage or that shedding away from the vascular wall (and attenuated local metabolism of Ang II) contributes to the increase in blood pressure is currently not known [70].

ACE2 constitutes a key enzymatic component of the RAS as a single catalytic step efficiently metabolizes Ang II to attenuate the Ang II-AT_1R pathway, and generates Ang-(1-7) that would activate the Ang-(1-7)-AT_7/MasR axis (Fig. 1). In isolated proximal tubules, Ang II is processed to Ang-(1-7), Ang-(1-4), Ang-(1-5), and Ang-(3-4) (Fig. 4) [44]. The selective neprilysin inhibitor SCH39370 essentially abolished Ang-(1-4) confirming a role for neprilysin to metabolize Ang II. Subsequent addition of an ACE inhibitor attenuates Ang-(1-5) and reveals higher levels of Ang-(1-7) that emphasize the importance of the ACE pathway to metabolize Ang-(1-7) in the kidney (Fig. 4a) [44]. Finally, Ang-(1-7) formation is essentially abolished by the ACE2 inhibitor MLN4760 (Fig. 4a). We also show that the MLN4760 (MLN)

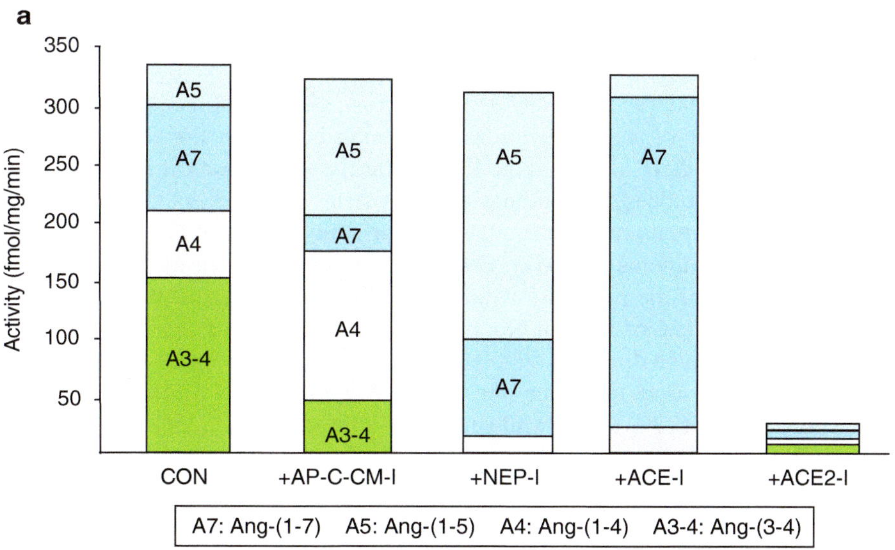

Fig. 4 Ang II processing to Ang-(1-7) and other metabolites in sheep proximal tubules. ^{125}I-Ang II (AII) was incubated with 50 μg of proximal tubules membranes for 30 minutes at 37 °C and the metabolites identified by HPLC. Panel **a**: Quantification of the peptidase activities for ^{125}I-AII metabolism from the sheep proximal tubule membranes expressed as the rate of metabolism products formed (fmol/mg/min). Conditions: Control (no inhibitors); +AP,CYS,CHM-I (inhibitors for aminopeptidase, carboxypeptidase, chymase, cysteine proteases); +NEP-I (addition of neprilysin inhibitor); +ACE-I (addition of ACE inhibitor); +ACE2-I (addition of ACE2 inhibitor). Panel **b**: Influence of ACE2 inhibition on half-life ($t_{1/2}$) of ^{125}I-Ang II (AII) in proximal tubules. Conditions: Control (no inhibitors); +MLN (only the ACE2 inhibitor). Data are mean values; $n = 5$; $*P < 0.05$ versus Control. Panel **c**: Metabolism pathway for Ang II to Ang-(1-7) and degradation products in sheep proximal tubules. (Adapted from Shaltout et al. [43])

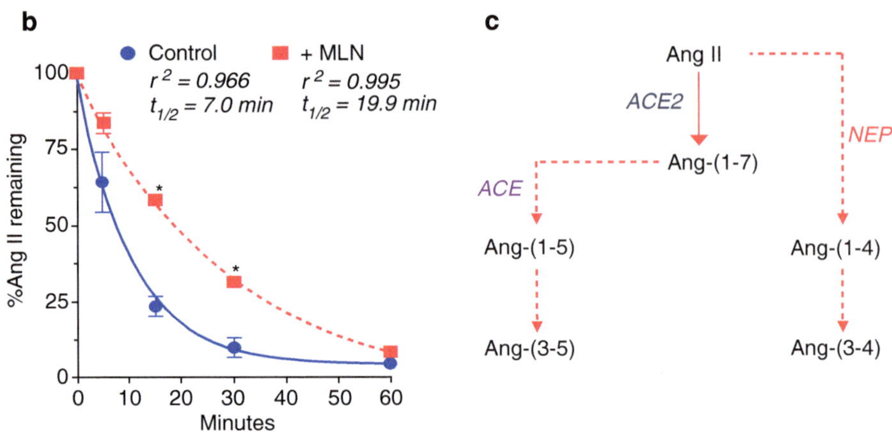

Fig. 4 (continued)

alone significantly attenuates Ang II metabolism by the isolated proximal tubules and increases the peptide's half-life by >2.5-fold, although it clearly does not abolish Ang II metabolism by other peptidases (Fig. 4b) [44]. Moreover, ACE2 protein expression and activity (Ang II to Ang-(1-7) conversion) is markedly reduced in the renal tubules, circulation, and brain medulla in a sheep model of fetal programming; the downregulation of this key peptidase may contribute to an altered ratio of Ang II to Ang-(1-7) in these important cardiovascular tissues that reflect fetal programming events [73]. Indeed, our findings are consistent with studies that demonstrate an exaggerated response to exogenous Ang II or in conditions of an activated RAS in ACE2-deficient animals [72–77]. Overexpression of ACE2 or administration of the soluble form of the peptidase which retains full enzymatic activity attenuates the Ang II-dependent increase in blood pressure and indices of target organ injury [78–86]. The premise of ACE2 supplementation is that sufficiently high levels of ACE2 are administered to reduce the Ang II: Ang-(1-7) ratio thereby attenuating the actions of Ang II-AT$_1$R axis while amplifying those of the Ang-(1-7)-AT$_7$/MasR effects (Fig. 5). Oudit and colleagues find that chronic administration of soluble ACE2 attenuated various indices of cardiac and renal injury, inflammation, and fibrosis in both type 1 and type 2 diabetic mice [79, 80]. Surprisingly, the administration of ACE2 reduced tissue levels of Ang II in the heart and kidney and increased the tissue contents of Ang-(1-7) [79]. Scholey and colleagues also report that ACE 2 given subcutaneously by an osmotic pump attenuated several indices of renal damage in the transgenic Col4A3-/- mouse, a model of Alport syndrome, as well as tended to lower blood pressure [87]. The renal protective effects of soluble ACE2 treatment were associated with a marked reduction in the ratio of Ang II: Ang-(1-7) in the kidney [87]. However, Wysocki et al. observed that neither the administration of ACE2 nor the chronic overexpression of the soluble peptidase by minicircle DNA conveyed any protective effects against diabetic nephropathy in diabetic mice [88]. In this study, plasma angiotensin peptides were quantified by UHPLC-MS

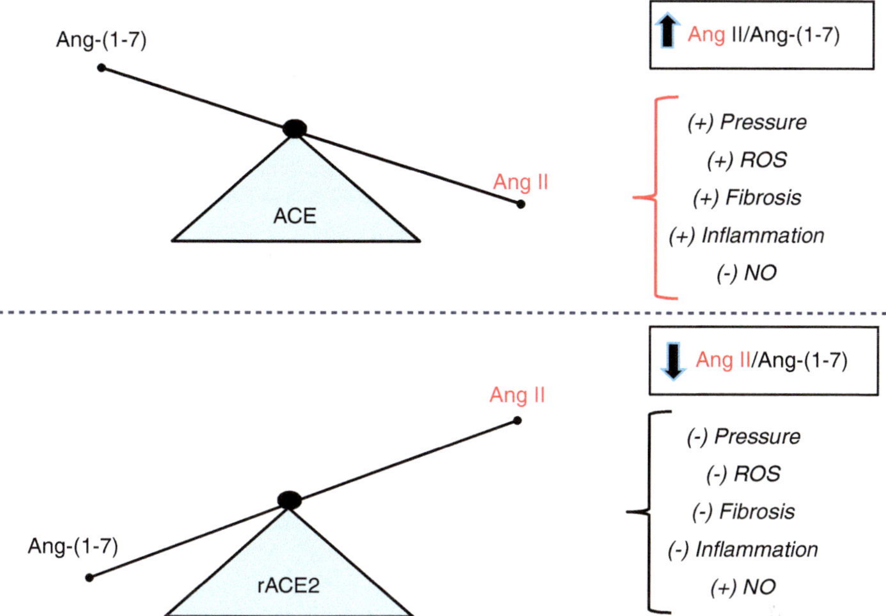

Fig. 5 Scheme depicting the influence of exogenous ACE2 on the ratio of Ang II to Ang-(1-7). **Upper panel:** Predominance of ACE activity results in a higher ratio of Ang II to Ang-(1-7) that may contribute to increased blood pressure (+BP), oxidative stress (ROS), fibrosis, and inflammation, but reduced levels of nitric oxide (-NO). **Lower panel:** Treatment with a soluble form of recombinant ACE2 (rACE2) or overexpression of the peptidase reduces the Ang II to Ang-(1-7) ratio that may lower BP, ROS, fibrosis and inflammation, but increase levels of NO

with sufficient sensitivity to detect peptides in the low pM range and resolve related N-terminal metabolites of Ang I, Ang II, and Ang-(1-7) that include Ang-(2-10), Ang-(2-8), Ang-(3-8), Ang-(2-7), and Ang-(3-7), respectively; these metabolites are not distinguished by direct RIAs or ELISAs unless coupled to HPLC/UHPLC separation prior to immunoassay. This analysis revealed that chronic ACE2 treatment reduced the total plasma ratio of Ang II: Ang-(1-7) approximately four-fold in the diabetic mice; however, the effect of ACE2 on renal peptide content was not addressed [88]. Moreover, it should be noted that the N-terminal metabolites Ang-(2-10), Ang-(2-8) or Ang III and Ang-(2-7) were the major components in plasma in this study that contrasts with the accepted profile of angiotensins in both the circulation and tissues [89].

In regard to the benefits of an activated ACE2 pathway, several compounds have been identified that may act as allosteric activators of ACE2 including xanthenone (XNT) and diminazene aceturate (DIZE) to promote a higher ratio of Ang-(1-7) to Ang II (89) [1]. Chronic treatment with DIZE ameliorated the extent of pulmonary hypertension and fibrosis, renal tissue injury, and myocardial infarction consistent with enhanced levels of Ang-(1-7) and a reduction in Ang II [90–93]. Interestingly, DIZE treatment was also associated with increased mRNA levels of

ACE2 suggesting that DIZE may exhibit actions apart from the direct activation of the peptidase. However, the stimulatory effects of DIZE on either ACE2 activity or expression have not been confirmed by others. Haber et al. found no effect of DIZE on soluble ACE2 activity or an influence on Ang II-dependent hypertension using similar doses of DIZE as previously reported [94]. Velkosa et al. also show no direct effect of various concentrations of DIZE on renal ACE2 activity ex vivo, as well as no in vivo effect on cardiac ACE2 activity or Ang-(1-7) content in the 5/6 nephrectomized rat following a 2-week administration of DIZE [95]. Indeed, this study reported that DIZE normalized the marked increase in cardiac ACE activity and Ang II suggesting that ACE may be a more relevant target than ACE2 to impact the cardiac RAS [95]. DIZE also failed to increase ACE2 activity or enhance the local vascular actions of Ang-(1-7) in a preparation of isolated pig coronary arteries [96].

Conceptually, the use of ACE2 as a therapeutic agent to chronically alter the balance of Ang II and Ang-(1-7) is challenging. ACE activity in the circulation and the vasculature surface is significant with a very high capacity to generate Ang II. This reflects not simply the abundance of ACE but the marked capability of angiotensinogen and renin to generate the ACE substrate Ang I. Moreover, reduced Ang II levels by exogenous ACE2 should stimulate the generation of Ang II that reflects inactivation of negative feedback mechanisms on renin. Therefore, it is difficult to conceive that sufficiently high levels of ACE2 can be achieved to chronically reduce Ang II *and* increase Ang-(1-7) except with the possible addition of an ACE inhibitor. In this case, supplementation of ACE2 may degrade residual levels of Ang II and the circulating levels of Ang-(1-7) may be augmented, particularly as the Ang-(1-7)-degrading pathway in the circulation is attenuated by the ACE inhibitor. In this regard, it is worth noting that Jin and colleagues have recently developed a Fc fusion protein to ACE2 that markedly prolongs the activity of the enzyme by reducing its clearance; however, the ACE2-fusion protein was assessed in an Ang II-infusion model that has suppressed endogenous Ang II, and other models of hypertension need to be examined [97].

It is presently unclear how increased circulating ACE2 augments Ang-(1-7) tissue levels as the intracellular mechanisms for Ang-(1-7) generation are not known. One possibility is that administered ACE2 increases circulating levels of Ang-(1-7) and the peptide is subsequently internalized by the MasR into a stable or protected intracellular compartment. Indeed, Gironacci and colleagues describe internalization of the Mas receptor following stimulation by Ang-(1-7); however, the assessment of intracellular Ang-(1-7) was not determined [98]. Alternatively, ACE2 treatment may alter that intracellular signaling milieu to attenuate oxidative stress or inflammation that impacts the local generation of Ang II and Ang-(1-7). Similar to neprilysin and ACE, ACE2 is primarily expressed as an ectocellular membrane-anchored peptidase that should readily access circulating or interstitial levels of Ang II. We observed that isolated nuclei from sheep renal cortex expressed ACE2 activity that converted Ang II to Ang-(1-7) and that the ACE2 inhibitor MLN4760 increased the oxidative stress response to Ang II suggesting that intracellular ACE2 may regulate the cellular balance of Ang II and Ang-(1-7) [99]. Tikellis et al. show that the renal content of Ang-(1-7) was reduced in ACE2$^{-/-}$ mice to a similar extent as that following chronic treatment with the potent ACE inhibitor perindopril [72]. Lavrentyev and colleagues also found that siRNA treatment against ACE2 reduced cellular levels of Ang-(1-7) in rat aortic smooth muscle cells [100]. Finally, Mompeon et al. recently demonstrate that estradiol increased the cellular levels of Ang-(1-7) that was associated with higher levels of both ACE and ACE2 by activation of the ERα subtype in human umbilical vein endothelial cells [101]. Overall, these studies suggest that intracellular expression of Ang-(1-7) is dependent in part on the processing of Ang II by ACE2 either intracellularly or extracellular conversion with the subsequent uptake of Ang-(1-7) by the Mas receptor.

Prolyl Carboxypeptidase

Prolyl carboxypeptidase is a monocarboxypeptidase [EC 3.4.16.2, Angiotensinase C, 58 kDa] with specificity for the C-terminal hydrolysis of the Pro-X bond where X is a hydrophobic residue [102]. The enzyme is capable of converting Ang II to Ang-(1-7), as well as degrading α-MSH, but activating the pre-kallikrein protease. In contrast to prolyl oligopeptidase and ACE2, the pH optima for prolyl carboxypeptidase is in the more acidic range of pH 4-5. Indeed, Grobe et al. find that prolyl carboxypeptidase was responsible for Ang II to Ang-(1-7) conversion in mouse renal cortex and urine at pH < 6.0 while ACE2 was predominant at pH >7.0 [103]. Velez et al. report that the mixed prolyl oligopeptidase/prolyl carboxypeptidase inhibitor ZPP partially blocked Ang II to Ang-(1-7) conversion, but had no effect on Ang I metabolism and concluded that the Ang-(1-7) forming enzymes from Ang II in human glomerular endothelial cells were prolyl carboxypeptidase and ACE2 [46]. Xu et al. find increased plasma levels of prolyl carboxypeptidase by ELISA in a cohort of obese diabetic patients; however, the circulating levels of Ang II or Ang-(1-7) were not evaluated in this study [104]. Interestingly, prolyl carboxypeptidase knockout mice exhibit higher blood pressure, increased oxidative stress, reduced vascular eNOS expression, renal damage, and cardiac dysfunction; however, neither circulating nor cardiac levels of Ang II and Ang-(1-7) were altered in this transgenic mouse as compared to wild-type [105, 106]. Finally, Jeong et al. report the prolyl carboxypeptidase knockout mice were protected against the metabolic effects of diet-induced obesity which runs counter to the expected effects of a higher Ang II: Ang-(1-7) ratio and suggest non-RAS targets of prolyl carboxypeptidase [107].

Angiotensin-(1-7) Degrading Pathways

Angiotensin-Converting Enzyme

The predominant pathway of the classical RAS for the conversion of Ang I to the bioactive peptide Ang II is catalyzed by the metallopeptidase ACE [EC 3.4.15.1], a dipeptidyl carboxypeptidase that cleaves two residues from the carboxy terminus of Ang I (Fig. 1) [108]. The peptidase is a membrane-bound, glycosylated protein (120–180 kDa) that is expressed in multiple tissues [108]. Soluble forms of the enzyme are present in the circulation, CSF, lymph fluid, and urine that retain peptidase activity [108]. The soluble form of ACE arises from the hydrolysis of the membrane-anchoring or stalk region of the protein that may reflect the processing by A Disintegrin and Metalloproteinase (ADAM) family of metallo-enzymes, although the precise role of ACE shedding in cardiovascular disease is presently unclear. Somatic ACE is characterized by two active sites termed N and C terminal domains that likely arose from the gene duplication of germinal or testicular ACE that contains only the single C terminal active site. In addition to forming Ang II, ACE degrades a number of other peptides that exhibit cardiovascular actions including bradykinin, substance P, and acetyl-SDKP [108]. Indeed, the cardioprotective effects of ACE inhibitors may reflect the protection of these peptides from metabolism, as well as the inhibitory effects on Ang II generation. In lieu of the increased circulating levels of Ang-(1-7) to ACE inhibitors, we postulated that Ang-(1-7) may be an endogenous substrate for ACE and demonstrated that ACE hydrolyzes Ang-(1-7) at the Ile^5-His^6 bond to yield the pentapeptide Ang-(1-5) and the dipeptide His-Pro (Fig. 1) [109]. Treatment with the potent ACE inhibitor lisinopril markedly reduced the clearance of the peptide and addition of the ACE inhibitor was required to demonstrate the accumulation of Ang-(1-7) derived from either Ang II or Ang I in isolated proximal tubules and following infusion of Ang I in SHR and WKY [44, 110]. Thus, the reduced metabolism of Ang-(1-7) likely contributes to the elevation in circulating levels of Ang-(1-7) following the chronic treatment with ACE inhibitors in experimental animals and in humans. These data suggest a pivotal role for ACE to regulate the balance of Ang II and Ang-(1-7) tone as the two peptides exhibit strikingly different actions from one another.

Dipeptidyl Peptidase 3

ACE clearly plays a role in the metabolism of Ang-(1-7), but there are other potential pathways that may regulate endogenous levels of the peptide [8]. Marshall and colleagues reported that ACE and a second peptidase activity in the sheep cerebrospinal fluid (CSF) degraded Ang-(1-7) [111–113]. Interestingly, the non-ACE degrading activity (subsequently identified as dipeptidyl peptidase 3, DPP3) accounted for a greater contribution of the metabolism of Ang-(1-7) than ACE [113]. Moreover, this activity was inversely correlated to CSF levels of Ang-(1-7) in control and betamethasone-exposed sheep, a model of fetal programming that exhibits elevated blood pressure and an attenuated baroreflex (Fig. 6). The Ang-(1-7)-degrading activity was also evident in sheep brain and kidney cortex, as well as in the human proximal tubule HK-2 cell line [52, 113]. The enzyme activity exhibited unusual characteristics as Ang I and other peptides equal to or greater than 10 residues were not substrates for the peptidase [52, 113]. Moreover, the inhibitor JMV-390, originally developed to block the metallo-endopeptidases neprilysin, thimet oligopeptidase and neurolysin, potently inhibited the Ang-(1-7)-degrading

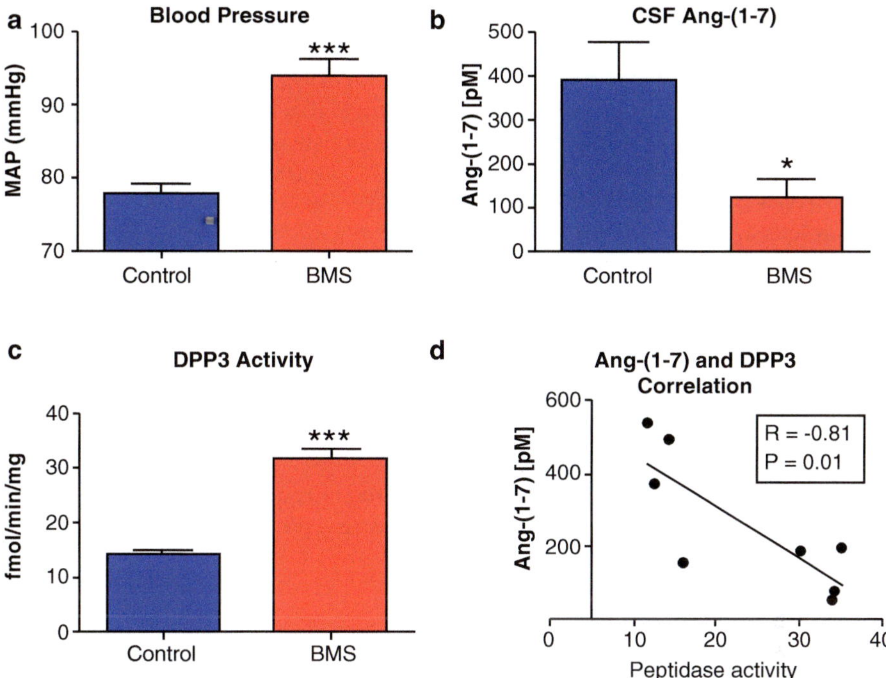

Fig. 6 Betamethasone-exposed (BMS) offspring exhibit higher mean arterial pressure (MAP) and greater CSF dipeptidyl aminopeptidase (DPP 3) activity than non-exposed sheep. MAP was higher in BMS animals at 6 months of age (**a**). CSF Ang-(1-7) peptide levels were lower in BMS animals (**b**). CSF DPP 3 was two-fold higher in BMS animals as compared to controls (**c**). Ang-(1-7) peptide levels negatively correlate with DPP 3 activity in the CSF ($r = -0.81$, $P = 0.01$) (**d**). Data are mean ± SEM; 4-5 per group; $*P < 0.05$ or $***P < 0.001$ vs. controls. (Adapted from Marshall et al. [113])

activity in the brain and kidney [$IC_{50} < 1$ nM]. Conversely, specific inhibitors against the three endopeptidases did not attenuate the Ang-(1-7) degrading activity [113]. Interestingly, the peptidase activity accounted for the sole degradative pathway in the cytosolic fraction and the media of the HK-2 cells while thimet oligopeptidase was responsible for intracellular generation of Ang-(1-7) from Ang I [52]. Utilizing the HK-2 cells as the source of the Ang-(1-7) degrading activity, we purified the peptidase from the cell cytosol by ion exchange and hydrophobic interaction chromatography and identified the enzyme as dipeptidyl peptidase 3 (EC 3.4.14.4, 85 kDa) [53].

DPP 3 belongs to a family of metallo-aminopeptidases that sequentially cleave 2 residues from the N-terminus of peptides ≤ 8 residues in length that explains our previous results that Ang I, apelin-13 and neurotensin were not substrates for the Ang-(1-7)-degrading activity in the CSF and brain [113]. We obtained a human recombinant form of DPP 3 that hydrolyzed Ang-(1-7) in two steps [55]. DPP 3 initially cleaved Ang-(1-7) at the Arg^2-Val^3 bond to form Ang-(3-7) and the dipeptide Arg^1-Asp^2. Ang-(3-7) is then very rapidly cut at Tyr^4-Ile^5 to form Ang-(5-7) and Val^3-Tyr^4. The kinetic analysis of DPP 3 hydrolysis revealed a higher efficiency constant [kcat/Km] for Ang-(3-7) than Ang-(1-7) [53]. The preferred hydrolysis of Ang-(3-7) by DPP 3 explains the inability to demonstrate the accumulation of Ang-(3-7) following the initial metabolism of Ang-(1-7). In regard to an in vivo role for DPP 3 to modulate Ang-(1-7), human HK-2 cells were treated with varying doses of the JMV-390 inhibitor and we assessed both the endogenous content of Ang-(1-7) and the intracellular DPP-3/Ang-(1-7) degrading activity in the cells. Treatment with 20 nM and 200 nM JMV-390 reduced DPP3 activity by >30% and > 80%, respectively, as compared to control suggesting that the inhibitor effectively penetrates the cells [53]. The lower dose of JMV increased the cellular content Ang-(1-7) approximately two-fold, although this did not reach statistical significance. The higher dose of JMV, however, significantly reduced the intracellular levels of the peptide [53]. The higher JMV dose may block other peptidases including thimet oligopeptidase that may be involved in the intracellular generation of Ang-(1-7) in the renal cells [53]. Thus, the blockade of Ang-(1-7)-forming enzymes by the high-dose JMV may override any protective effects of DPP 3 inhibition.

Aminopeptidase A

Both Ang II and Ang-(1-7) share the same N-terminal sequence and are likely substrates for N-terminal directed metabolism. Aminopeptidase A [EC 3.4.11.7, 50 kDa] was characterized as a classic angiotensinase that hydrolyzed the Asp^1-Arg^2 bond to Ang II to form Ang-(2-8) or Ang III [114]. Grobe and colleagues applied "in situ" MALDI to characterize both renal and cardiac metabolism of exogenous Ang II and Ang-(1-7) [115, 116]. Ang-(1-7) was the primary product from Ang II in the renal cortex while Ang III was the major metabolite in the medulla [115]. In the heart, Ang III and Ang-(1-7) were products of Ang II metabolism catalyzed by Aminopeptidase A and ACE2, respectively [116]. These data confirm earlier HPLC-based studies on the contribution of ACE2 to Ang-(1-7) formation in the mouse and human heart [68, 117]. In mouse podocytes, Aminopeptidase A contributed to the metabolism of Ang-(1-7) to Ang-(2-7) and the subsequent conversion to Ang-(3-7) by arginine aminopeptidase (EC 3.4.11.6) – an identical pathway for the metabolism of Ang II to Ang III and Ang-(3-8) [46, 114]. Aminopeptidase A is widely distributed in tissues predominantly in a membrane-bound form, although soluble forms are present in the circulation, urine, and CSF [114]. Aminopeptidase A knockout mice show an increase in blood pressure, an enhanced pressor response to Ang II and greater susceptibility to glomeruli injury that would be consistent with a role of the peptidase in the metabolism of Ang II; however, it is not known the extent that circulating endogenous levels of Ang-(1-7) are altered to potentially mitigate against the cardiovascular effects of higher Ang II [118, 119].

AGE-Induced Peptidase

The progression of epithelial to mesenchymal transition (EMT) in renal injury is an important process that leads to increased fibrosis and loss of epithelial function [120–122]. We assessed the role of Ang-(1-7) in EMT of renal epithelial NRK-52e cells provoked by either advanced glycation end products (AGEs) or TGF-β [123]. Treatment with Ang-(1-7) abolished EMT in the NRK-52e cells by the inhibition of the non-canonical ERK 1/2 signaling pathway stimulated by AGE [123]. In addition to stimulating EMT, AGE exposure reduced the intracellular levels of Ang-(1-7), but not Ang II (Fig. 7a, b, *respectively*). The intracellular processing of Ang I to Ang-(1-7) by thimet oligopeptidase tended to be reduced by AGE; however, Ang-(1-7) metabolism was significantly increased by AGE exposure (Fig. 7c, d, *respectively*). AGE-induced EMT may reflect lower Ang-(1-7) expression in the renal epithelial cells that may be permissive for the progression of EMT and increased fibrosis [120]. Although our data suggest that DPP 3 is not responsible for the AGE-induced metabolism of Ang-(1-7) in the NRK-52 cells, a distinct endopeptidase activity may participate in the cellular metabolism of the peptide [123].

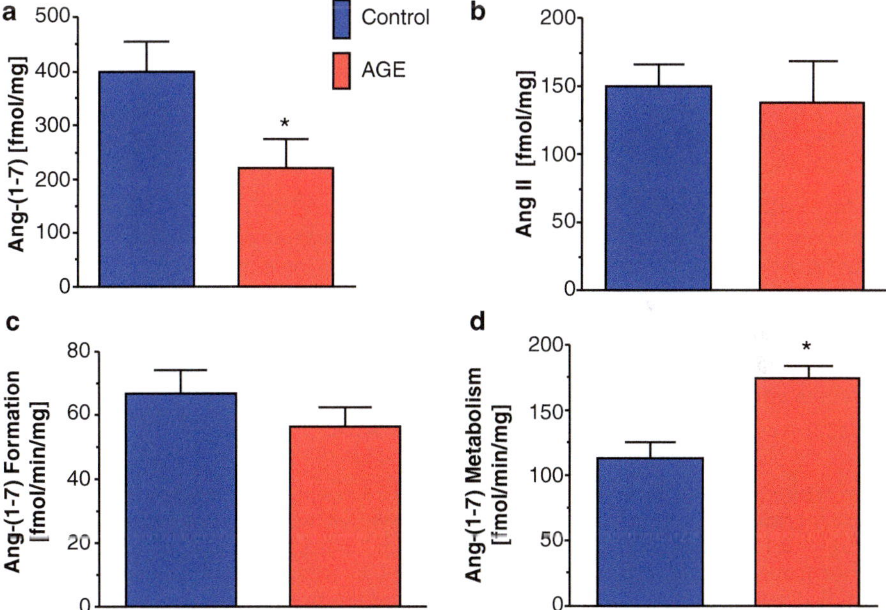

Fig. 7 Advanced glycation end products (AGEs) reduce cellular expression of Ang-(1-7) in NRK-52e renal epithelial cells. The AGE methyl glyoxal albumin (MGA) reduces Ang-(1-7) content (**a**) but not Ang II levels (**b**) in NRK-52e cells. MGA tended to reduce processing of Ang I to Ang-(1-7) (**c**), but significantly increased the metabolism of Ang-(1-7) (**d**) in the 100,000 xg cytosolic fraction of NRK-52e cells. Data are means ± SEM; $n = 5$. *$P < 0.05$ versus Control. Activities are expressed as fmol/minute/mg protein. (Adapted from Alzayadneh and Chappell [124])

Conclusion

Accumulating evidence from multiple laboratories clearly indicates that activation of the Ang-(1-7)-AT$_7$/MasR axis appears to be a potentially important therapeutic target in the treatment of cardiovascular disease and other pathologies [125]. In lieu of the described peptidases that degrade Ang-(1-7) and the apparent reliance on Ang II as an endogenous substrate for the generation of the peptide in several tissues, the development of selective non-peptide analogs of Ang-(1-7) that target both the Mas receptor and the newly described Mas-related receptor D (MRG-D) [125] may be of particular benefit in terms of their oral availability, greater resistance to peptidase metabolism, improved selectivity, their potential ability to accumulate within the cell and to cross the blood-brain barrier.

A functional intracellular AT$_7$/MasR was first identified on isolated nuclei from the ovine kidney that was linked to NO generation following exposure to low pM doses of Ang-(1-7) [124]. These findings were recently corroborated in the brain that revealed evidence of the MasR on nuclei and mitochondria, as well as Ang-(1-7)-dependent stimulation of NO [126]. Moreover, AT$_7$/MasR binding on renal nuclei and the Ang-(1-7)-evoked NO response were attenuated in aged sheep and in adult sheep exposed to betamethasone in utero as compared to the younger non-exposed animals [9, 99]. Conversely, intracellular levels of the AT$_1$R that were associated with stimulation of oxidative stress on isolated nuclei were increased in the older animals and in the betamethasone-exposed sheep [9, 98]. Abadir et al. originally reported an increased ratio of AT$_1$R: AT$_2$R receptors in isolated renal mitochondrial of aging mice [127]. Valenzuela et al. also find reduced an increased AT$_1$R:AT$_2$R ratio in brain mitochondria of aging mice that may lead to higher cellular levels of oxidative stress [128, 129]. Overall, these novel findings of nuclear and mitochondrial angiotensin receptors support extensive evidence for an intracellular RAS within various tissues although the exact role of an activated intracellular RAS to influence cardiovascular disease and other pathologies is not resolved and requires further study. Nevertheless, peptidase-resistant and cell-permeable agonists of the Ang-(1-7) axis may provide additional cardioprotective effects to conventional approaches to block the RAS and may expand the targeted therapies required to combat cardiovascular disease.

Acknowledgments I am most fortunate to have shared a moment in time with a number of very talented graduate students and postdoctoral fellows including Karl Pendergrass, Liliya Yamaleyeva, Hossam Shaltout, TanYa Gwathmey, Sarah Lindsey, Jonathan Cohen, Shea Gilliam, Allyson Marshall, Ebaa Al-Zayadneh, Bryan Wilson, and Nildris Cruz-Diaz, as well as Brian Westwood and Nancy Pirro whose research efforts to the studies in this review I gratefully acknowledge. Portions of this work were supported by grants from the National Institute of Health grants (HL-56973, HL-51952, HD-084227, HD-047584, HL-091797, HL-146818, GM102773 and HD-017644) and the American Heart Association (AHA-151521, AHA-355741 and AHA-18TPA34170522). An unrestricted grant from the Farley-Hudson Foundation (Jacksonville, NC), Groskert Heart Fund, and the Wake Forest Venture Fund is also acknowledged.

Literature Cited

1. Hernandez Prada JA, Ferreira AJ, Katovich MJ, Shenoy V, Qi Y, Santos RA, et al. Structure-based identification of small-molecule angiotensin-converting enzyme 2 activators as novel antihypertensive agents. Hypertension (Dallas, TX: 1979). 2008;51(5):1312–7. https://doi.org/10.1161/hypertensionaha.107.108944.

2. Aroor AR, Demarco VG, Jia G, Sun Z, Nistala R, Meininger GA, et al. The role of tissue renin-angiotensin-aldosterone system in the development of endothelial dysfunction and arterial stiffness. Front Endocrinol. 2013;4:161. https://doi.org/10.3389/fendo.2013.00161.

3. Ferrario CM, Strawn WB. Role of the renin-angiotensin-aldosterone system and proinflammatory mediators in cardiovascular disease. Am J Cardiol. 2006;98(1):121–8. https://doi.org/10.1016/j.amjcard.2006.01.059.

4. Te Riet L, van Esch JH, Roks AJ, van den Meiracker AH, Danser AH. Hypertension: renin-angiotensin-aldosterone system alterations. Circ Res. 2015;116(6):960–75. https://doi.org/10.1161/circresaha.116.303587.

5. Kobori H, Nangaku M, Navar LG, Nishiyama A. The intrarenal renin-angiotensin system: from physiology to the pathobiology of hypertension and kidney disease. Pharmacol Rev. 2007;59(3):251–87. https://doi.org/10.1124/pr.59.3.3.

6. Navar LG, Prieto MC, Satou R, Kobori H. Intrarenal angiotensin II and its contribution to the genesis of chronic hypertension. Curr Opin Pharmacol. 2011;11(2):180–6. https://doi.org/10.1016/j.coph.2011.01.009.

7. Chappell MC, Brosnihan KB, Diz DI, Ferrario CM. Identification of angiotensin-(1-7) in rat brain. Evidence for differential processing of angiotensin peptides. J Biol Chem. 1989;264(28):16518–23.

8. Santos RA. Angiotensin-(1-7). Hypertension (Dallas, TX: 1979). 2014;63(6):1138–47. https://doi.org/10.1161/hypertensionaha.113.01274.

9. Chappell MC, Marshall AC, Alzayadneh EM, Shaltout HA, Diz DI. Update on the angiotensin converting enzyme 2-angiotensin (1-7)-MAS receptor axis: fetal programing, sex differences, and intracellular pathways. Front Endocrinol. 2014;4:201. https://doi.org/10.3389/fendo.2013.00201.

10. Carey RM, Padia SH. Role of angiotensin AT(2) receptors in natriuresis: Intrarenal mechanisms and therapeutic potential. Clin Exp Pharmacol Physiol. 2013;40(8):527–34. https://doi.org/10.1111/1440-1681.12059.

11. Sumners C, de Kloet AD, Krause EG, Unger T, Steckelings UM. Angiotensin type 2 receptors: blood pressure regulation and end organ damage. Curr Opin Pharmacol. 2015;21:115–21. https://doi.org/10.1016/j.coph.2015.01.004.

12. Gonzalez AA, Lara LS, Luffman C, Seth DM, Prieto MC. Soluble form of the (pro)renin receptor is augmented in the collecting duct and urine of chronic angiotensin II-dependent hypertensive rats. Hypertension (Dallas, TX: 1979). 2011;57(4):859–64. https://doi.org/10.1161/hypertensionaha.110.167957.

13. Navar LG, Kobori H, Prieto MC, Gonzalez-Villalobos RA. Intratubular renin-angiotensin system in hypertension. Hypertension (Dallas, TX: 1979). 2011;57(3):355–62. https://doi.org/10.1161/hypertensionaha.110.163519.

14. Gonzalez AA, Liu L, Lara LS, Seth DM, Navar LG, Prieto MC. Angiotensin II stimulates renin in inner medullary collecting duct cells via protein kinase C and independent of epithelial sodium channel and mineralocorticoid receptor activity. Hypertension (Dallas, TX: 1979). 2011;57(3):594–9. https://doi.org/10.1161/hypertensionaha.110.165902.

15. Peters J, Clausmeyer S. Intracellular sorting of renin: cell type specific differences and their consequences. J Mol Cell Cardiol. 2002;34(12):1561–8.

16. Lavoie JL, Liu X, Bianco RA, Beltz TG, Johnson AK, Sigmund CD. Evidence supporting a functional role for intracellular renin in the brain. Hypertension (Dallas, TX: 1979). 2006;47(3):461–6. https://doi.org/10.1161/01.HYP.0000203308.52919.dc.

17. Ishigami T, Kino T, Chen L, Minegishi S, Araki N, Umemura M, et al. Identification of bona fide alternative renin transcripts expressed along cortical tubules and potential roles in promoting insulin resistance in vivo without significant plasma renin activity elevation. Hypertension (Dallas, TX: 1979). 2014;64(1):125–33. https://doi.org/10.1161/hypertensionaha.114.03394.

18. Peters J, Wanka H, Peters B, Hoffmann S. A renin transcript lacking exon 1 encodes for a non-secretory intracellular renin that increases aldosterone production in transgenic rats. J Cell Mol Med. 2008;12(4):1229–37. https://doi.org/10.1111/j.1582-4934.2008.00132.x.

19. Clausmeyer S, Sturzebecher R, Peters J. An alternative transcript of the rat renin gene can result in a truncated prorenin that is transported into adrenal mitochondria. Circ Res. 1999;84(3):337–44.

20. Wanka H, Kessler N, Ellmer J, Endlich N, Peters BS, Clausmeyer S, et al. Cytosolic renin is targeted to mitochondria and induces apoptosis in H9c2 rat cardiomyoblasts. J Cell Mol Med. 2009;13(9a):2926–37. https://doi.org/10.1111/j.1582-4934.2008.00448.x.

21. Wanka H, Staar D, Lutze P, Peters B, Hildebrandt J, Beck T, et al. Anti-necrotic and cardio-protective effects of a cytosolic renin isoform under ischemia-related conditions. J Mol Med (Berlin, Germany). 2016;94(1):61–9. https://doi.org/10.1007/s00109-015-1321-z.

22. Wilson BA, Nautiyal M, Gwathmey TM, Rose JC, Chappell MC. Evidence for a mitochondrial angiotensin-(1-7) system in the kidney. Am J Physiol Renal Physiol. 2016;310(7):F637–f45. https://doi.org/10.1152/ajprenal.00479.2015.

23. Alzayadneh EM, Chappell MC. Nuclear expression of renin-angiotensin system components in NRK-52E renal epithelial cells. J Renin Angiotensin Aldosterone Syst: JRAAS. 2015;16(4):1135–48. https://doi.org/10.1177/1470320313515039.

24. Li XC, Cook JL, Rubera I, Tauc M, Zhang F, Zhuo JL. Intrarenal transfer of an intracellular fluorescent fusion of angiotensin II selectively in proximal tubules increases blood pressure in rats and mice. Am J Physiol Renal Physiol. 2011;300(5):F1076–88. https://doi.org/10.1152/ajprenal.00329.2010.

25. Crowley SD, Gurley SB, Herrera MJ, Ruiz P, Griffiths R, Kumar AP, et al. Angiotensin II causes hypertension and cardiac hypertrophy through its receptors in the kidney. Proc Natl Acad Sci U S A. 2006;103(47):17985–90. https://doi.org/10.1073/pnas.0605545103.

26. Li XC, Hopfer U, Zhuo JL. Novel signaling mechanisms of intracellular angiotensin II-induced NHE3 expression and activation in mouse proximal tubule cells. Am J Physiol Renal Physiol. 2012;303(12):F1617–28. https://doi.org/10.1152/ajprenal.00219.2012.

27. Li XC, Zhuo JL. Proximal tubule-dominant transfer of AT(1a) receptors induces blood pressure responses to intracellular angiotensin II in AT(1a) receptor-deficient mice. Am J Physiol Regul Integr Comp Physiol. 2013;304(8):R588–98. https://doi.org/10.1152/ajpregu.00338.2012.

28. Davisson RL, Ding Y, Stec DE, Catterall JF, Sigmund CD. Novel mechanism of hypertension revealed by cell-specific targeting of human angiotensinogen in transgenic mice. Physiol Genomics. 1999;1(1):3–9. https://doi.org/10.1152/physiolgenomics.1999.1.1.3.

29. Campbell DJ. Angiotensin II generation in vivo: does it involve enzymes other than renin and angiotensin-converting enzyme? J Renin Angiotensin Aldosterone Syst: JRAAS. 2012;13(2):314–6. https://doi.org/10.1177/1470320312447162.

30. Nagata S, Kato J, Sasaki K, Minamino N, Eto T, Kitamura K. Isolation and identification of proangiotensin-12, a possible component of the renin-angiotensin system. Biochem Biophys Res Commun. 2006;350(4):1026–31. https://doi.org/10.1016/j.bbrc.2006.09.146.

31. Westwood BM, Chappell MC. Divergent pathways for the angiotensin-(1-12) metabolism in the rat circulation and kidney. Peptides. 2012;35(2):190–5. https://doi.org/10.1016/j.peptides.2012.03.025.

32. Turner AJ. Neprilysin. In: Barrett A, Woessner J, Rawlings N, editors. Handbook of proteolytic enzymes. 2nd ed. London: Elsevier Academic Press; 2004. p. 419–26.

33. McKinnie SM, Fischer C, Tran KM, Wang W, Mosquera F, Oudit GY, et al. The metalloprotease neprilysin degrades and inactivates apelin peptides. Chembiochem: Eur J Chem Biol. 2016;17(16):1495–8. https://doi.org/10.1002/cbic.201600244.

34. Cataliotti A, Boerrigter G, Chen HH, Jougasaki M, Costello LC, Tsuruda T, et al. Differential actions of vasopeptidase inhibition versus angiotensin-converting enzyme inhibition on diuretic therapy in experimental congestive heart failure. Circulation. 2002;105(5):639–44.

35. Tikkanen I, Tikkanen T, Cao Z, Allen TJ, Davis BJ, Lassila M, et al. Combined inhibition of neutral endopeptidase with angiotensin converting enzyme or endothelin converting enzyme in experimental diabetes. J Hypertens. 2002;20(4):707–14.

36. Kubota E, Dean RG, Hubner RA, Casley DJ, Johnston CI, Burrell LM. Differential tissue and enzyme inhibitory effects of the vasopeptidase inhibitor omapatrilat in the rat. Clin Sci (London, England: 1979). 2003;105(3):339–45. https://doi.org/10.1042/cs20030097.

37. Hubers SA, Brown NJ. Combined angiotensin receptor antagonism and neprilysin inhibition. Circulation. 2016;133(11):1115–24. https://doi.org/10.1161/circulationaha.115.018622.

38. Rice GI, Thomas DA, Grant PJ, Turner AJ, Hooper NM. Evaluation of angiotensin-converting enzyme (ACE), its homologue ACE2 and neprilysin in angiotensin peptide metabolism. Biochem J. 2004;383(Pt 1):45–51. https://doi.org/10.1042/bj20040634.

39. Yamamoto K, Chappell MC, Brosnihan KB, Ferrario CM. In vivo metabolism of angiotensin I by neutral endopeptidase (EC 3.4.24.11) in spontaneously hypertensive rats. Hypertension (Dallas, TX: 1979). 1992;19(6 Pt 2):692–6.

40. Iyer SN, Chappell MC, Averill DB, Diz DI, Ferrario CM. Vasodepressor actions of angiotensin-(1-7) unmasked during combined treatment with lisinopril and losartan. Hypertension (Dallas, TX: 1979). 1998;31(2):699–705.

41. Chappell MC, Gomez MN, Pirro NT, Ferrario CM. Release of angiotensin-(1-7) from the rat hindlimb: influence of angiotensin-converting enzyme inhibition. Hypertension (Dallas, TX: 1979). 2000;35(1 Pt 2):348–52.

42. Campbell DJ, Anastasopoulos F, Duncan AM, James GM, Kladis A, Briscoe TA. Effects of neutral endopeptidase inhibition and combined angiotensin converting enzyme and neutral endopeptidase inhibition on angiotensin and bradykinin peptides in rats. J Pharmacol Exp Ther. 1998;287(2):567–77.

43. Domenig O, Manzel A, Grobe N, Konigshausen E, Kaltenecker CC, Kovarik JJ, et al. Neprilysin is a mediator of alternative renin-angiotensin-system activation in the murine and human kidney. Sci Rep. 2016;6:33678. https://doi.org/10.1038/srep33678.

44. Shaltout HA, Westwood BM, Averill DB, Ferrario CM, Figueroa JP, Diz DI, et al. Angiotensin metabolism in renal proximal tubules, urine, and serum of sheep: evidence for ACE2-dependent processing of angiotensin II. Am J Physiol Renal Physiol. 2007;292(1):F82–91. https://doi.org/10.1152/ajprenal.00139.2006.

45. Allred AJ, Diz DI, Ferrario CM, Chappell MC. Pathways for angiotensin-(1---7) metabolism in pulmonary and renal tissues. Am J Physiol Renal Physiol. 2000;279(5):F841–50. https://doi.org/10.1152/ajprenal.2000.279.5.F841.

46. Velez JC, Ierardi JL, Bland AM, Morinelli TA, Arthur JM, Raymond JR, et al. Enzymatic processing of angiotensin peptides by human glomerular endothelial cells. Am J Physiol Renal Physiol. 2012;302(12):F1583–94. https://doi.org/10.1152/ajprenal.00087.2012.

47. Schwacke JH, Spainhour JC, Ierardi JL, Chaves JM, Arthur JM, Janech MG, et al. Network modeling reveals steps in angiotensin peptide processing. Hypertension (Dallas, TX: 1979). 2013;61(3):690–700. https://doi.org/10.1161/hypertensionaha.111.00318.

48. Velez JC, Ryan KJ, Harbeson CE, Bland AM, Budisavljevic MN, Arthur JM, et al. Angiotensin I is largely converted to angiotensin (1-7) and angiotensin (2-10) by isolated rat glomeruli. Hypertension (Dallas, TX: 1979). 2009;53(5):790–7. https://doi.org/10.1161/hypertensionaha.109.128819.

49. Modrall JG, Sadjadi J, Brosnihan KB, Gallagher PE, Yu CH, Kramer GL, et al. Depletion of tissue angiotensin-converting enzyme differentially influences the intrarenal and urinary expression of angiotensin peptides. Hypertension (Dallas, TX: 1979). 2004;43(4):849–53. https://doi.org/10.1161/01.HYP.0000121462.27393.f6.

50. Barrett AJ, Chen JM. Thimet oligopeptidase. In: Barrett A, Woessner J, Rawlings N, editors. Handbook of Proteolytic enzymes. 2nd ed. London: Elsevier Academic Press; 2004. p. 352–6.

51. Pereira MG, Souza LL, Becari C, Duarte DA, Camacho FR, Oliveira JA, et al. Angiotensin II-independent angiotensin-(1-7) formation in rat hippocampus: involvement of thimet oligopeptidase. Hypertension (Dallas, TX: 1979). 2013;62(5):879–85. https://doi.org/10.1161/hypertensionaha.113.01613.

52. Wilson BA, Cruz-Diaz N, Marshall AC, Pirro NT, Su Y, Gwathmey TM, et al. An angiotensin-(1-7) peptidase in the kidney cortex, proximal tubules, and human HK-2 epithelial cells that is distinct from insulin-degrading enzyme. Am J Physiol Renal Physiol. 2015;308(6):F594–601. https://doi.org/10.1152/ajprenal.00609.2014.

53. Cruz-Diaz N, Wilson BA, Pirro NT, Brosnihan KB, Marshall AC, Chappell MC. Identification of dipeptidyl peptidase 3 as the angiotensin-(1-7) degrading peptidase in human HK-2 renal epithelial cells. Peptides. 2016;83:29–37. https://doi.org/10.1016/j.peptides.2016.06.005.

54. Suski M, Gebska A, Olszanecki R, Stachowicz A, Uracz D, Madej J, et al. Influence of atorvastatin on angiotensin I metabolism in resting and TNF-alpha-activated rat vascular smooth muscle cells. J Renin Angiotensin Aldosterone Syst: JRAAS. 2014;15(4):378–83. https://doi.org/10.1177/1470320313475907.

55. Chappell MC, Tallant EA, Brosnihan KB, Ferrario CM. Conversion of angiotensin I to angiotensin-(1-7) by thimet oligopeptidase (EC 3.4. 24.15) in vascular smooth muscle cells. J Vasc Med Biol. 1994;5:129–37.

56. Chappell MC. Emerging evidence for a functional angiotensin-converting enzyme 2-angiotensin-(1-7)-MAS receptor axis: more than regulation of blood pressure? Hypertension (Dallas, TX: 1979). 2007;50(4):596–9. https://doi.org/10.1161/hypertensionaha.106.076216.

57. Gwathmey TM, Alzayadneh EM, Pendergrass KD, Chappell MC. Novel roles of nuclear angiotensin receptors and signaling mechanisms. Am J Physiol Regul Integr Comp Physiol. 2012;302(5):R518–30. https://doi.org/10.1152/ajpregu.00525.2011.

58. Ellis B, Li XC, Miguel-Qin E, Gu V, Zhuo JL. Evidence for a functional intracellular angiotensin system in the proximal tubule of the kidney. Am J Physiol Regul Integr Comp Physiol. 2012;302(5):R494–509. https://doi.org/10.1152/ajpregu.00487.2011.

59. Kumar R, Thomas CM, Yong QC, Chen W, Baker KM. The intracrine renin-angiotensin system. Clin Sci (London, England: 1979). 2012;123(5):273–84. https://doi.org/10.1042/cs20120089.

60. Cook JL, Re RN. Lessons from in vitro studies and a related intracellular angiotensin II transgenic mouse model. Am J Physiol Regul Integr Comp Physiol. 2012;302(5):R482–93. https://doi.org/10.1152/ajpregu.00493.2011.

61. Abadir PM, Walston JD, Carey RM. Subcellular characteristics of functional intracellular renin-angiotensin systems. Peptides. 2012;38(2):437–45. https://doi.org/10.1016/j.peptides.2012.09.016.

62. Myohanen TT, Garcia-Horsman JA, Tenorio-Laranga J, Mannisto PT. Issues about the physiological functions of prolyl oligopeptidase based on its discordant spatial association with substrates and inconsistencies among mRNA, protein levels, and enzymatic activity. J Histochem Cytochem: Off J Histochem Soc. 2009;57(9):831–48. https://doi.org/10.1369/jhc.2009.953711.

63. Santos RA, Brosnihan KB, Jacobsen DW, DiCorleto PE, Ferrario CM. Production of angiotensin-(1-7) by human vascular endothelium. Hypertension (Dallas, TX: 1979). 1992;19(2 Suppl):Ii56–61.

64. Penttinen A, Tenorio-Laranga J, Siikanen A, Morawski M, Rossner S, Garcia-Horsman JA. Prolyl oligopeptidase: a rising star on the stage of neuroinflammation research. CNS Neurol Disord Drug Targets. 2011;10(3):340–8.

65. Lawandi J, Gerber-Lemaire S, Juillerat-Jeanneret L, Moitessier N. Inhibitors of prolyl oligopeptidases for the therapy of human diseases: defining diseases and inhibitors. J Med Chem. 2010;53(9):3423–38. https://doi.org/10.1021/jm901104g.

66. Donoghue M, Hsieh F, Baronas E, Godbout K, Gosselin M, Stagliano N, et al. A novel angiotensin-converting enzyme-related carboxypeptidase (ACE2) converts angiotensin I to angiotensin 1-9. Circ Res. 2000;87(5):E1–9.

67. Vickers C, Hales P, Kaushik V, Dick L, Gavin J, Tang J, et al. Hydrolysis of biological peptides by human angiotensin-converting enzyme-related carboxypeptidase. J Biol Chem. 2002;277(17):14838–43. https://doi.org/10.1074/jbc.M200581200.

68. Garabelli PJ, Modrall JG, Penninger JM, Ferrario CM, Chappell MC. Distinct roles for angiotensin-converting enzyme 2 and carboxypeptidase A in the processing of angiotensins within the murine heart. Exp Physiol. 2008;93(5):613–21. https://doi.org/10.1113/expphysiol.2007.040246.

69. Rice GI, Jones AL, Grant PJ, Carter AM, Turner AJ, Hooper NM. Circulating activities of angiotensin-converting enzyme, its homolog, angiotensin-converting enzyme 2, and neprilysin in a family study. Hypertension (Dallas, TX: 1979). 2006;48(5):914–20. https://doi.org/10.1161/01.hyp.0000244543.91937.79.

70. Yamaleyeva LM, Gilliam-Davis S, Almeida I, Brosnihan KB, Lindsey SH, Chappell MC. Differential regulation of circulating and renal ACE2 and ACE in hypertensive mRen2. Lewis rats with early-onset diabetes. Am J Physiol Renal Physiol. 2012;302(11):F1374–84. https://doi.org/10.1152/ajprenal.00656.2011.

71. Epelman S, Shrestha K, Troughton RW, Francis GS, Sen S, Klein AL, et al. Soluble angiotensin-converting enzyme 2 in human heart failure: relation with myocardial function and clinical outcomes. J Card Fail. 2009;15(7):565–71. https://doi.org/10.1016/j.cardfail.2009.01.014.

72. Tikellis C, Bialkowski K, Pete J, Sheehy K, Su Q, Johnston C, et al. ACE2 deficiency modifies renoprotection afforded by ACE inhibition in experimental diabetes. Diabetes. 2008;57(4):1018–25. https://doi.org/10.2337/db07-1212.

73. South AM, Shaltout HA, Washburn LK, Hendricks AS, Diz DI, Chappell MC. Fetal programming and the angiotensin-(1-7) axis: a review of the experimental and clinical data. Clin Sci (Lond). 2019;113(1):55–74. https://doi.org/10.1042/CS20171550.

74. Soler MJ, Wysocki J, Ye M, Lloveras J, Kanwar Y, Batlle D. ACE2 inhibition worsens glomerular injury in association with increased ACE expression in streptozotocin-induced diabetic mice. Kidney Int. 2007;72(5):614–23. https://doi.org/10.1038/sj.ki.5002373.

75. Oudit GY, Herzenberg AM, Kassiri Z, Wong D, Reich H, Khokha R, et al. Loss of angiotensin-converting enzyme-2 leads to the late development of angiotensin II-dependent glomerulosclerosis. Am J Pathol. 2006;168(6):1808–20. https://doi.org/10.2353/ajpath.2006.051091.

76. Crackower MA, Sarao R, Oudit GY, Yagil C, Kozieradzki I, Scanga SE, et al. Angiotensin-converting enzyme 2 is an essential regulator of heart function. Nature. 2002;417(6891):822–8. https://doi.org/10.1038/nature00786.

77. Oudit GY, Kassiri Z, Patel MP, Chappell M, Butany J, Backx PH, et al. Angiotensin II-mediated oxidative stress and inflammation mediate the age-dependent cardiomyopathy in ACE2 null mice. Cardiovasc Res. 2007;75(1):29–39. https://doi.org/10.1016/j.cardiores.2007.04.007.

78. Chhabra KH, Xia H, Pedersen KB, Speth RC, Lazartigues E. Pancreatic angiotensin-converting enzyme 2 improves glycemia in angiotensin II-infused mice. Am J Physiol Endocrinol Metab. 2013;304(8):E874–84. https://doi.org/10.1152/ajpendo.00490.2012.

79. Oudit GY, Liu GC, Zhong J, Basu R, Chow FL, Zhou J, et al. Human recombinant ACE2 reduces the progression of diabetic nephropathy. Diabetes. 2010;59(2):529–38. https://doi.org/10.2337/db09-1218.

80. Zhong J, Basu R, Guo D, Chow FL, Byrns S, Schuster M et al. Angiotensin-converting enzyme 2 suppresses pathological hypertrophy, myocardial fibrosis, and cardiac dysfunction. Circulation. 2010;122(7):717-728, 18 p following 28. doi:https://doi.org/10.1161/circulationaha.110.955369.

81. Wysocki J, Ye M, Rodriguez E, Gonzalez-Pacheco FR, Barrios C, Evora K, et al. Targeting the degradation of angiotensin II with recombinant angiotensin-converting enzyme 2: prevention of angiotensin II-dependent hypertension. Hypertension (Dallas, TX: 1979). 2010;55(1):90–8. https://doi.org/10.1161/hypertensionaha.109.138420.

82. Ye M, Wysocki J, Gonzalez-Pacheco FR, Salem M, Evora K, Garcia-Halpin L, et al. Murine recombinant angiotensin-converting enzyme 2: effect on angiotensin II-dependent hypertension and distinctive angiotensin-converting enzyme 2 inhibitor characteristics on rodent and human angiotensin-converting enzyme 2. Hypertension (Dallas, TX: 1979). 2012;60(3):730–40. https://doi.org/10.1161/hypertensionaha.112.198622.

83. Huentelman MJ, Grobe JL, Vazquez J, Stewart JM, Mecca AP, Katovich MJ, et al. Protection from angiotensin II-induced cardiac hypertrophy and fibrosis by systemic

lentiviral delivery of ACE2 in rats. Exp Physiol. 2005;90(5):783–90. https://doi.org/10.1113/expphysiol.2005.031096.

84. Diez-Freire C, Vazquez J, Correa de Adjounian MF, Ferrari MF, Yuan L, Silver X, et al. ACE2 gene transfer attenuates hypertension-linked pathophysiological changes in the SHR. Physiol Genomics. 2006;27(1):12–9. https://doi.org/10.1152/physiolgenomics.00312.2005.

85. Shenoy V, Ferreira AJ, Qi Y, Fraga-Silva RA, Diez-Freire C, Dooies A, et al. The angiotensin-converting enzyme 2/angiogenesis-(1-7)/Mas axis confers cardiopulmonary protection against lung fibrosis and pulmonary hypertension. Am J Respir Crit Care Med. 2010;182(8):1065–72. https://doi.org/10.1164/rccm.200912-1840OC.

86. Yamazato M, Yamazato Y, Sun C, Diez-Freire C, Raizada MK. Overexpression of angiotensin-converting enzyme 2 in the rostral ventrolateral medulla causes long-term decrease in blood pressure in the spontaneously hypertensive rats. Hypertension (Dallas, TX: 1979). 2007;49(4):926–31. https://doi.org/10.1161/01.hyp.0000259942.38108.20.

87. Bae EH, Fang F, Williams VR, Konvalinka A, Zhou X, Patel VB, et al. Murine recombinant angiotensin-converting enzyme 2 attenuates kidney injury in experimental Alport syndrome. Kidney Int. 2017;91(6):1347–61. https://doi.org/10.1016/j.kint.2016.12.022.

88. Wysocki J, Ye M, Khattab AM, Fogo A, Martin A, David NV, et al. Angiotensin-converting enzyme 2 amplification limited to the circulation does not protect mice from development of diabetic nephropathy. Kidney Int. 2017;91(6):1336–46. https://doi.org/10.1016/j.kint.2016.09.032.

89. Chappell MC. Biochemical evaluation of the renin-angiotensin system: the good, bad, and absolute? Am J Physiol Heart Circ Physiol. 2016;310(2):H137–52. https://doi.org/10.1152/ajpheart.00618.2015.

90. Qi Y, Zhang J, Cole-Jeffrey CT, Shenoy V, Espejo A, Hanna M, et al. Diminazene aceturate enhances angiotensin-converting enzyme 2 activity and attenuates ischemia-induced cardiac pathophysiology. Hypertension (Dallas, TX: 1979). 2013;62(4):746–52. https://doi.org/10.1161/hypertensionaha.113.01337.

91. De Maria ML, Araujo LD, Fraga-Silva RA, Pereira LA, Ribeiro HJ, Menezes GB, et al. Anti-hypertensive effects of diminazene aceturate: an angiotensin- converting enzyme 2 activator in rats. Protein Pept Lett. 2016;23(1):9–16.

92. Macedo LM, Souza AP, De Maria ML, Borges CL, Soares CM, Pedrino GR, et al. Cardioprotective effects of diminazene aceturate in pressure-overloaded rat hearts. Life Sci. 2016;155:63–9. https://doi.org/10.1016/j.lfs.2016.04.036.

93. Shenoy V, Gjymishka A, Jarajapu YP, Qi Y, Afzal A, Rigatto K, et al. Diminazene attenuates pulmonary hypertension and improves angiogenic progenitor cell functions in experimental models. Am J Respir Crit Care Med. 2013;187(6):648–57. https://doi.org/10.1164/rccm.201205-0880OC.

94. Haber PK, Ye M, Wysocki J, Maier C, Haque SK, Batlle D. Angiotensin-converting enzyme 2-independent action of presumed angiotensin-converting enzyme 2 activators: studies in vivo, ex vivo, and in vitro. Hypertension (Dallas, TX: 1979). 2014;63(4):774–82. https://doi.org/10.1161/hypertensionaha.113.02856.

95. Velkoska E, Patel SK, Griggs K, Burrell LM. Diminazene aceturate improves cardiac fibrosis and diastolic dysfunction in rats with kidney disease. PLoS One. 2016;11(8):e161760. https://doi.org/10.1371/journal.pone.0161760.

96. Raffai G, Khang G, Vanhoutte PM. Angiotensin-(1-7) augments endothelium-dependent relaxations of porcine coronary arteries to bradykinin by inhibiting angiotensin-converting enzyme 1. J Cardiovasc Pharmacol. 2014;63(5):453–60. https://doi.org/10.1097/fjc.0000000000000069.

97. Liu P, Wysocki J, Souma T, Ye M, Ramirez V, Zhou B, et al. Novel ACE2-Fc chimeric fusion provides long-lasting hypertension control and organ protection in mouse models of systemic renin angiotensin system activation. Kidney Int. 2018; https://doi.org/10.1016/j.kint.2018.01.029.

98. Cerniello FM, Carretero OA, Longo Carbajosa NA, Cerrato BD, Santos RA, Grecco HE, et al. MAS1 receptor trafficking involves ERK1/2 activation through a beta-arrestin2-dependent pathway. Hypertension (Dallas, TX: 1979). 2017;70(5):982–9. https://doi.org/10.1161/hypertensionaha.117.09789.

99. Gwathmey TM, Shaltout HA, Rose JC, Diz DI, Chappell MC. Glucocorticoid-induced fetal programming alters the functional complement of angiotensin receptor subtypes within the kidney. Hypertension (Dallas, TX: 1979). 2011;57(3):620–6. https://doi.org/10.1161/hypertensionaha.110.164970.
100. Lavrentyev EN, Estes AM, Malik KU. Mechanism of high glucose induced angiotensin II production in rat vascular smooth muscle cells. Circ Res. 2007;101(5):455–64. https://doi.org/10.1161/circresaha.107.151852.
101. Mompeon A, Lazaro-Franco M, Bueno-Beti C, Perez-Cremades D, Vidal-Gomez X, Monsalve E, et al. Estradiol, acting through ER alpha, induces endothelial non-classic renin-angiotensin system increasing angiotensin 1-7 production. Mol Cell Endocrinol. 2016;422:1–8. https://doi.org/10.1016/j.mce.2015.11.004.
102. Tan F, Erdos EG. Lysosomal Pro-X carboxypeptidase. In: Barrett A, Woessner J, Rawlings N, editors. Handbook of proteolytic enzymes. 2nd ed. London: Elsevier Academic Press; 2004. p. 1936–8.
103. Grobe N, Weir NM, Leiva O, Ong FS, Bernstein KE, Schmaier AH, et al. Identification of prolyl carboxypeptidase as an alternative enzyme for processing of renal angiotensin II using mass spectrometry. Am J Physiol Cell Physiol. 2013;304(10):C945–53. https://doi.org/10.1152/ajpcell.00346.2012.
104. Xu S, Lind L, Zhao L, Lindahl B, Venge P. Plasma prolylcarboxypeptidase (angiotensinase C) is increased in obesity and diabetes mellitus and related to cardiovascular dysfunction. Clin Chem. 2012;58(7):1110–5. https://doi.org/10.1373/clinchem.2011.179291.
105. Adams GN, LaRusch GA, Stavrou E, Zhou Y, Nieman MT, Jacobs GH, et al. Murine prolylcarboxypeptidase depletion induces vascular dysfunction with hypertension and faster arterial thrombosis. Blood. 2011;117(14):3929–37. https://doi.org/10.1182/blood-2010-11-318527.
106. Maier C, Schadock I, Haber PK, Wysocki J, Ye M, Kanwar Y, et al. Prolylcarboxypeptidase deficiency is associated with increased blood pressure, glomerular lesions, and cardiac dysfunction independent of altered circulating and cardiac angiotensin II. J Mol Med (Berlin, Germany). 2017;95(5):473–86. https://doi.org/10.1007/s00109-017-1513-9.
107. Jeong JK, Szabo G, Raso GM, Meli R, Diano S. Deletion of prolyl carboxypeptidase attenuates the metabolic effects of diet-induced obesity. Am J Physiol Endocrinol Metab. 2012;302(12):E1502–10. https://doi.org/10.1152/ajpendo.00544.2011.
108. Bernstein KE, Ong FS, Blackwell WL, Shah KH, Giani JF, Gonzalez-Villalobos RA, et al. A modern understanding of the traditional and nontraditional biological functions of angiotensin-converting enzyme. Pharmacol Rev. 2013;65(1):1–46. https://doi.org/10.1124/pr.112.006809.
109. Chappell MC, Pirro NT, Sykes A, Ferrario CM. Metabolism of angiotensin-(1-7) by angiotensin-converting enzyme. Hypertension (Dallas, TX: 1979). 1998;31(1 Pt 2):362–7.
110. Yamada K, Iyer SN, Chappell MC, Ganten D, Ferrario CM. Converting enzyme determines plasma clearance of angiotensin-(1-7). Hypertension (Dallas, TX: 1979). 1998;32(3):496–502.
111. Marshall AC, Shaltout HA, Pirro NT, Rose JC, Diz DI, Chappell MC. Antenatal betamethasone exposure is associated with lower ANG-(1-7) and increased ACE in the CSF of adult sheep. Am J Physiol Regul Integr Comp Physiol. 2013;305(7):R679–88. https://doi.org/10.1152/ajpregu.00321.2013.
112. Marshall AC, Shaltout HA, Pirro NT, Rose JC, Diz DI, Chappell MC. Enhanced activity of an angiotensin-(1-7) neuropeptidase in glucocorticoid-induced fetal programming. Peptides. 2014;52:74–81. https://doi.org/10.1016/j.peptides.2013.12.006.
113. Marshall AC, Pirro NT, Rose JC, Diz DI, Chappell MC. Evidence for an angiotensin-(1-7) neuropeptidase expressed in the brain medulla and CSF of sheep. J Neurochem. 2014;130(2):313–23. https://doi.org/10.1111/jnc.12720.
114. Wang JO, Cooper MD. Aminopeptidase A. In: Barrett A, Woessner J, Rawlings N, editors. Handbook of proteolytic enzymes. 2nd ed. London: Elsevier Academic Press; 2004. p. 299–301.

115. Grobe N, Elased KM, Cool DR, Morris M. Mass spectrometry for the molecular imaging of angiotensin metabolism in kidney. Am J Physiol Endocrinol Metab. 2012;302(8):E1016–24. https://doi.org/10.1152/ajpendo.00515.2011.
116. Alghamri MS, Morris M, Meszaros JG, Elased KM, Grobe N. Novel role of aminopeptidase-A in angiotensin-(1-7) metabolism post myocardial infarction. Am J Physiol Heart Circ Physiol. 2014;306(7):H1032–40. https://doi.org/10.1152/ajpheart.00911.2013.
117. Zisman LS, Keller RS, Weaver B, Lin Q, Speth R, Bristow MR, et al. Increased angiotensin-(1-7)-forming activity in failing human heart ventricles: evidence for upregulation of the angiotensin-converting enzyme homologue ACE2. Circulation. 2003;108(14):1707–12. https://doi.org/10.1161/01.cir.0000094734.67990.99.
118. Velez JCQ, Arif E, Rodgers J, Hicks MP, Arthur JM, Nihalani D, et al. Deficiency of the angiotensinase aminopeptidase A increases susceptibility to glomerular injury. J Am Soc Nephrol: JASN. 2017;28(7):2119–32. https://doi.org/10.1681/asn.2016111166.
119. Mitsui T, Nomura S, Okada M, Ohno Y, Kobayashi H, Nakashima Y, et al. Hypertension and angiotensin II hypersensitivity in aminopeptidase A-deficient mice. Mol Med (Cambridge, Mass). 2003;9(1-2):57–62.
120. Chappell MC, Al Zayadneh EM. Angiotensin-(1-7) and the regulation of anti-fibrotic signaling pathways. J Cell Signal. 2017;2(1):1–3.
121. Inoue T, Umezawa A, Takenaka T, Suzuki H, Okada H. The contribution of epithelial-mesenchymal transition to renal fibrosis differs among kidney disease models. Kidney Int. 2015;87(1):233–8. https://doi.org/10.1038/ki.2014.235.
122. Tennakoon AH, Izawa T, Kuwamura M, Yamate J. Pathogenesis of type 2 epithelial to mesenchymal transition (EMT) in renal and hepatic fibrosis. J Clin Med. 2015;5(1) https://doi.org/10.3390/jcm5010004.
123. Alzayadneh EM, Chappell MC. Angiotensin-(1-7) abolishes AGE-induced cellular hypertrophy and myofibroblast transformation via inhibition of ERK1/2. Cell Signal. 2014;26(12):3027–35. https://doi.org/10.1016/j.cellsig.2014.09.010.
124. Gwathmey TM, Westwood BM, Pirro NT, Tang L, Rose JC, Diz DI, et al. Nuclear angiotensin-(1-7) receptor is functionally coupled to the formation of nitric oxide. Am J Physiol Renal Physiol. 2010;299(5):F983–90. https://doi.org/10.1152/ajprenal.00371.2010.
125. Santos RAS, Sampaio WO, Alzamora AC, Motta-Santos D, Alenina N, Bader M, et al. The ACE2/angiotensin-(1-7)/MAS Axis of the renin-angiotensin system: focus on angiotensin-(1-7). Physiol Rev. 2018;98(1):505–53. https://doi.org/10.1152/physrev.00023.2016.
126. Costa-Besada MA, Valenzuela R, Garrido-Gil P, Villar-Cheda B, Parga JA, Lanciego JL, Labandeira-Garcia JL. Paracrine and intracrine angiotensin 1-7/Mas receptor axis in the substantia nigra of rodents, monkeys, and humans. Mol Neurobiol. 2017; https://doi.org/10.1007/s12035-017-0805-y. [Epub ahead of print].
127. Abadir PM, Foster DB, Crow M, Cooke CA, Rucker JJ, Jain A, et al. Identification and characterization of a functional mitochondrial angiotensin system. Proc Natl Acad Sci U S A. 2011;108(36):14849–54. https://doi.org/10.1073/pnas.1101507108.
128. Valenzuela R, Costa-Besada MA, Iglesias-Gonzalez J, Perez-Costas E, Villar-Cheda B, Garrido-Gil P, et al. Mitochondrial angiotensin receptors in dopaminergic neurons. Role in cell protection and aging-related vulnerability to neurodegeneration. Cell Death Dis. 2016;7(10):e2427. https://doi.org/10.1038/cddis.2016.327.
129. Villar-Cheda B, Costa-Besada MA, Valenzuela R, Perez-Costas E, Melendez-Ferro M, Labandeira-Garcia JL. The intracellular angiotensin system buffers deleterious effects of the extracellular paracrine system. Cell Death Dis. 2017;8(9):e3044. https://doi.org/10.1038/cddis.2017.439.

Part I

Tools for Studying Angiotensin-(1-7)

Tools for Studying Angiotensin-(1-7)

Robson Augusto Souza Santos

Pharmacological Tools

The main pharmacological tools used to study the ACE2/ANG-(1–7)/MAS axis are MAS agonists that stimulate NO production/release [ANG-(1–7), AVE 0991, CGEN 861, CGEN 856, CGEN 856S, cyclic ANG-(1–7), NorLeu^3-A(1–7)] [1, 2, 9, 10, 21, 29, 32, 33, 36, 37, 39, 44, 46]. Two antagonists [D-Ala^7-ANG-(1–7) (A-779) and D-Pro^7-ANG-(1–7)] are also available. Interestingly, other MAS ligands, the non-peptides AR234960 (agonist) and AR244555 (inverse agonist) and neuropeptide FF (NPFF) appear to act through a different signaling pathway in which ANG-(1–7) is ineffective [20, 47]. This opens the possibility that some of the non-peptide or peptidic compounds act as biased MAS agonists [42].

Chronic administration of ANG-(1–7) has been tested in animals using an inclusion compound, hydroxypropyl-β-cyclodextrin/ANG-(1–7) [HPβCD-ANG-(1–7)]. This compound protects ANG-(1–7) from inactivation by digestive tract enzymes and permits its oral administration [16, 22]. Since cyclodextrins are metabolized by bacteria in the colon [12], only ANG-(1–7) enters the bloodstream. Therefore, the inclusion compound can be considered a long-lasting releasing system. This approach has led to the description of many beneficial cardiovascular and metabolic effects of ANG-(1–7), including antithrombogenesis [14], attenuation of cardiac remodeling induced by isoproterenol treatment [25], reduction of the lesion area, and attenuation of acute and chronic postinfarction cardiac dysfunction [26]. There have also been reports of antihypertensive effects [5] and improvements of erectile

R. A. S. Santos (✉)
National Institute of Science and Technology in Nano Biopharmaceutics –
Department of Physiology and Biophysics, Biological Sciences Institute,
Federal University of Minas Gerais, Belo Horizonte, MG, Brazil

Department of Physiology and Pharmacology, Biological Sciences Institute,
Federal University of Minas Gerais, Belo Horizonte, MG, Brazil
e-mail: santos@icb.ufmg.br

© Springer Nature Switzerland AG 2019
R. A. S. Santos (ed.), *Angiotensin (1-7)*,
https://doi.org/10.1007/978-3-030-22696-1_2

dysfunction [15], muscular dystrophy [3, 34], and type II diabetes mellitus [35]. Recently, a beneficial effect of this compound was reported in subjects undergoing a model of skeletal muscle lesion [4].

In addition to cyclodextrins, cyclic ANG-(1–7) is also undergoing preclinical testing (146, 281). It is more resistant to enzymatic hydrolysis than ANG-(1–7). Interestingly, the vasorelaxation produced by cyclic ANG-(1–7) in aortic rings from Sprague-Dawley rats is only partially blocked by the MAS antagonist A-779 (281), whereas this effect is completely blocked by the ANG-(1–7) analog D-Pro^7-ANG-(1–7), an ANG-(1–7)/alamandine antagonist (297). This pharmacological profile suggests that cyclic ANG-(1–7) could be a dual MAS/MrgD agonist sharing ANG-(1–7) and alamandine characteristics [21].

In addition to MAS agonists, recombinant human ACE2 (hACE2) is currently being used in recent studies [18]. Another interesting possibility is the use of ACE2 activators to alter the balance between the ACE/ANG II/ AT1R and the ACE2/ANG-(1–7)/MAS axes. The group of Prof. Raizada identified small-molecule ACE2 activators ([19]). The first compound they found was the 1-[(2-dimethylamino)ethylamino]-4-(hydroxymethyl)-7-[(4-methylphenyl) sulfonyloxy]-9H-xanthene-9-one, or XNT. Acute administration of XNT induced a dose-dependent hypotensive response in spontaneously hypertensive rats (SHR), while long-term treatment with this compound improved cardiac function and reversed the cardiac and renal fibrosis in these animals [19]. Oral administration of XNT was able to attenuate diabetes-induced heart dysfunction [28]. XNT and prevented the increase in right ventricular systolic pressure and hypertrophy in a monocrotaline-induced pulmonary hypertension model [11] and attenuated thrombus formation in SHR [17]. Numerous protective effects have been reported with another putative ACE2 activator, DIZE [6–8, 13, 15, 23, 24, 27, 30, 31, 40, 41, 43, 45]. It should be pointed out, however, that the effects of these small-molecule activators could be ACE2-independent (234). Moreover, ACE2 can cleave other substrates (140), a fact which should be taken into account when interpreting results obtained with methods involving the gain or loss of ACE2 functions.

More recently, oral delivery of ACE2 and ANG-(1–7) bioencapsulated in plant cells has been reported to attenuate pulmonary hypertension [38].

References

1. Abdallah WF, Louie SG, Zhang Y, Rodgers KE, Sivok E, S diZerega G, Humayun MS. NorLeu3A(1-7) accelerates clear corneal full thickness wound healing. Invest Ophthalmol Vis Sci. 2016;57:2187–94. https://doi.org/10.1167/iovs.15-18515.
2. Aboye T, Meeks CJ, Majumder S, Shekhtman A, Rodgers K, Camarero JA. Design of a MCoTI-based cyclotide with angiotensin (1-7)-like activity. Molecules. 2016;21:152. https://doi.org/10.3390/molecules21020152.
3. Acuna MJ, Pessina P, Olguin H, Cabrera D, Vio CP, Bader M, Munoz-Canoves P, Santos RA, Cabello-Verrugio C, Brandan E. Restoration of muscle strength in dystrophic muscle by

angiotensin-1-7 through inhibition of TGF-beta signalling. Hum Mol Genet. 2014;23:1237–49. https://doi.org/10.1093/hmg/ddt514.

4. Becker LK, Totou N, Moura S, Kangussu L, Millán RDS, Campagnole-Santos MJ, Coelho D, Motta-Santos D, Santos RAS. Eccentric overload muscle damage is attenuated by a novel angiotensin- (1-7) treatment. Int J Sports Med. 2018;39:743–8. https://doi.org/10.1055/a-0633-8892.

5. Bertagnolli M, Casali KR, De Sousa FB, Rigatto K, Becker L, Santos SH, Dias LD, Pinto G, Dartora DR, Schaan BD, Milan RD, Irigoyen MC, Santos RA. An orally active angiotensin-(1-7) inclusion compound and exercise training produce similar cardiovascular effects in spontaneously hypertensive rats. Peptides. 2014;51:65–73. https://doi.org/10.1016/j.peptides.2013.11.006.

6. Coutinho DC, Monnerat-Cahli G, Ferreira AJ, Medei E. Activation of angiotensin-converting enzyme 2 improves cardiac electrical changes in ventricular repolarization in streptozotocin-induced hyperglycaemic rats. Europace. 2014;16:1689–96. https://doi.org/10.1093/europace/euu070.

7. de Macedo SM, Guimarares TA, Andrade JM, Guimaraes AL, Batista de Paula AM, Ferreira AJ, Sousa Santos SH. Angiotensin converting enzyme 2 activator (DIZE) modulates metabolic profiles in mice, decreasing lipogenesis. Protein Pept Lett. 2015;22:332–40.

8. De Maria ML, Araújo LD, Fraga-Silva RA, Pereira LA, Ribeiro HJ, Menezes GB, Shenoy V, Raizada MK, Ferreira AJ. Anti-hypertensive effects of diminazene aceturate: an angiotensin-converting enzyme 2 activator in rats. Protein Pept Lett. 2016;23:9–16.

9. Ebermann L, Spillmann F, Sidiropoulos M, Escher F, Heringer-Walther S, Schultheiss HP, Tschöpe C, Walther T. The angiotensin-(1-7) receptor agonist AVE0991 is cardioprotective in diabetic rats. Eur J Pharmacol. 2008;590:276–80. https://doi.org/10.1016/j.ejphar.2008.05.024.

10. Ferreira AJ, Jacoby BA, Araújo CA, Macedo FA, Silva GA, Almeida AP, Caliari MV, Santos RA. The nonpeptide angiotensin-(1-7) receptor Mas agonist AVE-0991 attenuates heart failure induced by myocardial infarction. Am J Physiol Heart Circ Physiol. 2007;292:H1113–9. https://doi.org/10.1152/ajpheart.00828.2006.

11. Ferreira AJ, Shenoy V, Yamazato Y, Sriramula S, Francis J, Yuan L, Castellano RK, Ostrov DA, Oh SP, Katovich MJ, Raizada MK. Evidence for angiotensin-converting enzyme 2 as a therapeutic target for the prevention of pulmonary hypertension. Am J Respir Crit Care Med. 2009;179:1048–54. https://doi.org/10.1164/rccm.200811-1678OC.

12. Fetzner A, Böhm S, Schreder S, Schubert R. Degradation of raw or film-incorporated β-cyclodextrin by enzymes and colonic bacteria. Eur J Pharm Biopharm. 2004;58:91–7. https://doi.org/10.1016/j.ejpb.2004.02.001.

13. Foureaux G, Nogueira JC, Nogueira BS, Fulgêncio GO, Menezes GB, Fernandes SO, Cardoso VN, Fernandes RS, Oliveira GP, Franca JR, Faraco AA, Raizada MK, Ferreira AJ. Antiglaucomatous effects of the activation of intrinsic angiotensin-converting enzyme 2. Invest Ophthalmol Vis Sci. 2013;54:4296–306. https://doi.org/10.1167/iovs.12-11427.

14. Fraga-Silva RA, Costa-Fraga FP, De Sousa FB, Alenina N, Bader M, Sinisterra RD, Santos RA. An orally active formulation of angiotensin-(1-7) produces an antithrombotic effect. Clin Sao Paulo. 2011;66:837–41.

15. Fraga-Silva RA, Costa-Fraga FP, Montecucco F, Sturny M, Faye Y, Mach F, Pelli G, Shenoy V, da Silva RF, Raizada MK, Santos RA, Stergiopulos N. Diminazene protects corpus cavernosum against hypercholesterolemia-induced injury. J Sex Med. 2015;12:289–302. https://doi.org/10.1111/jsm.12757.

16. Fraga-Silva RA, Da Silva DG, Montecucco F, Mach F, Stergiopulos N, da Silva RF, Santos RA. The angiotensin-converting enzyme 2/angiotensin-(1-7)/Mas receptor axis: a potential target for treating thrombotic diseases. Thromb Haemost. 2012;108:1089–96. https://doi.org/10.1160/TH12-06-0396.

17. Fraga-Silva RA, Sorg BS, Wankhede M, Dedeugd C, Jun JY, Baker MB, Li Y, Castellano RK, Katovich MJ, Raizada MK, Ferreira AJ. ACE2 activation promotes antithrombotic activity. Mol Med. 2010;16:210–5. https://doi.org/10.2119/molmed.2009.00160.

18. Haschke M, Schuster M, Poglitsch M, Loibner H, Salzberg M, Bruggisser M, Penninger J, Krähenbühl S. Pharmacokinetics and pharmacodynamics of recombinant human angiotensin-converting enzyme 2 in healthy human subjects. Clin Pharmacokinet. 2013;52:783–92. https://doi.org/10.1007/s40262-013-0072-7.

19. Hernández Prada JA, Ferreira AJ, Katovich MJ, Shenoy V, Qi Y, Santos RAS, Castellano RK, Lampkins AJ, Gubala V, Ostrov DA, Raizada MK. 2008. Structure-based identification of small-molecule angiotensin-converting enzyme 2 activators as novel antihypertensive agents. Hypertens. Dallas Tex. 1979;51:1312–7. https://doi.org/10.1161/HYPERTENSIONAHA.107.108944.

20. Karnik SS, Unal H, Kemp JR, Tirupula KC, Eguchi S, Vanderheyden PML, Thomas WG. International Union of Basic and Clinical Pharmacology. XCIX. Angiotensin receptors: interpreters of pathophysiological angiotensinergic stimuli. Pharmacol Rev. 2015;67:754–819. https://doi.org/10.1124/pr.114.010454.

21. Kluskens LD, Nelemans SA, Rink R, de Vries L, Meter-Arkema A, Wang Y, Walther T, Kuipers A, Moll GN, Haas M. Angiotensin-(1-7) with thioether bridge: an angiotensin-converting enzyme-resistant, potent angiotensin-(1-7) analog. J Pharmacol Exp Ther. 2009;328:849–54. https://doi.org/10.1124/jpet.108.146431.

22. Lula I, Denadai AL, Resende JM, de Sousa FB, de Lima GF, Pilo-Veloso D, Heine T, Duarte HA, Santos RA, Sinisterra RD. Study of angiotensin-(1-7) vasoactive peptide and its beta-cyclodextrin inclusion complexes: complete sequence-specific NMR assignments and structural studies. Peptides. 2007;28:2199–210. https://doi.org/10.1016/j.peptides.2007.08.011.

23. Macedo LM, Souza ÁP, De Maria ML, Borges CL, Soares CM, Pedrino GR, Colugnati DB, Santos RA, Mendes EP, Ferreira AJ, Castro CH. Cardioprotective effects of diminazene aceturate in pressure-overloaded rat hearts. Life Sci. 2016;155:63–9. https://doi.org/10.1016/j.lfs.2016.04.036.

24. Malek M, Nematbakhsh M. The preventive effects of diminazene aceturate in renal ischemia/reperfusion injury in male and female rats. Adv Prev Med. 2014;2014(740647) https://doi.org/10.1155/2014/740647.

25. Marques FD, Ferreira AJ, Sinisterra RD, Jacoby BA, Sousa FB, Caliari MV, Silva GA, Melo MB, Nadu AP, Souza LE, Irigoyen MC, Almeida AP, Santos RA. An oral formulation of angiotensin-(1-7) produces cardioprotective effects in infarcted and isoproterenol-treated rats. Hypertension. 2011;57:477–83. https://doi.org/10.1161/HYPERTENSIONAHA.110.167346.

26. Marques FD, Melo MB, Souza LE, Irigoyen MC, Sinisterra RD, de Sousa FB, Savergnini SQ, Braga VB, Ferreira AJ, Santos RA. Beneficial effects of long-term administration of an oral formulation of angiotensin-(1-7) in infarcted rats. Int J Hypertens. 2012;2012(795452) https://doi.org/10.1155/2012/795452.

27. Mecca AP, Regenhardt RW, O'Connor TE, Joseph JP, Raizada MK, Katovich MJ, Sumners C. Cerebroprotection by angiotensin-(1-7) in endothelin-1-induced ischaemic stroke. Exp Physiol. 2011;96:1084–96. https://doi.org/10.1113/expphysiol.2011.058578.

28. Murça TM, Moraes PL, Capuruço CAB, Santos SHS, Melo MB, Santos RAS, Shenoy V, Katovich MJ, Raizada MK, Ferreira AJ. Oral administration of an angiotensin-converting enzyme 2 activator ameliorates diabetes-induced cardiac dysfunction. Regul Pept. 2012;177:107–15. https://doi.org/10.1016/j.regpep.2012.05.093.

29. Pinheiro SV, Simões e Silva AC, Sampaio WO, de Paula RD, Mendes EP, Bontempo ED, Pesquero JB, Walther T, Alenina N, Bader M, Bleich M, Santos RA. Nonpeptide AVE 0991 is an angiotensin-(1-7) receptor Mas agonist in the mouse kidney. Hypertension. 2004;44:490–6. https://doi.org/10.1161/01.HYP.0000141438.64887.42.

30. Qiu Y, Shil PK, Zhu P, Yang H, Verma A, Lei B, Li Q. Angiotensin-converting enzyme 2 (ACE2) activator diminazene aceturate ameliorates endotoxin-induced uveitis in mice. Invest Ophthalmol Vis Sci. 2014;55:3809–18. https://doi.org/10.1167/iovs.14-13883.

31. Rigatto K, Casali KR, Shenoy V, Katovich MJ, Raizada MK. Diminazene aceturate improves autonomic modulation in pulmonary hypertension. Eur J Pharmacol. 2013;713:89–93. https://doi.org/10.1016/j.ejphar.2013.04.017.

32. Rodgers KE, Espinoza T, Felix J, Roda N, Maldonado S, diZerega G. Acceleration of healing, reduction of fibrotic scar, and normalization of tissue architecture by an angiotensin analogue, NorLeu3-A(1-7). Plast Reconstr Surg. 2003;111:1195–206. https://doi.org/10.1097/01.PRS.0000047403.23105.66.

33. Rodrigues-Machado MG, Magalhães GS, Cardoso JA, Kangussu LM, Murari A, Caliari MV, Oliveira ML, Cara DC, Noviello ML, Marques FD, Pereira JM, Lautner RQ, Santos RA, Campagnole-Santos MJ. AVE 0991, a non-peptide mimic of angiotensin-(1-7) effects, attenuates pulmonary remodelling in a model of chronic asthma. Br J Pharmacol. 2013;170:835–46. https://doi.org/10.1111/bph.12318.

34. Sabharwal R, Cicha MZ, Sinisterra RD, De Sousa FB, Santos RA, Chapleau MW. Chronic oral administration of Ang-(1-7) improves skeletal muscle, autonomic and locomotor phenotypes in muscular dystrophy. Clin Sci. 2014;127:101–9. https://doi.org/10.1042/CS20130602.

35. Santos SH, Giani JF, Burghi V, Miquet JG, Qadri F, Braga JF, Todiras M, Kotnik K, Alenina N, Dominici FP, Santos RA, Bader M. Oral administration of angiotensin-(1-7) ameliorates type 2 diabetes in rats. J Mol Med Berl. 2014;92:255–65.. https://doi.org/10.1007/s00109-013-1087-0

36. Savergnini SQ, Beiman M, Lautner RQ, de Paula-Carvalho V, Allahdadi K, Pessoa DC, Costa-Fraga FP, Fraga-Silva RA, Cojocaru G, Cohen Y, Bader M, de Almeida AP, Rotman G, Santos RA. Vascular relaxation, antihypertensive effect, and cardioprotection of a novel peptide agonist of the MAS receptor. Hypertension. 2010;56:112–20. https://doi.org/10.1161/HYPERTENSIONAHA.110.152942.

37. Savergnini SQ, Ianzer D, Carvalho MBL, Ferreira AJ, Silva GAB, Marques FD, Peluso AAB, Beiman M, Cojocaru G, Cohen Y, Almeida AP, Rotman G, Santos RAS. The novel Mas agonist, CGEN-856S, attenuates isoproterenol-induced cardiac remodeling and myocardial infarction injury in rats. PLoS One. 2013;8:e57757. https://doi.org/10.1371/journal.pone.0057757.

38. Shenoy V, Ferreira AJ, Qi Y, Fraga-Silva RA, Díez-Freire C, Dooies A, Jun JY, Sriramula S, Mariappan N, Pourang D, Venugopal CS, Francis J, Reudelhuber T, Santos RA, Patel JM, Raizada MK, Katovich MJ. The angiotensin-converting enzyme 2/angiogenesis-(1-7)/Mas axis confers cardiopulmonary protection against lung fibrosis and pulmonary hypertension. Am J Respir Crit Care Med. 2010;182:1065–72. https://doi.org/10.1164/rccm.200912-1840OC.

39. Singh K, Sharma K, Singh M, Sharma P. Possible mechanism of the cardio-renal protective effects of AVE-0991, a non-peptide Mas-receptor agonist, in diabetic rats. J Renin-Angiotensin-Aldosterone Syst. 2012;13:334–40. https://doi.org/10.1177/1470320311435534.

40. Singh N, Joshi S, Guo L, Baker MB, Li Y, Castellano RK, Raizada MK, Jarajapu YPR. ACE2/Ang-(1-7)/Mas axis stimulates vascular repair-relevant functions of CD34+ cells. Am J Physiol Heart Circ Physiol. 2015;309:H1697–707. https://doi.org/10.1152/ajpheart.00854.2014.

41. Souza LKM, Nicolau LAD, Sousa NA, Araújo TSL, Sousa FBM, Costa DS, Souza FM, Pacífico DM, Martins CS, Silva RO, Souza MHLP, Cerqueira GS, Medeiros JVR. Diminazene aceturate, an angiotensin-converting enzyme II activator, prevents gastric mucosal damage in mice: role of the angiotensin-(1-7)/Mas receptor axis. Biochem Pharmacol. 2016;112:50–9. https://doi.org/10.1016/j.bcp.2016.05.010.

42. Strachan RT, Sun J, Rominger DH, Violin JD, Ahn S, Thomsen ARB, Zhu X, Kleist A, Costa T, Lefkowitz RJ. Divergent transducer-specific molecular efficacies generate biased agonism at a G protein-coupled receptor (GPCR). J Biol Chem. 2014;289:14211–24. https://doi.org/10.1074/jbc.M114.548131.

43. Tao L, Qiu Y, Fu X, Lin R, Lei C, Wang J, Lei B. Angiotensin-converting enzyme 2 activator diminazene aceturate prevents lipopolysaccharide-induced inflammation by inhibiting MAPK and NF-κB pathways in human retinal pigment epithelium. J Neuroinflammation. 2016;13(35) https://doi.org/10.1186/s12974-016-0489-7.

44. Toton-Zuranska J, Gajda M, Pyka-Fosciak G, Kus K, Pawlowska M, Niepsuj A, Wolkow P, Olszanecki R, Jawien J, Korbut R. AVE 0991-angiotensin-(1-7) receptor agonist, inhibits atherogenesis in apoE-knockout mice. J Physiol Pharmacol. 2010;61:181–3.
45. Velkoska E, Patel SK, Burrell LM. Angiotensin converting enzyme 2 and diminazene: role in cardiovascular and blood pressure regulation. Curr Opin Nephrol Hypertens. 2016;25:384–95. https://doi.org/10.1097/MNH.0000000000000254.
46. Wiemer G, Dobrucki LW, Louka FR, Malinski T, Heitsch H. AVE 0991, a nonpeptide mimic of the effects of angiotensin-(1-7) on the endothelium. Hypertension. 2002;40:847–52.
47. Zhang T, Li Z, Dang H, Chen R, Liaw C, Tran T-A, Boatman PD, Connolly DT, Adams JW. Inhibition of mas G-protein signaling improves coronary blood flow, reduces myocardial infarct size, and provides long-term cardioprotection. Am J Physiol Heart Circ Physiol. 2012;302:H299–311. https://doi.org/10.1152/ajpheart.00723.2011.

Genetic Models

Natalia Alenina and Michael Bader

Introduction

Transgenic and knockout animal models are the most effective tools to study cardiovascular hormone systems, since they reveal effects of changes in single components of these systems on the whole physiology. In particular, studies on the renin-angiotensin systems (RAS) have profited from this technology in recent decades [3, 5, 58]. Therefore, it was warranted to establish such models also for the novel RAS consisting of ACE2, Ang-(1-7), and Mas (Table 1). Despite that these three components comprise a common axis, distinct phenotypes of models with one of the components altered are expected since each of the three components has distinct additional functions independent from the two other molecules. ACE2, in particular, is a protein with several functions, a carboxypeptidase metabolizing a multitude of peptides, such as AngII and apelins, thereby either activating or inactivating them [104], a protein with a collectrin domain, which is involved in amino acid uptake in the gut [34, 95], and the receptor for the severe acute respiratory syndrome (SARS) coronavirus [45]. Moreover, also Ang-(1-7) may interact with other receptors than Mas and Mas may have other ligands or exert ligand-independent effects [4, 86].

N. Alenina
Max-Delbrück-Center for Molecular Medicine (MDC), Berlin, Germany

DZHK (German Center for Cardiovascular Research), Partner Site Berlin, Berlin, Germany

M. Bader (✉)
Max-Delbrück-Center for Molecular Medicine (MDC), Berlin, Germany

DZHK (German Center for Cardiovascular Research), Partner Site Berlin, Berlin, Germany

Berlin Institute of Health (BIH), Berlin, Germany

Charité - University Medicine, Berlin, Germany

Institute for Biology, University of Lübeck, Lübeck, Germany
e-mail: mbader@mdc-berlin.de

© Springer Nature Switzerland AG 2019
P. A. S. Santos (ed.), *Angiotensin (1-7)*,
https://doi.org/10.1007/978-3-030-22696-1_3

Table 1 Independently generated transgenic and knockout mouse and rat models for the ACE2/Ang-(1-7)/Mas axis of the RAS

Gene	Method	Species	Promoter	Expressing tissue	Reference
ACE2	ESC-Knockout	Mouse	–	–	[17]
ACE2	ESC-Knockout	Mouse	–	–	[31]
ACE2	ESC-Knockout	Mouse	–	–	[125]
ACE2	TALEN-Knockout	Mouse	–	–	[47]
ACE2	CRISPR-Knockout	Mouse	–	–	[47, 129]
ACE2 S680D	CRISPR-Knockin	Mouse	–	–	[47]
ACE2	TALEN-Knockout	Rat	–	–	[130]
ACE2 (mouse)	Transgene, stopflox	Mouse	Rosa26	Ubiquitous, inducible	[75, 107]
ACE2 (human)	Transgene	Mouse	ACE2	Ubiquitous	[126]
ACE2 (human)	Transgene	Mouse	CMV	Ubiquitous	[102]
ACE2 (human)	Transgene	Mouse	Cytokeratin 18	Airways	[53]
ACE2 (human)	Transgene	Mouse	Cardiac α-MHC	Heart	[21]
ACE2 (human)	Transgene	Mouse	Nephrin	Podocytes	[61]
ACE2 (human)	Transgene	Mouse	Synapsin	Neurons	[26]
ACE2 (human)	Transgene, floxed	Mouse	Synapsin	Neurons	[117]
ACE2 (human)	Transgene	Rat	SM-MHC	Smooth muscle	[79]
Mas	ESC-Knockout	Mouse	–	–	[105]
Mas	ESC-Knockout	Mouse	–	–	[20, 113]
Mas	ZFN-Knockout	Rat	–	–	https://rgd.mcw.edu/rgdweb/report/gene/main.html?id=3049
Mas (rat)	Transgene	Mouse	Opsin	Retina	[122]
Ang-(1-7)	Transgene	Mouse	Cardiac α-MHC	Heart	[54]
Ang-(1-7)	Transgene	Rat	Cardiac α-MHC	Heart	[27]
Ang-(1-7)	Transgene	Rat	CMV	Testis	[28]

CMV cytomegalovirus, *ESC* embryonic stem cell, *MHC* myosin-heavy chain, *SM* smooth muscle, *TALEN* transcription activator-like effector nuclease, *ZFN* zinc-finger nuclease

ACE2 Models

ACE2 Knockout Mice

Since the ACE2 gene is localized on the X-chromosome, male mice with ACE2 gene deletion (ACE2$^{-/y}$) are already deficient in the enzyme in the hemizygous state. Based on the pleiotropic actions of this protein, mice lacking ACE2 are expected to exhibit increased levels of AngII, decreased levels of Ang-(1-7) and tryptophan, as well as alterations in other peptide levels, which all may contribute to observed phenotypes. ACE2$^{-/y}$ mice and also heterozygous female ACE2$^{+/-}$ mice were more susceptible to cardiac injury induced by pressure overload, AngII infusion, or diabetes [69, 109, 115, 125] and ACE2$^{-/y}$ mice developed cardiac abnormalities at older age [17] probably due to an increased level of Ang II [68]. However, the spontaneous appearance of cardiac alterations could not be confirmed by another group and therefore remains controversial [31, 32, 125]. However, obesity-induced epicardial inflammation was worsened and caused cardiac dysfunction in ACE2$^{-/y}$ mice [70]. Furthermore, in heart and skeletal muscle, ACE2 was involved in training-induced physiological hypertrophy [59].

There were also inconsistencies in the reports about hypertension in ACE2$^{-/y}$ mice, but it is now accepted that this phenotype is depending on the strain of mice appearing in C57BL/6 and FVB/N but not in 129 mice [32, 38, 77, 101]. ACE2-deficient mice on C57BL/6 background even developed a pre-ecclampsia-like syndrome when pregnant [8] and placental hypoxia and uterine artery dysfuncion caused fetal growth restriction in these animals [124]. In ACE2$^{-/-}$ female mice, estrogen cannot inhibit obesity-induced hypertension in contrast to wild-type controls [114]. We and others have described AngII-dependent endothelial dysfunction in ACE2-deficient mice [49, 77], which probably mediated the prohypertensive phenotype. However, an increased sympathetic outflow may have also contributed [119]. On the other hand, ACE2 also degrades the vasodilator apelin peptides which consequently accumulate in ACE2$^{-/y}$ mice and counteract the effects on the RAS [110]. Nevertheless, there were several other vascular effects of genetic ACE2-deletion such as a worsening of atherosclerosis and aortic aneurysm in apolipoprotein E (ApoE)- and low-density lipoprotein receptor-deficient mice [63, 81, 99, 100] and an increased neointima formation after vascular injury [81], to which the endothelial dysfunction was a major contributor.

In double knockout mice for ACE2 and ApoE, also the renal injury induced by atherosclerosis was aggravated [38]. Moreover, ACE2$^{-/y}$ mice spontaneously developed glomerulosclerosis in older age [67] and were more susceptible to renal ischemia/reperfusion injury due to increased cytokine expression, inflammation, and oxidative stress [23]. Accordingly, genetic ACE2 deficiency led to accelerated nephropathy in streptozotocin (STZ)-induced and Akita diabetic mice [91, 115]. Furthermore, knockout mice for ACE2 infused with AngII showed enhanced collagen I deposition in renal glomeruli and expression of genes related to fibrosis, such as smooth muscle actin, transforming growth factor β (TGF-β), and procollagen I, probably through activation of ERK1/2 and enhancement of protein kinase C levels [133]. ACE2-deficient mice also showed a worse outcome in shock-induced kidney injury [127], chronic hepatic injury [66], liver steatosis [12, 64], and cerulein-induced pancreatitis [48].

The lung is a major site of ACE2 expression. Accordingly, ACE2$^{-/y}$ mice exhibited an aggravated pathogenesis of lung injury induced by cigarette smoke, air pollution, bleomycin, influenza virus or respiratory syncytial virus [30, 36, 46, 80, 134], of pulmonary hypertension [129], and of acute respiratory distress syndrome [37]. In most of these injury models, the increased oxidative stress observed in kidneys [116], livers [12, 64], and vessels [71] of ACE2$^{-/y}$ mice contributed to the exacerbation.

ACE2 in the gut with its collectrin domain is part of the amino acid uptake system and, therefore, mice lacking this protein showed reduced tryptophan in the blood, an altered gut microflora, and intestinal inflammation [34, 95]. These results were recently confirmed in a novel ACE2-deficient mouse model on an outbred genetic background generated by transcription-activator-like effector nucleases (TALEN) [47]. Whether the collectrin-domain-dependent effects contributed to the metabolic alterations shown in ACE2-deficient mice, such as insulin resistance and impaired glucose homeostasis [12, 63] in particular under a high-fat diet [15, 50, 90, 92, 123] needs still to elucidated [7]. However, in the liver, the carboxypeptidase function of ACE2 was more relevant for these metabolic effects since they could be ameliorated by Ang-(1-7) infusion [12].

ACE2 in the brain also influences behavior since ACE2-deficient mice showed impaired performance in cognition and memory tests [111].

ACE2 S680D Knockin Mouse

Recently, it was discovered that serine 680 of mouse ACE2 is phosphorylated by AMP kinase, leading to increased stability of the protein. When this phosphorylation was mimicked (S680D) in knockin mice by CRISPR/Cas9 technology, the resulting animals were partially resistant to a pulmonary hypertension model [129].

ACE2 Knockout Rats

ACE2 knockout rats have recently been established using TALEN technology [130]. These animals exhibited cardiac hypertrophy and impaired heart function; however, their blood pressure was not reported. Therefore, it remains unclear whether the cardiac effects are direct or caused by hypertension.

Inducible Mouse ACE2 Overexpression in Mouse

In order to allow tissue-specific activation of ACE2 expression, the mouse ACE2 coding region was knocked into the Rosa26 locus of mice with a Stop-lox cassette in front of it, which inhibits transcription. This cassette can be removed by Cre-recombinase expression and then ACE2 gets highly expressed in the cells expressing Cre-recombinase. When Cre-recombinase was expressed in the germline, ubiquitously ACE2 overexpressing mice resulted, which were protected from post-infarction cardiac dysfunction [75] and exhibited less anxiety-related behavior [107]. The same behavioral effects were also observed when the gene was only activated in CRH (corticotropin-releasing hormone) expressing cells using the corresponding Cre-recombinase-expressing mouse for breeding with the ACE2/Rosa26 animals [108].

Human ACE2 Overexpression in Mouse

Human ACE2 is hijacked by the SARS virus as a receptor to enter cells. In order to create a model for this disease, mice were "humanized" by several groups by inserting human ACE2 transgenes in their genome either using the ACE2 promoter itself [126], the ubiquitously active cytomegalovirus (CMV) promoter [102, 128], or the airway-specific cytokeratin 18 promoter [53, 62]. These animals were also suitable for studies on the role of ACE2 in other diseases and therefore the first model was tested in a kidney injury model and showed a protected phenotype [127]. Moreover, it was shown to be protected from AngII-induced hypertension and myocardial fibrosis [109].

Human ACE2 Overexpression in Mouse Heart

When human ACE2 was overexpressed in hearts of transgenic mice, surprisingly ventricular tachycardia and sudden death was observed accompanied by a dysregulation of connexin expression [21]. Apelin, which is also a substrate for ACE2 [104], may in this case be lacking and this deficiency may have caused the cardiac dysfunction [41].

Human ACE2 Overexpression in Mouse Podocytes

When human ACE2 was overexpressed in kidneys of transgenic mice, particularly in podocytes using the nephrin promoter, the animals became protected from diabetes-induced renal injury [61]. The authors provided evidence that the relative amounts of AngII and Ang-(1-7) are critical for the phenotype by increased AngII upregulating TGF-β.

Human ACE2 Overexpression in Mouse Brain

When human ACE2 was overexpressed in brains of transgenic mice using the synapsin promoter, a protective phenotype is observed for several cardiovascular diseases. This included hypertension induced by peripheral infusions of AngII [26] and by desoxycorticosterone acetate (DOCA)/salt treatment [118], cardiac hypertrophy elicited by AngII [25], coronary ligation-induced chronic heart failure [120], and stroke triggered by middle cerebral artery occlusion [14, 132]. In another model, the ACE2 transgene was flanked by loxP sites and it could therefore be specifically deleted in distinct brain regions by the local injection of Cre-recombinase-expressing adeno-associated viruses to assess the relevance of these areas for the blood pressure increase after DOCA/salt treatment. Such experiments revealed the paraventricular nucleus of the hypothalamus and the subfornical organ as important but not exclusive contributors to hypertension development [117]. The shift in the balance between Ang-(1-7) and AngII in brain regions important for cardiovascular control modulated local NO and ROS production as well as cyclooxygenase-mediated neuroinflammation [97] and likely caused the beneficial effects of ACE2 in the brain. Accordingly, the AngII-dependent deleterious effects on brain tissues observed in double transgenic mice expressing human angiotensinogen and human renin were mitigated in triple transgenic animals additionally expressing human ACE2 [14, 131].

Human ACE2 Overexpression in Rat Vascular Smooth Muscle

When we overexpressed human ACE2 in vascular smooth muscle of transgenic rats of the spontaneously hypertensive stroke-prone (SHRSP) strain using the smooth muscle myosin heavy chain promoter, blood pressure was significantly reduced [79]. This confirmed a study postulating that reduced ACE2 is an important genetic determinant for hypertension in this strain [17]. Reduced blood pressure was accompanied by decreased oxidative stress and improved endothelial function [79].

Mas Models

Mas Knockout Mice

When we generated Mas-deficient (Mas$^{-/-}$) mice, it was not yet known that it is the receptor for Ang-(1-7) [105]. Therefore, phenotyping concentrated on the brain as major Mas-expressing organ. Male (but not female [106]) Mas-deficient mice showed increased anxiety-like behavior and long-term potentiation (LTP) in the hippocampus [105]. Surprisingly, despite the improved LTP, object recognition memory was impaired [43]. However, Mas$^{-/-}$ mice showed delayed extinction of fear memory [42] and were protected from cognitive impairments induced by ischemia but only in the presence of the AngII AT2 receptor [35] supporting a role of the dimerization of both receptors in brain function [44].

After our discovery that Mas is the receptor for Ang-(1-7) [85], we performed comprehensive cardiovascular phenotyping. Mas-deficient mice on the C57BL/6 background exhibited spontaneous cardiac fibrosis and dysfunction [13, 72, 83, 113]. Increased oxidative stress and endothelial dysfunction were observed on all genetic backgrounds studied (C57BL/6 and FVB/N) [33, 78, 121], but only resulted in hypertension in FVB/N mice. Possibly, an autonomic dysbalance in Mas$^{-/-}$ mice also contributed to the increased blood pressure [76]. Moreover, regional blood flow and local vascular resistance were differentially altered in different tissues of Mas$^{-/-}$ mice [10], which may also be the cause for the increased vascular resistance in the *corpus cavernosum* and the resulting erectile dysfunction observed in these mice [29].

Mas$^{-/-}$ mice showed an impaired renal function with increased urinary volume and proteinuria [74]. However, Esteban and coworkers found that Mas knockout mice presented an attenuation of renal damage in the unilateral ureteral obstruction and in the renal ischemia/reperfusion model [22]. The authors reported that Ang-(1-7) infusion led to NF-κB activation and inflammation via Mas. In contrast, Kim et al. showed protective effects of Ang-(1-7) infusion in the same model [40] and no aggravation of renal injury produced by kidney ischemia/reperfusion was observed in Mas$^{-/-}$ mice [6]. Moreover, Mas$^{-/-}$ mice were protected from adriamycin-induced renal injury, again confirming the protective actions of the ACE2/Ang-(1-7)/Mas axis of the RAS in the kidney [94]. The discrepancy between the studies remained unresolved, but anti-inflammatory and protective actions of Mas have repeatedly been described also in other organs: Ang-(1-7) protected from intracranial aneurysm only in wild-type but not in Mas$^{-/-}$ mice [73]. Mas deficiency promoted atherosclerosis and autoimmune encephalitis by affecting macrophage polarization and migration [33] and by increasing vascular intima proliferation [2]. The effects on macrophages and other leukocytes were probably also the reason for the higher susceptibility of Mas$^{-/-}$ mice in an endotoxic shock model [65, 96]. Moreover, Mas$^{-/-}$ mice presented aggravated inflammatory pain [16] and allergic pulmonary inflammation [51].

Mas$^{-/-}$ mice are also a model for metabolic syndrome since they developed metabolic abnormalities, such as type 2 diabetes mellitus and dyslipidemia [88], besides their hypertensive phenotype. On the mechanistic level, this was accompanied by decreased PPARγ expression in fat tissue [52] and a change in the relative amounts of α and β cells in pancreatic islets [24]. Ang-(1-7), mainly via Mas, stimulated insulin secretion from β cells [82]. Furthermore, Mas$^{-/-}$ mice developed liver steatosis when bred with ApoE-deficient mice [93] and Mas$^{-/-}$ female mice were more susceptible to obesity-induced hypertension [113]. Ang-(1-7) and Mas were involved in vascular repair, which is deficient in diabetes, and hindlimb ischemia-induced progenitor cell mobilization was absent in Mas$^{-/-}$ mice [103].

In skeletal muscle, Ang-(1-7) and Mas protected from atrophy since Mas$^{-/-}$ mice were more susceptible to a Duchenne muscular dystrophy model (mdx) [1] and to immobilization-induced atrophy [56].

Mas Knockout Rats

Mas knockout rats have been established using Zinc-finger nuclease technology but their phenotype is only partially reported on the Rat Genome Database website (https://rgd.mcw.edu/rgdweb/report/gene/main.html?id=3049).

Mas Overexpression in Retina

Transgenic mice overexpressing Mas in the retina under the control of the opsin promoter developed degeneration of photoreceptors [122]. This surprising phenotype may have been caused by the ligand-independent constitutive activity of Mas [4] causing proliferative effects in cells when the gene is overexpressed.

Ang-(1-7) Models

Transgenic Rats Overexpressing Ang-(1-7)

The group of Timothy Reudelhuber invented a method to express and secrete peptides from an artificial protein without the need of specific proteases in transgenic animals [54, 55]. Using this method, Ang-(1-7) was overexpressed in transgenic rats (TGR(A1-7)3292) using the CMV promoter [28]. These animals mainly expressed the peptide in the testis, which nevertheless significantly increased plasma levels of Ang-(1-7). As a consequence, total peripheral resistance was decreased together with increases in the blood flow to several organs. Nonetheless, the animals remained normotensive, probably since they exhibited an improved pumping function of the heart [11]. These cardiac effects also protected the heart from pressure and ischemia-induced damage [84] as well as from DOCA-induced diastolic dysfunction [19]. A part of these effects may be due to alterations in autonomic regulation observed in these rats [18]. The increased levels of plasma Ang-(1-7) exerted antinatriuretic actions in the kidney resulting in reduced urinary flow and increased urinary osmolality [28]. Furthermore, TGR(A1-7)3292 rats exhibited metabolic improvements such as decreased plasma lipid levels, improved glucose tolerance, less fat tissue, decreased lipogenesis, and less cafeteria-diet-induced obesity [9, 57, 87, 89]. Moreover, these rats presented a reduction in anxiety-like behavior [39] and in the response to stress [60].

Transgenic Mice and Rats Overexpressing Ang-(1-7) in the Heart

We also generated transgenic mice and rats expressing the Ang-(1-7) release protein specifically in the heart using the α cardiac myosin heavy chain promoter. Both lines showed a slightly improved heart function at baseline and were protected from cardiac hypertrophy [27, 54], but, interestingly, not from myocardial infarction [112].

Conclusions

As summarized in this chapter, several genetically altered rat and mouse models have been generated changing the expression of components of the ACE2/Ang-(1-7)/Mas axis of the RAS (Table 1). With the help of these models, physiological and pathophysiological functions of this axis have been elucidated. Nevertheless, novel models are warranted with cell-type-specific deficiency of ACE2 or Mas to further delineate their tissue-specific effects. The already collected findings are the basis for the development of novel therapeutic strategies for cardiovascular and metabolic diseases by targeting ACE2 or Mas [86, 98].

References

1. Acuna MJ, Pessina P, Olguin H, Cabrera D, Vio CP, Bader M, Munoz-Canoves P, Santos RA, Cabello-Verrugio C, Brandan E. Restoration of muscle strength in dystrophic muscle by angiotensin-1-7 through inhibition of TGF-beta signalling. Hum Mol Genet. 2014;23:1237–49.
2. Alsaadon H, Kruzliak P, Smardencas A, Hayes A, Bader M, Angus P, Herath C, Zulli A. Increased aortic intimal proliferation due to MasR deletion in vitro. Int J Exp Pathol. 2015;96:183.
3. Bader M. Rat models of cardiovascular diseases. Methods Mol Biol. 2010;597:403–14.
4. Bader M, Alenina N, Andrade-Navarro MA, and Santos RA. Mas and its related G protein-coupled receptors. Pharmacol Rev. 2014;66:1080–105.
5. Bader M, Bohnemeier H, Zollmann FS, Lockley-Jones OE, Ganten D. Transgenic animals in cardiovascular disease research. Exp Physiol. 2000;85:713–31.
6. Barroso LC, Silveira KD, Lima CX, Borges V, Bader M, Rachid M, Santos RA, Souza DG. Simoes e Silva AC, and Teixeira MM. Renoprotective effects of AVE0991, a nonpeptide Mas receptor agonist, in experimental acute renal injury. *Int.* J Hypertens. 2012;2012:808726.
7. Bernardi S, Tikellis C, Candido R, Tsorotes D, Pickering RJ, Bossi F, Carretta R, Fabris B, Cooper ME, Thomas MC. ACE2 deficiency shifts energy metabolism towards glucose utilization. Metabolism. 2015;64:406–15.
8. Bharadwaj MS, Strawn WB, Groban L, Yamaleyeva LM, Chappell MC, Horta C, Atkins K, Firmes L, Gurley SB, Brosnihan KB. Angiotensin-converting enzyme 2 deficiency is associated with impaired gestational weight gain and fetal growth restriction. Hypertension. 2011;58:852–8.
9. Bilman V, Mares-Guia L, Nadu AP, Bader M, Campagnole-Santos MJ, Santos RA, Santos SH. Decreased hepatic gluconeogenesis in transgenic rats with increased circulating angiotensin-(1-7). Peptides. 2012;37:247–51.
10. Botelho-Santos GA, Bader M, Alenina N, Santos RA. Altered regional blood flow distribution in Mas-deficient mice. Ther Adv Cardiovasc Dis. 2012;6:201–11.
11. Botelho-Santos GA, Sampaio WO, Reudelhuber TL, Bader M, Campagnole-Santos MJ, Santos RA. Expression of an angiotensin-(1-7)-producing fusion protein in rats induced marked changes in regional vascular resistance. Am J Phys. 2007;292:H2485–90.
12. Cao X, Yang FY, Xin Z, Xie RR, Yang JK. The ACE2/Ang-(1-7)/Mas axis can inhibit hepatic insulin resistance. Mol Cell Endocrinol. 2014;393:30–8.
13. Castro CH, Santos RA, Ferreira AJ, Bader M, Alenina N, Almeida AP. Effects of genetic deletion of angiotensin-(1-7) receptor Mas on cardiac function during ischemia/reperfusion in the isolated perfused mouse heart. Life Sci. 2006;80:264–8.
14. Chen J, Zhao Y, Chen S, Wang J, Xiao X, Ma X, Penchikala M, Xia H, Lazartigues E, Zhao B, Chen Y. Neuronal over-expression of ACE2 protects brain from ischemia-induced damage. Neuropharmacology. 2014;79:550–8.

15. Chodavarapu H, Chhabra KH, Xia H, Shenoy V, Yue X, Lazartigues E. High-fat diet-induced glucose dysregulation is independent of changes in islet ACE2 in mice. Am J Physiol Regul Integr Comp Physiol. 2016;311:R1223–33.
16. Costa AC, Romero TR, Pacheco DF, Perez AC, Savernini A, Santos RR, Duarte ID. Participation of AT1 and Mas receptors in the modulation of inflammatory pain. Peptides. 2014;61:17–22.
17. Crackower MA, Sarao R, Oudit GY, Yagil C, Kozieradzki I, Scanga SE, Oliveira-dos-Santos AJ, da Costa J, Zhang L, Pei Y, Scholey J, Ferrario CM, Manoukian AS, Chappell MC, Backx PH, Yagil Y, Penninger JM. Angiotensin-converting enzyme 2 is an essential regulator of heart function. Nature. 2002;417:822–8.
18. Dartora DR, Irigoyen MC, Casali KR, Moraes-Silva IC, Bertagnolli M, Bader M, Santos RAS. Improved cardiovascular autonomic modulation in transgenic rats expressing an Ang-(1-7)-producing fusion protein. Can J Physiol Pharmacol. 2017;95:993–8.
19. de Almeida PW, Melo MB, Lima RF, Gavioli M, Santiago NM, Greco L, Jesus IC, Nocchi E, Parreira A, Alves MN, Mitraud L, Resende RR, Campagnole-Santos MJ, Dos Santos RA, Guatimosim S. Beneficial effects of angiotensin-(1-7) against deoxycorticosterone acetate-induced diastolic dysfunction occur independently of changes in blood pressure. Hypertension. 2015;66:389–95.
20. Dickinson ME, Flenniken AM, Ji X, Teboul L, Wong MD, White JK, Meehan TF, Weninger WJ, Westerberg H, Adissu H, Baker CN, Bower L, Brown JM, Caddle LB, Chiani F, Clary D, Cleak J, Daly MJ, Denegre JM, Doe B, Dolan ME, Edie SM, Fuchs H, Gailus-Durner V, Galli A, Gambadoro A, Gallegos J, Guo S, Horner NR, Hsu CW, Johnson SJ, Kalaga S, Keith LC, Lanoue L, Lawson TN, Lek M, Mark M, Marschall S, Mason J, ML ME, Newbigging S, Nutter LM, Peterson KA, Ramirez-Solis R, Rowland DJ, Ryder E, Samocha KE, Seavitt JR, Selloum M, Szoke-Kovacs Z, Tamura M, Trainor AG, Tudose I, Wakana S, Warren J, Wendling O, West DB, Wong L, Yoshiki A, MacArthur DG, Tocchini-Valentini GP, Gao X, Flicek P, Bradley A, Skarnes WC, Justice MJ, Parkinson HE, Moore M, Wells S, Braun RE, Svenson KL, de Angelis MH, Herault Y, Mohun T, Mallon AM, Henkelman RM, Brown SD, Adams DJ, Lloyd KC, McKerlie C, Beaudet AL, Bucan M, Murray SA. High-throughput discovery of novel developmental phenotypes. Nature. 2016;537:508–14.
21. Donoghue M, Wakimoto H, Maguire CT, Acton S, Hales P, Stagliano N, Fairchild-Huntress V, Xu J, Lorenz JN, Kadambi V, Berul CI, Breitbart RE. Heart block, ventricular tachycardia, and sudden death in ACE2 transgenic mice with downregulated connexins. J Mol Cell Cardiol. 2003;35:1043–53.
22. Esteban V, Heringer-Walther S, Sterner-Kock A, de BR, van den Engel S, Wang Y, Mezzano S, Egido J, Schultheiss HP, Ruiz-Ortega M, Walther T. Angiotensin-(1-7) and the g protein-coupled receptor MAS are key players in renal inflammation. PLoS ONE. 2009;4:e5406.
23. Fang F, Liu GC, Zhou X, Yang S, Reich HN, Williams V, Hu A, Pan J, Konvalinka A, Oudit GY, Scholey JW, John R. Loss of ACE2 exacerbates murine renal ischemia-reperfusion injury. PLoS One. 2013;8:e71433.
24. Felix BJ, Ravizzoni DD, Alenina N, Bader M, Santos RA. Glucagon-producing cells are increased in Mas-deficient mice. Endocr Connect. 2017;6:27–32.
25. Feng Y, Hans C, McIlwain E, Varner KJ, Lazartigues E. Angiotensin-converting enzyme 2 over-expression in the central nervous system reduces angiotensin-II-mediated cardiac hypertrophy. PLoS One. 2012;7:e48910.
26. Feng Y, Xia H, Cai Y, Halabi CM, Becker LK, Santos RA, Speth RC, Sigmund CD, Lazartigues E. Brain-selective overexpression of human angiotensin-converting enzyme type 2 attenuates neurogenic hypertension. Circ Res. 2010;106:373–82.
27. Ferreira AJ, Castro CH, Guatimosim S, Almeida PW, Gomes ER, Dias-Peixoto MF, Alves MN, Fagundes-Moura CR, Rentzsch B, Gava E, Almeida AP, Guimaraes AM, Kitten GT, Reudelhuber T, Bader M, Santos RA. Attenuation of isoproterenol-induced cardiac fibrosis in transgenic rats harboring an angiotensin-(1-7)-producing fusion protein in the heart. Ther Adv Cardiovasc Dis. 2010;4:83–96.
28. Ferreira AJ, Pinheiro SVB, Castro CH, Silva GAB, Simoes e Silva AC, Almeida AP, Bader M, Rentzsch B, Reudelhuber TL, Santos RA. Renal functions in transgenic rats expressing an angiotensin-(1-7)-producing fusion protein. Regul Pept. 2006;137:128–33.

29. Goncalves ACC, Leite R, Silva RAF, Pinheiro SVB, Sampaio WO, Reis AB, Reis FM, Thouyz RM, Webb R, Alenina N, Bader M, Santos RA. The vasodilator angiotensin-(1-7)-Mas axis plays an essential role in erectile function. Am J Phys. 2007;293:2588–96.

30. Gu H, Xie Z, Li T, Zhang S, Lai C, Zhu P, Wang K, Han L, Duan Y, Zhao Z, Yang X, Xing L, Zhang P, Wang Z, Li R, Yu JJ, Wang X, Yang P. Angiotensin-converting enzyme 2 inhibits lung injury induced by respiratory syncytial virus. Sci Rep. 2016;6:19840.

31. Gurley SB, Allred A, Le TH, Griffiths R, Mao L, Philip N, Haystead TA, Donoghue M, Breitbart RE, Acton SL, Rockman HA, Coffman TM. Altered blood pressure responses and normal cardiac phenotype in ACE2-null mice. J Clin Invest. 2006;116:2218–25.

32. Gurley SB, Coffman TM. Angiotensin-converting enzyme 2 gene targeting studies in mice: mixed messages. Exp Physiol. 2008;93:538–42.

33. Hammer A, Yang G, Friedrich J, Kovacs A, Lee DH, Grave K, Jorg S, Alenina N, Grosch J, Winkler J, Gold R, Bader M, Manzel A, Rump LC, Muller DN, Linker RA, Stegbauer J. Role of the receptor Mas in macrophage-mediated inflammation in vivo. Proc Natl Acad Sci U S A. 2016;113:14109–14.

34. Hashimoto T, Perlot T, Rehman A, Trichereau J, Ishiguro H, Paolino M, Sigl V, Hanada T, Hanada R, Lipinski S, Wild B, Camargo SM, Singer D, Richter A, Kuba K, Fukamizu A, Schreiber S, Clevers H, Verrey F, Rosenstiel P, Penninger JM. ACE2 links amino acid malnutrition to microbial ecology and intestinal inflammation. Nature. 2012;487:477–81.

35. Higaki A, Mogi M, Iwanami J, Min LJ, Bai HY, Shan BS, Kukida M, Yamauchi T, Tsukuda K, Kan-No H, Ikeda S, Higaki J, Horiuchi M. Beneficial Effect of Mas Receptor Deficiency on Vascular Cognitive Impairment in the Presence of Angiotensin II Type 2 Receptor. J Am Heart Assoc. 2018;7.

36. Hung YH, Hsieh WY, Hsieh JS, Liu FC, Tsai CH, Lu LC, Huang CY, Wu CL, Lin CS. Alternative roles of STAT3 and MAPK signaling pathways in the MMPs activation and progression of lung injury induced by cigarette smoke exposure in ACE2 knockout mice. Int J Biol Sci. 2016;12:454–65.

37. Imai Y, Kuba K, Rao S, Huan Y, Guo F, Guan B, Yang P, Sarao R, Wada T, Leong-Poi H, Crackower MA, Fukamizu A, Hui CC, Hein L, Uhlig S, Slutsky AS, Jiang C, Penninger JM. Angiotensin-converting enzyme 2 protects from severe acute lung failure. Nature. 2005;436:112–6.

38. Jin HY, Chen LJ, Zhang ZZ, Xu YL, Song B, Xu R, Oudit GY, Gao PJ, Zhu DL, Zhong JC. Deletion of angiotensin-converting enzyme 2 exacerbates renal inflammation and injury in apolipoprotein E-deficient mice through modulation of the nephrin and TNF-alpha-TNFRSF1A signaling. J Transl Med. 2015;13:255.

39. Kangussu LM, Almeida-Santos AF, Moreira FA, Fontes MAP, Santos RAS, Aguiar DC, Campagnole-Santos MJ. Reduced anxiety-like behavior in transgenic rats with chronically overproduction of angiotensin-(1-7): role of the Mas receptor. Behav Brain Res. 2017;331:193–8.

40. Kim CS, Kim IJ, Bae EH, Ma SK, Lee J, Kim SW. Angiotensin-(1-7) attenuates kidney injury due to obstructive nephropathy in rats. PLoS One. 2015;10:e0142664.

41. Kuba K, Zhang L, Imai Y, Arab S, Chen M, Maekawa Y, Leschnik M, Leibbrandt A, Markovic M, Schwaighofer J, Beetz N, Musialek R, Neely GG, Komnenovic V, Kolm U, Metzler B, Ricci R, Hara H, Meixner A, Nghiem M, Chen X, Dawood F, Wong KM, Sarao R, Cukerman E, Kimura A, Hein L, Thalhammer J, Liu PP, Penninger JM. Impaired heart contractility in Apelin gene-deficient mice associated with aging and pressure overload. Circ Res. 2007;101:e32–42.

42. Lazaroni TL, Bastos CP, Moraes MF, Santos RS, Pereira GS. Angiotensin-(1-7)/Mas axis modulates fear memory and extinction in mice. Neurobiol Learn Mem. 2016;127:27–33.

43. Lazaroni TL, Raslan AC, Fontes WR, de Oliveira ML, Bader M, Alenina N, Moraes MF, Dos Santos RA, Pereira GS. Angiotensin-(1-7)/Mas axis integrity is required for the expression of object recognition memory. Neurobiol Learn Mem. 2012;97:113–23.

44. Leonhardt J, Villela DC, Teichmann A, Munter LM, Mayer MC, Mardahl M, Kirsch S, Namsolleck P, Lucht K, Benz V, Alenina N, Daniell N, Horiuchi M, Iwai M, Multhaup G, Schulein R, Bader M, Santos RA, Unger T, Steckelings UM. Evidence for Heterodimerization and functional interaction of the angiotensin type 2 receptor and the receptor MAS. Hypertension. 2017;69:1128–35.

45. Li W, Moore MJ, Vasilieva N, Sui J, Wong SK, Berne MA, Somasundaran M, Sullivan JL, Luzuriaga K, Greenough TC, Choe H, Farzan M. Angiotensin-converting enzyme 2 is a functional receptor for the SARS coronavirus. Nature. 2003;426:450–4.
46. Lin CI, Tsai CH, Sun YL, Hsieh WY, Lin YC, Chen CY, Lin CS. Instillation of particulate matter 2.5 induced acute lung injury and attenuated the injury recovery in ACE2 knockout mice. Int J Biol Sci. 2018;14:253–65.
47. Liu C, Xiao L, Li F, Zhang H, Li Q, Liu H, Fu S, Li C, Zhang X, Wang J, Staunstrup NH, Li Y, Yang H. Generation of outbred Ace2 knockout mice by RNA transfection of TALENs displaying colitis reminiscent pathophysiology and inflammation. Transgenic Res. 2015;24:433–46.
48. Liu R, Qi H, Wang J, Wang Y, Cui L, Wen Y, Yin C. Angiotensin-converting enzyme (ACE and ACE2) imbalance correlates with the severity of cerulein-induced acute pancreatitis in mice. Exp Physiol. 2014;99:651–63.
49. Lovren F, Pan Y, Quan A, Teoh H, Wang G, Shukla PC, Levitt KS, Oudit GY, Al-Omran M, Stewart DJ, Slutsky AS, Peterson MD, Backx PH, Penninger JM, Verma S. Angiotensin converting enzyme-2 confers endothelial protection and attenuates atherosclerosis. Am J Physiol Heart Circ Physiol. 2008;295:H1377–84.
50. Lu CL, Wang Y, Yuan L, Li Y, Li XY. The angiotensin-converting enzyme 2/angiotensin (1-7)/ Mas axis protects the function of pancreatic beta cells by improving the function of islet microvascular endothelial cells. Int J Mol Med. 2014;34:1293–300.
51. Magalhaes GS, Rodrigues-Machado MD, Motta-Santos D, Alenina N, Bader M, Santos RA, Barcelos LS, Campagnole-Santos MJ. Chronic allergic pulmonary inflammation is aggravated in angiotensin-(1-7) Mas receptor knockout mice. Am J Physiol Lung Cell Mol Physiol ajplung. 2016;311:L1141.
52. Mario EG, Santos SH, Ferreira AV, Bader M, Santos RA, Botion LM. Angiotensin-(1-7) Mas-receptor deficiency decreases peroxisome proliferator-activated receptor gamma expression in adipocytes. Peptides. 2012;33:174–7.
53. McCray PB Jr, Pewe L, Wohlford-Lenane C, Hickey M, Manzel L, Shi L, Netland J, Jia HP, Halabi C, Sigmund CD, Meyerholz DK, Kirby P, Look DC, Perlman S. Lethal infection of K18-hACE2 mice infected with severe acute respiratory syndrome coronavirus. J Virol. 2007;81:813–21.
54. Mercure C, Yogi A, Callera GE, Aranha AB, Bader M, Ferreira AJ, Santos RA, Walther T, Thouyz RM, Reudelhuber TL. Angiotensin 1-7 blunts hypertensive cardiac remodeling by a direct effect on the heart. Circ Res. 2008;103:1319–26.
55. Methot D, Lapointe MC, Touyz RM, Yang XP, Carretero OA, Deschepper CF, Schiffrin EL, Thibault G, Reudelhuber TL. Tissue targeting of angiotensin peptides. J Biol Chem. 1997;272:12994–9.
56. Morales MG, Abrigo J, Acuna MJ, Santos RA, Bader M, Brandan E, Simon F, Olguin H, Cabrera D, Cabello-Verrugio C. Angiotensin-(1-7) attenuates disuse skeletal muscle atrophy in mice via its receptor, Mas. Dis Model Mech. 2016;9:441–9.
57. Moreira CCL, Lourenco FC, Mario EG, Santos RAS, Botion LM, Chaves VE. Long-term effects of angiotensin-(1-7) on lipid metabolism in the adipose tissue and liver. Peptides. 2017;92:16–22.
58. Mori MAS, Bader M, Pesquero JB. Genetically altered animals in the study of the metabolic functions of peptide hormone systems. Curr Opin Nephrol Hypertens. 2008;17:11–7.
59. Motta-Santos D, Dos Santos RA, Oliveira M, Qadri F, Poglitsch M, Mosienko V, Kappes BL, Campagnole-Santos MJ, Penninger M, Alenina N, Bader M. Effects of ACE2 deficiency on physical performance and physiological adaptations of cardiac and skeletal muscle to exercise. Hypertens Res. 2016;39:506–12.
60. Moura SD, Ribeiro MF, Limborco-Filho M, de Oliveira ML, Hamamoto D, Xavier CH, Moreira FA, Santos RA, Campagnole-Santos MJ, Peliky Fontes MA. Chronic overexpression of angiotensin-(1-7) in rats reduces cardiac reactivity to acute stress and dampens anxious behavior. Stress. 2017;20:189–96.

61. Nadarajah R, Milagres R, Dilauro M, Gutsol A, Xiao F, Zimpelmann J, Kennedy C, Wysocki J, Batlle D, Burns KD. Podocyte-specific overexpression of human angiotensin-converting enzyme 2 attenuates diabetic nephropathy in mice. Kidney Int. 2012;82:292–303.

62. Netland J, Meyerholz DK, Moore S, Cassell M, Perlman S. Severe acute respiratory syndrome coronavirus infection causes neuronal death in the absence of encephalitis in mice transgenic for human ACE2. J Virol. 2008;82:7264–75.

63. Niu MJ, Yang JK, Lin SS, Ji XJ, Guo LM. Loss of angiotensin-converting enzyme 2 leads to impaired glucose homeostasis in mice. Endocrine. 2008;34:56–61.

64. Nunes-Souza V, Alenina N, Qadri F, Penninger JM, Santos RA, Bader M, Rabelo LA. CD36/ Sirtuin 1 Axis impairment contributes to hepatic steatosis in ACE2-deficient mice. Oxidative Med Cell Longev. 2016;2016:6487509.

65. Oliveira-Lima OC, Pinto MC, Duchene J, Qadri F, Souza LL, Alenina N, Bader M, Santos RA, Carvalho-Tavares J. Mas receptor deficiency exacerbates lipopolysaccharide-induced cerebral and systemic inflammation in mice. Immunobiology. 2015;220:1311–21.

66. Osterreicher CH, Taura K, De MS, Seki E, Penz-Osterreicher M, Kodama Y, Kluwe J, Schuster M, Oudit GY, Penninger JM, Brenner DA. Angiotensin-converting-enzyme 2 inhibits liver fibrosis in mice. Hepatology (Baltimore, Md). 2009;50:929–38.

67. Oudit GY, Herzenberg AM, Kassiri Z, Wong D, Reich H, Khokha R, Crackower MA, Backx PH, Penninger JM, Scholey JW. Loss of angiotensin-converting enzyme-2 leads to the late development of angiotensin II-dependent glomerulosclerosis. Am J Pathol. 2006;168:1808–20.

68. Oudit GY, Kassiri Z, Patel MP, Chappell M, Butany J, Backx PH, Tsushima RG, Scholey JW, Khokha R, Penninger JM. Angiotensin II-mediated oxidative stress and inflammation mediate the age-dependent cardiomyopathy in ACE2 null mice. Cardiovasc Res. 2007;75:29–39.

69. Patel VB, Bodiga S, Fan D, Das SK, Wang Z, Wang W, Basu R, Zhong J, Kassiri Z, Oudit GY. Cardioprotective effects mediated by angiotensin II type 1 receptor blockade and enhancing angiotensin 1-7 in experimental heart failure in angiotensin-converting enzyme 2-null mice. Hypertension. 2012;59:1195–203.

70. Patel VB, Mori J, McLean BA, Basu R, Das SK, Ramprasath T, Parajuli N, Penninger JM, Grant MB, Lopaschuk GD, Oudit GY. ACE2 deficiency worsens Epicardial adipose tissue inflammation and cardiac dysfunction in response to diet-induced obesity. Diabetes. 2016;65:85–95.

71. Patel VB, Zhong JC, Fan D, Basu R, Morton JS, Parajuli N, McMurtry MS, Davidge ST, Kassiri Z, Oudit GY. Angiotensin-converting enzyme 2 is a critical determinant of angiotensin II-induced loss of vascular smooth muscle cells and adverse vascular remodeling. Hypertension. 2014;64:157–64.

72. Peiro C, Vallejo S, Gembardt F, Azcutia V, Heringer-Walther S, Rodriguez-Manas L, Schultheiss HP, Sanchez-Ferrer CF, Walther T. Endothelial dysfunction through genetic deletion or inhibition of the G protein-coupled receptor Mas: a new target to improve endothelial function. J Hypertens. 2007;25:2421–5.

73. Pena Silva RA, Kung DK, Mitchell IJ, Alenina N, Bader M, Santos RA, Faraci FM, Heistad DD, Hasan DM. Angiotensin 1-7 reduces mortality and rupture of intracranial aneurysms in mice. Hypertension. 2014;64:362–8.

74. Pinheiro SVB, Ferreira AJ, Kitten GT, da Silveira KD, da Silva DA, Santos SHS, Gava E, Castro CH, Magalhaes JA, da Mota RK, Botelho-Santos GA, Bader M, Alenina N, Santos RA, Simoes e Silva AC. Genetic deletion of the angiotensin(1-7) receptor Mas leads to glomerular hyperfiltration and microalbuminuria. Kidney Int. 2009;75:1184–93.

75. Qi YF, Zhang J, Wang L, Shenoy V, Krause E, Oh SP, Pepine CJ, Katovich MJ, Raizada MK. Angiotensin-converting enzyme 2 inhibits high-mobility group box 1 and attenuates cardiac dysfunction post-myocardial ischemia. J Mol Med (Berl). 2016;94:37–49.

76. Rabello CK, Ravizzoni DD, Moura M, Bertagnolli M, Bader M, Haibara A, Alenina N, Irigoyen MC, Santos RA. Increased vascular sympathetic modulation in mice with Mas receptor deficiency. J Renin-Angiotensin-Aldosterone Syst. 2016;17:1470320316643643.

77. Rabelo LA, Todiras M, Nunes-Souza V, Qadri F, Szijarto IA, Gollasch M, Penninger JM, Bader M, Santos RA, Alenina N. Genetic deletion of ACE2 induces vascular dysfunction in C57BL/6 mice: role of nitric oxide imbalance and oxidative stress. PLoS One. 2016;11:e0150255.
78. Rabelo LA, Xu P, Todiras M, Sampaio WO, Buttgereit J, Bader M, Santos RA, Alenina N. Ablation of angiotensin (1-7) receptor Mas in C57Bl/6 mice causes endothelial dysfunction. J Am Soc Hypertens. 2008;2:418–24.
79. Rentzsch B, Todiras M, Iliescu R, Popova E, Campos LA, Oliveira ML, Baltatu OC, Santos RA, Bader M. Transgenic ACE2 overexpression in vessels of SHRSP rats reduces blood pressure and improves endothelial function. Hypertension. 2008;52:967–73.
80. Rey-Parra GJ, Vadivel A, Coltan L, Hall A, Eaton F, Schuster M, Loibner H, Penninger JM, Kassiri Z, Oudit GY, Thebaud B. Angiotensin converting enzyme 2 abrogates bleomycin-induced lung injury. J Mol Med (Berl). 2012;90:637.
81. Sahara M, Ikutomi M, Morita T, Minami Y, Nakajima T, Hirata Y, Nagai R, Sata M. Deletion of angiotensin-converting enzyme 2 promotes the development of atherosclerosis and arterial neointima formation. Cardiovasc Res. 2014;101:236–46.
82. Sahr A, Wolke C, Maczewsky J, Krippeit-Drews P, Tetzner A, Drews G, Venz S, Gurtler S, van den Brandt J, Berg S, Doring P, Dombrowski F, Walther T, Lendeckel U. The angiotensin-(1-7)/Mas Axis improves pancreatic beta-cell function in vitro and in vivo. Endocrinology. 2016;157:4677–90.
83. Santos RA, Castro CH, Gava E, Pinheiro SVB, Almeida AP, Paula DR, Cruz JS, Ramos AS, Rosa KT, Irigoyen MC, Bader M, Alenina N, Ferreira AJ. Impairment of in vitro and in vivo heart function in angiotensin-(1-7) receptor Mas knockout mice. Hypertension. 2006;47:996–1002.
84. Santos RA, Ferreira AJ, Nadu AP, Braga AN, Almeida AP, Campagnole-Santos MJ, Baltatu O, Iliescu R, Reudelhuber TL, Bader M. Expression of an angiotensin-(1-7)-producing fusion protein produces cardioprotective effects in rats. Physiol Genomics. 2004;17:292–9.
85. Santos RA, Simoes e Silva AC, Maric C, DMR S, Machado RP, de Buhr I, Heringer-Walther S, SVB P, Lopes MT, Bader M, Mendes EP, Lemos VS, Campagnole-Santos MJ, Schultheiss HP, Speth R, Walther T. Angiotensin-(1-7) is an endogenous ligand for the G-protein coupled receptor Mas. Proc Natl Acad Sci U S A. 2003;100:8258–63.
86. Santos RAS, Sampaio WO, Alzamora AC, Motta-Santos D, Alenina N, Bader M, Campagnole-Santos MJ. The ACE2/angiotensin-(1-7)/MAS Axis of the renin-angiotensin system: focus on angiotensin-(1-7). Physiol Rev. 2018;98:505–53.
87. Santos SHS, Braga JF, Mario EG, Porto LCJ, Botion LM, Alenina N, Bader M, Santos RA. Improved lipid and glucose metabolism in transgenic rats with increased circulating angiotensin-(1-7). Arterioscler Thromb Vasc Biol. 2010;30:953–61.
88. Santos SHS, Fernandes LR, Mario EG, Ferreira AVM, Porto LCJ, Alvarez-Leite JI, Botion LM, Bader M, Alenina N, Santos RA. Mas deficiency in FVB/N in mice produces marked changes in lipid and glycemic metabolism. Diabetes. 2008;57:340–7.
89. Schuchard J, Winkler M, Stolting I, Schuster F, Vogt FM, Barkhausen J, Thorns C, Santos RA, Bader M, Raasch W. Lack of weight gain after angiotensin AT1 receptor blockade in diet-induced obesity is partly mediated by an angiotensin-(1-7)/Mas-dependent pathway. Br J Pharmacol. 2015;172:3764–78.
90. Shi TT, Yang FY, Liu C, Cao X, Lu J, Zhang XL, Yuan MX, Chen C, Yang JK. Angiotensin-converting enzyme 2 regulates mitochondrial function in pancreatic beta-cells. Biochem Biophys Res Commun. 2018;495:860–6.
91. Shiota A, Yamamoto K, Ohishi M, Tatara Y, Ohnishi M, Maekawa Y, Iwamoto Y, Takeda M, Rakugi H. Loss of ACE2 accelerates time-dependent glomerular and tubulointerstitial damage in streptozotocin-induced diabetic mice. Hypertens Res. 2010;33:298–307.
92. Shoemaker R, Yiannikouris F, Thatcher S, Cassis L. ACE2 deficiency reduces beta-cell mass and impairs beta-cell proliferation in obese C57BL/6 mice. Am J Physiol Endocrinol Metab. 2015;309:E621–31.
93. Silva AR, Aguilar EC, Alvarez-Leite JI, da Silva RF, Arantes RM, Bader M, Alenina N, Pelli G, Lenglet S, Galan K, Montecucco F, Mach F, Santos SH, Santos RA. Mas receptor deficiency is associated with worsening of lipid profile and severe hepatic steatosis in ApoE knockout mice. Am J Physiol Regul Integr Comp Physiol. 2013;305:R1323–30.

94. Silveira KD, Barroso LC, Vieira AT, Cisalpino D, Lima CX, Bader M, Arantes RM, Dos Santos RA, Simoes-E-Silva AC, Teixeira MM. Beneficial effects of the activation of the Angiotensin-(1-7) MAS receptor in a murine model of adriamycin-induced nephropathy. PLoS One. 2013;8:e66082.

95. Singer D, Camargo SM, Ramadan T, Schafer M, Mariotta L, Herzog B, Huggel K, Wolfer D, Werner S, Penninger JM, Verrey F. Defective intestinal amino acid absorption in Ace2 null mice. Am J Physiol Gastrointest Liver Physiol. 2012;303:G686–95.

96. Souza LL, Duchene J, Todiras M, Azevedo LCP, Costa-Neto CM, Alenina N, Santos RA, Bader M. Receptor Mas protects mice against hypothermia and mortality induced by endotoxemia. Shock. 2014;41:331–6.

97. Sriramula S, Xia H, Xu P, Lazartigues E. Brain-targeted angiotensin-converting enzyme 2 overexpression attenuates neurogenic hypertension by inhibiting cyclooxygenase-mediated inflammation. Hypertension. 2015;65:577–86.

98. Steckelings UM, Paulis L, Unger T, Bader M. Emerging drugs which target the renin-angiotensin-aldosterone system. Expert Opin Emerg Drugs. 2011;16:619–30.

99. Thatcher SE, Zhang X, Howatt DA, Yiannikouris F, Gurley SB, Ennis T, Curci JA, Daugherty A, Cassis LA. Angiotensin-converting enzyme 2 decreases formation and severity of angiotensin II-induced abdominal aortic aneurysms. Arterioscler Thromb Vasc Biol. 2014;34:2617–23.

100. Thomas MC, Pickering RJ, Tsorotes D, Koitka A, Sheehy K, Bernardi S, Toffoli B, Nguyen-Huu TP, Head GA, Fu Y, Chin-Dusting J, Cooper ME, Tikellis C. Genetic Ace2 deficiency accentuates vascular inflammation and atherosclerosis in the ApoE knockout mouse. Circ Res. 2010;107:888–97.

101. Tikellis C, Brown R, Head GA, Cooper ME, Thomas MC. Angiotensin-converting enzyme 2 mediates hyperfiltration associated with diabetes. Am J Physiol Renal Physiol. 2014;306:F773–80.

102. Tseng CT, Huang C, Newman P, Wang N, Narayanan K, Watts DM, Makino S, Packard MM, Zaki SR, Chan TS, Peters CJ. Severe acute respiratory syndrome coronavirus infection of mice transgenic for the human angiotensin-converting enzyme 2 virus receptor. J Virol. 2007;81:1162–73.

103. Vasam G, Joshi S, Thatcher SE, Bartelmez SH, Cassis LA, Jarajapu YP. Reversal of bone marrow Mobilopathy and enhanced vascular repair by Angiotensin-(1-7) in diabetes. Diabetes. 2017;66:505–18.

104. Vickers C, Hales P, Kaushik V, Dick L, Gavin J, Tang J, Godbout K, Parsons T, Baronas E, Hsieh F, Acton S, Patane M, Nichols A, Tummino P. Hydrolysis of biological peptides by human angiotensin-converting enzyme-related carboxypeptidase. J Biol Chem. 2002;277:14838–43.

105. Walther T, Balschun D, Voigt JP, Fink H, Zuschratter W, Birchmeier C, Ganten D, Bader M. Sustained long term potentiation and anxiety in mice lacking the Mas protooncogene. J Biol Chem. 1998;273:11867–73.

106. Walther T, Voigt JP, Fink H, Bader M. Sex specific behavioural alterations in Mas-deficient mice. Behav Brain Res. 2000;107:105–9.

107. Wang L, de Kloet AD, Pati D, Hiller H, Smith JA, Pioquinto DJ, Ludin JA, Oh SP, Katovich MJ, Frazier CJ, Raizada MK, Krause EG. Increasing brain angiotensin converting enzyme 2 activity decreases anxiety-like behavior in male mice by activating central Mas receptors. Neuropharmacology. 2016;105:114–23.

108. Wang LA, de Kloet AD, Smeltzer MD, Cahill KM, Hiller H, Bruce EB, Pioquinto DJ, Ludin JA, Katovich MJ, Raizada MK, Krause EG. Coupling corticotropin-releasing-hormone and angiotensin converting enzyme 2 dampens stress responsiveness in male mice. Neuropharmacology. 2018;133:85–93.

109. Wang LP, Fan SJ, Li SM, Wang XJ, Gao JL, Yang XH. Protective role of ACE2-Ang-(1-7)-Mas in myocardial fibrosis by downregulating KCa3.1 channel via ERK1/2 pathway. Pflugers Arch. 2016;468:2041–51.

110. Wang W, McKinnie SM, Farhan M, Paul M, McDonald T, McLean B, Llorens-Cortes C, Hazra S, Murray AG, Vederas JC, Oudit GY. Angiotensin-converting enzyme 2 metabolizes and

partially inactivates Pyr-Apelin-13 and Apelin-17: physiological effects in the cardiovascular system. Hypertension. 2016;68:365–77.

111. Wang XL, Iwanami J, Min LJ, Tsukuda K, Nakaoka H, Bai HY, Shan BS, Kan-No H, Kukida M, Chisaka T, Yamauchi T, Higaki A, Mogi M, Horiuchi M. Deficiency of angiotensin-converting enzyme 2 causes deterioration of cognitive function. NPJ Aging Mech Dis. 2016;2:16024.

112. Wang Y, Qian C, Roks AJ, Westermann D, Schumacher SM, Escher F, Schoemaker RG, Reudelhuber TL, Van Gilst WH, Schultheiss HP, Tschope C, Walther T. Circulating rather than cardiac angiotensin-(1-7) stimulates cardioprotection after myocardial infarction. Circ Heart Fail. 2010;3:286–93.

113. Wang Y, Shoemaker R, Powell D, Su W, Thatcher S, Cassis L. Differential effects of Mas receptor deficiency on cardiac function and blood pressure in obese male and female mice. Am J Physiol Heart Circ Physiol. 2017;312:H459–68.

114. Wang Y, Shoemaker R, Thatcher SE, Batifoulier-Yiannikouris F, English VL, Cassis LA. Administration of 17beta-estradiol to ovariectomized obese female mice reverses obesity-hypertension through an ACE2-dependent mechanism. Am J Physiol Endocrinol Metab. 2015;308:E1066–75.

115. Wong DW, Oudit GY, Reich H, Kassiri Z, Zhou J, Liu QC, Backx PH, Penninger JM, Herzenberg AM, Scholey JW. Loss of angiotensin-converting enzyme-2 (Ace2) accelerates diabetic kidney injury. Am J Pathol. 2007;171:438–51.

116. Wysocki J, Ortiz-Melo DI, Mattocks NK, Xu K, Prescott J, Evora K, Ye M, Sparks MA, Haque SK, Batlle D, Gurley SB. ACE2 deficiency increases NADPH-mediated oxidative stress in the kidney. Physiol Rep. 2014;2:e00264.

117. Xia H, de Queiroz TM, Sriramula S, Feng Y, Johnson T, Mungrue IN, Lazartigues E. Brain ACE2 overexpression reduces DOCA-salt hypertension independently of endoplasmic reticulum stress. Am J Physiol Regul Integr Comp Physiol. 2015;308:R370–8.

118. Xia H, Sriramula S, Chhabra KH, Lazartigues E. Brain angiotensin-converting enzyme type 2 shedding contributes to the development of neurogenic hypertension. Circ Res. 2013;113:1087–96.

119. Xia H, Suda S, Bindom S, Feng Y, Gurley SB, Seth D, Navar LG, Lazartigues E. ACE2-mediated reduction of oxidative stress in the central nervous system is associated with improvement of autonomic function. PLoS One. 2011;6:e22682.

120. Xiao L, Gao L, Lazartigues E, Zucker IH. Brain-selective overexpression of angiotensin-converting enzyme 2 attenuates sympathetic nerve activity and enhances baroreflex function in chronic heart failure. Hypertension. 2011;58:1057–65.

121. Xu P, Goncalves ACC, Todiras M, Rabelo LA, Sampaio WO, Moura MM, Santos SS, Luft FC, Bader M, Gross V, Alenina N, Santos RA. Endothelial dysfunction and elevated blood pressure in Mas gene-deleted mice. Hypertension. 2008;51:574–80.

122. Xu X, Quiambao AB, Roveri L, Pardue MT, Marx JL, Rohlich P, Peachey NS, Al Ubaidi MR. Degeneration of cone photoreceptors induced by expression of the Mas1 protooncogene. Exp Neurol. 2000;163:207–19.

123. Xuan X, Gao F, Ma X, Huang C, Wang Y, Deng H, Wang S, Li W, Yuan L. Activation of ACE2/angiotensin (1-7) attenuates pancreatic beta cell dedifferentiation in a high-fat-diet mouse model. Metabolism. 2018;81:83–96.

124. Yamaleyeva LM, Pulgar VM, Lindsey SH, Yamane L, Varagic J, McGee C, da Silva M, Lopes BP, Gurley SB, Brosnihan KB. Uterine artery dysfunction in pregnant ACE2 knockout mice is associated with placental hypoxia and reduced umbilical blood flow velocity. Am J Physiol Endocrinol Metab. 2015;309:E84–94.

125. Yamamoto K, Ohishi M, Katsuya T, Ito N, Ikushima M, Kaibe M, Tatara Y, Shiota A, Sugano S, Takeda S, Rakugi H, Ogihara T. Deletion of angiotensin-converting enzyme 2 accelerates pressure overload-induced cardiac dysfunction by increasing local angiotensin II. Hypertension. 2006;47:718–26.

126. Yang XH, Deng W, Tong Z, Liu YX, Zhang LF, Zhu H, Gao H, Huang L, Liu YL, Ma CM, Xu YF, Ding MX, Deng HK, Qin C. Mice transgenic for human angiotensin-converting enzyme 2 provide a model for SARS coronavirus infection. Comp Med. 2007;57:450–9.
127. Yang XH, Wang YH, Wang JJ, Liu YC, Deng W, Qin C, Gao JL, Zhang LY. Role of angiotensin-converting enzyme (ACE and ACE2) imbalance on tourniquet-induced remote kidney injury in a mouse hindlimb ischemia-reperfusion model. Peptides. 2012;36:60–70.
128. Yoshikawa N, Yoshikawa T, Hill T, Huang C, Watts DM, Makino S, Milligan G, Chan T, Peters CJ, Tseng CT. Differential virological and immunological outcome of severe acute respiratory syndrome coronavirus infection in susceptible and resistant transgenic mice expressing human angiotensin-converting enzyme 2. J Virol. 2009;83:5451–65.
129. Zhang J, Dong J, Martin M, He M, Gongol B, Marin TL, Chen L, Shi X, Yin Y, Shang F, Wu Y, Huang HY, Zhang J, Zhang Y, Kang J, Moya EA, Huang HD, Powell FL, Chen Z, Thistlethwaite PA, Yuan ZY, Shyy JY. AMPK phosphorylation of ACE2 in endothelium mitigates pulmonary hypertension. Am J Respir Crit Care Med. 2018;198:509.
130. Zhang ZZ, Cheng YW, Jin HY, Chang Q, Shang QH, Xu YL, Chen LX, Xu R, Song B, Zhong JC. The sirtuin 6 prevents angiotensin II-mediated myocardial fibrosis and injury by targeting AMPK-ACE2 signaling. Oncotarget. 2017;8:72302–14.
131. Zheng J, Li G, Chen S, Bihl J, Buck J, Zhu Y, Xia H, Lazartigues E, Chen Y, Olson JE. Activation of the ACE2/Ang-(1-7)/Mas pathway reduces oxygen-glucose deprivation-induced tissue swelling, ROS production, and cell death in mouse brain with angiotensin II overproduction. Neuroscience. 2014;273:39–51.
132. Zheng JL, Li GZ, Chen SZ, Wang JJ, Olson JE, Xia HJ, Lazartigues E, Zhu YL, Chen YF. Angiotensin converting enzyme 2/Ang-(1-7)/Mas axis protects brain from ischemic injury with a tendency of age-dependence. CNS Neurosci Ther. 2014;20:452–9.
133. Zhong J, Guo D, Chen CB, Wang W, Schuster M, Loibner H, Penninger JM, Scholey JW, Kassiri Z, Oudit GY. Prevention of angiotensin II-mediated renal oxidative stress, inflammation, and fibrosis by angiotensin-converting enzyme 2. Hypertension. 2011;57:314–22.
134. Zou Z, Yan Y, Shu Y, Gao R, Sun Y, Li X, Ju X, Liang Z, Liu Q, Zhao Y, Guo F, Bai T, Han Z, Zhu J, Zhou H, Huang F, Li C, Lu H, Li N, Li D, Jin N, Penninger JM, Jiang C. Angiotensin-converting enzyme 2 protects from lethal avian influenza a H5N1 infections. Nat Commun. 2014;5:3594.

Part II

Actions of Angiotensin-(1-7)

Brain

Maria Jose Campagnole-Santos, Mariela M. Gironacci, and Marco Antônio Peliky Fontes

Ang-(1-7) Metabolism in the Brain

All components of the renin-angiotensin system (RAS) are present in the brain [47, 73]. The first evidence that Ang-(1-7) could be generated in areas of the central nervous system (CNS) was obtained from studies of the hydrolysis of [^{125}I]-Ang I in brain homogenates [109]. In this study, homogenate of micropunctures of medullary areas produced Ang-(1-7) from [^{125}I]-Ang I or [^{125}I]-Ang II, both in the absence or presence of ACE inhibitor showing that Ang-(1-7) could not be just a pathway of Ang II metabolism. Currently, it is accepted that Ang-(1-7) is formed from Ang I through cleavage at the Pro^7-Phe^8 peptide bound by several endopeptidases, such as prolyl endopeptidase, thimet oligopeptidase, and neutral endopeptidase (neprylisin) [73, 103]. The carboxypeptidase ACE2, which removes the carboxyterminal phenylalanine from Ang II [35, 36, 123], seems to be the main Ang-(1-7)-forming enzyme. The catalytic efficiency of ACE2 to generate Ang-(1-7) from Ang II is almost 500-fold greater than that shown for conversion of Ang I to Ang-(1–9) and 10- or 600-fold higher than that described for two other Ang-(1-7)-forming enzymes: prolyl endopeptidase and prolyl carboxypeptidase, respectively [103, 104, 125]. The exact site for the generation of Ang-(1-7) in the brain, if intracellular or extracellular, is still uncertain, as well as, it is not known

M. J. Campagnole-Santos (✉) · M. A. P. Fontes
Departamento de Fisiologia e Biofísica, INCT-Nanobiofar, Instituto de Ciências Biológicas, Universidade Federal de Minas Gerais, Belo Horizonte, MG, Brazil
e-mail: mjcs@icb.ufmg.br

M. M. Gironacci
Departamento de Química Biológica, IQUIFIB (UBA-CONICET), Facultad de Farmacia y Bioquímica, Universidad de Buenos Aires, Buenos Aires, Argentina

© Springer Nature Switzerland AG 2019
R. A. S. Santos (ed.), *Angiotensin (1-7)*,
https://doi.org/10.1007/978-3-030-22696-1_4

whether the generation of Ang peptides is solely dependent on brain Aogen, ACE, ACE2, and renin, taking into account the discrepancies in the location of RAS components [17, 56].

Ang-(1-7) Location in the Brain

Immunostaining for Ang-(1-7) was initially shown mainly in paraventricular (PVN), supraoptic (SON), and suprachiasmatic nuclei of hypothalamus, bed nucleus of the stria terminalis, substantia innominata, median eminence, and neurohypophysis [15, 18, 76]. Consistent with this observation, Ang-(1-7) immunoreactivity was reported in neurons from hypothalamus, brainstem [83], and PVN [10] of rats. In extracts from the rat hypothalamus, approximately equimolar amounts of Ang-(1-7), Ang II, and Ang I were detected [25]. A similar profile was observed in the medulla oblongata and amygdala, although the content of these three peptides was 40–70% lower than that determined in the hypothalamus [25]. Recently, it has been shown that Ang-(1-7) is a preferential peptide formed from Ang I through thimet oligopeptidase [98]. Further, increase in thimet oligopeptidase and Ang-(1-7) was observed in the hippocampus of epileptic rats [98], which is in agreement with the reported increase in Ang-(1-7) in the hippocampus of rats during the acute and silent phases of pilocarpine-induced epilepsy [55].

MAS Location in the Brain

There is still a debate on the true ligand of AT_1 receptor in the brain, whether it is Ang II or Ang III [48]; however, the effects described for Ang-(1-7) appear to be related to MAS receptor (MasR; [111]). Soon after the initial description of biological actions of Ang-(1-7), several studies raised the hypothesis of the existence of a selective receptor for mediating the physiological effects of Ang-(1-7) in the brain. First, Ang-(1-7) was shown to display opposite effect on baroreflex control. Ang-(1-7), given ICV, facilitated, while Ang II attenuated the baroreflex control of heart rate [20]. Additional studies strengthened this hypothesis by showing that Ang-(1-7) did not bind to AT1 or AT2 receptors [11, 105], that Ang-(1-7) presented different effects from those induced by Ang II, both centrally and peripherally [110], and, finally, the description of a selective antagonist for Ang-(1-7) actions, A779 [108], which did not interfere with Ang II at AT1 or AT2 receptor. Other studies using PD123319 raised the hypothesis that some of Ang-(1-7) actions could be mediated by AT2 receptor [51, 52]. However, PD123319 was recently described as MrgD receptor antagonist, which mediates alamandine effects [77]. Alamandine [Ala^1-$DesAsp^1$-Ang-(1-7)] can be formed endogenously from Ang-(1-7) or Ang A and expresses actions at the central nervous system (CNS) [77]. These data together with the observation that Ang-(1-7) actions could be blocked by PD123319 in AT2KO mice [114] indicate that future studies will have to show whether in fact some of Ang-(1-7) actions could be mediated by AT2 in the brain.

In 2003, with the identification of the G-coupled Mas protein as an Ang-(1-7) receptor [111], Mas was identified in different areas of the brain, especially those related to cardiovascular control [12]. There was a strong staining in the nucleus tractus solitarii (NTS), caudal and rostral ventrolateral medulla (CVLM and RVLM), inferior olive, and parvo and magnocellular portions of the PVN, SON, and lateral preoptic area (LPA) of normotensive SD rats. However, other areas were also stained for Mas, such as the hippocampal nucleus, different subregions of the frontal cortex, anterodorsal thalamic nucleus, basomedial and basolateral amygdaloid nucleus, and hypoglossal nucleus [12, 46]. Recently, Mas immunostaining was shown mainly in the soma of neurons and microglia of adult rat cerebral cortex and not in astroglia, and to exist in both non-nuclear and nuclear compartments [88, 101]. Further, MasR protein expression was greater in neurons from hypothalamus of spontaneously hypertensive rats (SHR) than from normotensive Wistar-Kyoto rats [84].

Ang-(1-7) Actions in the Brain

Historically, it is important to mention that the first biological actions described for Ang-(1-7) were related to the CNS. Schiavone et al. [112] found that Ang-(1-7) was equipotent to Ang II in stimulating arginine–vasopressin release from rat hypothalamo-neurohypophysial preparation in vitro. In the same year, Ang-(1-7)-immunoreactivity was found in the neurohypophysis and hypothalamic regions [15]. Subsequent study showed that Ang-(1-7) was able to induce fall in blood pressure after in vivo microinjection into the nucleus tractus solitarii (NTS) [19]. At present, we recognize that Ang-(1-7) acts as an important neuromodulator, especially in areas related to tonic and reflex control of arterial pressure, in the hypothalamus and in the dorsomedial and ventrolateral medulla. At these sites, cardiovascular effects induced by Ang-(1-7) are blocked by A-779, which is recognized as a selective Mas receptor antagonist [108], suggesting Ang-(1-7) actions in the brain are mainly mediated by interaction with MasR.

No alteration in blood pressure or drinking behavior was observed after short-term (up to few hours) infusion of Ang-(1-7) into the lateral ventricle [20] or microinjection into the PVN [100], which contrasts with the classical stimulatory effect mediated by AT1 in the brain [86]. However, when long-term infusion (14–28 days) of Ang-(1-7) was performed, an attenuation in high arterial pressure was observed in DOCA-salt [58, 138], TGR(mREN2)L-27 hypertensive rats [49, 51, 70, 92] or Ang II hypertension [137]. Lowering blood pressure effect of Ang-(1-7) in DOCA-salt hypertensive rats was related to improvement in baroreflex bradycardia, restoration of the baroreflex control of renal sympathetic activity, and regaining of the balance of cardiac autonomic tone [58, 70]. Enhancement of the baroreflex control of the RSNA was also observed after 4 days of ICV infusion of Ang-(1-7) in rabbits subjected to chronic heart failure [71]. Similar effects were also observed in transgenic mice overexpressing human ACE2 selectively in the brain [32, 40, 118, 131, 133]. On the other hand, increased ACE2 membrane shedding by ADAM17 in the brain contributes to the high blood pressure of mice subjected to the DOCA-salt model [132]. Additionally, it was shown that mice lacking MasR presented important imbalance in the neural control of blood pressure, with blunted sensitivity of not only the baroreflex but also the chemo- and Bezold–Jarisch reflexes [31]. More recently, studying rats that develop metabolic syndrome after chronic fructose intake, it was shown that fructose-fed rats receiving Ang-(1-7) infusion into the lateral ventricle had normalized baseline MAP, baroreflex control of HR, and reduced cardiac sympathetic tone [57]. Strikingly, along with these cardiovascular improvements, these rats presented normalized glucose tolerance, glycemia, insulinemia, and HOMA score [57]. These data suggest that activation of Ang-(1-7)/Mas pathway in the brain may play an important beneficial role against cardiovascular and metabolic disorders. In agreement, the overexpression of ACE2 selectively in the RVLM [139] or PVN [118] induces a significant decrease in blood pressure of SHR [139], attenuation of the sympathetic activity, and improvement of the baroreflex function in animals with congestive heart failure [71].

Cardiovascular Effects of Angiotensin-(1-7) at Specific Medullary Sites

Among the central actions, the more consistent effects of Ang-(1-7) are related to the modulation of the baroreflex, especially improving the bradycardic component of the baroreflex control of heart rate in normotensive [20] or hypertensive animals [16, 62, 93]. The facilitatory effect of Ang-(1-7) on baroreflex control of heart rate is also consistently observed after microinjection into the NTS, a key region in the brain stem controlling cardiovascular reflex function. Microinjection of Ang-(1-7) induces facilitation, while injection of Ang II produces attenuation of the baroreflex bradycardia, whereas A-779 and losartan, the selective Mas and AT1 receptor antagonists, blocked the effect of their respective peptide in normotensive or hypertensive rats [9, 26, 29, 33, 65, 106, 107].

At the CVLM, inhibitory area of the baroreflex arch, Ang-(1-7) induces Mas-related hypotensive responses similar to Ang II [44, 99, 115]. In addition, it was shown to release glutamate and decrease taurine [128]. At the RVLM, the main relay area for the sympathetic control of the cardiovascular system, microinjections of Ang-(1-7) induces pressor response [43, 44, 115, 145], an effect accompanied by increase in renal sympathetic nerve activity in normotensive [99, 145] and renovascular hypertensive animals [43, 44, 79, 82, 91, 116, 145]. Conversely, selective blockade of endogenous Ang-(1-7) by A-779 results in a decrease in BP in normotensive and hypertensive rats [43, 44, 82, 91, 116, 117, 128, 145]. Ang-(1-7) pressor effect at the RVLM is increased after hemorrhage [81] and in hypertensive rats [24, 41, 90, 91, 124]. In keeping with the excitatory role of Ang-(1-7) after acute injections, A779 or inhibition of ACE2 with the compound DX600 induces greater decreases in blood pressure in SHR [91]. Although Ang-(1-7)/Mas at the RVLM does not alter baroreflex [6], it increases cardiac sympathetic afferent reflex (CSAR), which contributes to sympathetic excitation in hypertension [79, 91, 145]. The facilitation of CSAR seems to be mediated by MasR activation of cAMP-protein kinase A and increases in NAD(P)H oxidase activity and superoxide anion level [79]. Peripheral mechanisms triggered by Ang-(1-7) at the CVLM and RVLM involve modulation of sympathetic tonus and a nitroxidergic pathway sensible to NOS inhibitors – LNAME and 7-NI [5, 22, 94].

Stress-induced hypertension induces increase in the expression of ACE and AT1, along with decrease in ACE2 and hyperresponsiveness of Ang II at the RVLM [37]. Moderate physical exercise during SHR development (7–23 weeks old) attenuated hypertension, prevented increase in TNFα, IL-1β, ACE, and AT1 expression in the RVLM, and upregulated IL-10, ACE2, and Mas at this site [1]. In addition, these changes were associated with reduced plasma Ang II levels, reduced neuronal activity, reduced NADPH-oxidase subunit gp91(phox) and inducible NO synthase in trained SHRs, indicating reduced oxidative stress [1]. Exercise training rescues ACE2 overexpression in the RVLM of animals subjected to heart failure [72]. A779 induced a pressor response in exercise-trained normotensive rats [13], suggesting that endogenous Ang-(1-7) triggers inhibition of pressor neurons at the RVLM. Moreover, long-term ACE2 expression in the RVLM in SHR produced significant and long-term reduction in blood pressure [139], suggesting that increasing RVLM Ang-(1-7) contributes to the anti-hypertensive effect of exercise training.

Cardiovascular Effects of Angiotensin-(1-7) at Specific Hypothalamic Sites

Ang-(1-7)-immunoreactivity is present in different hypothalamic regions, including the paraventricular, supraoptic, and suprachiasmatic nuclei and in the median eminence [15, 76]. MasR are also present in the parvo and magnocellular portions of the paraventricular hypothalamic nucleus (PVN), supraoptic nucleus, and lateral preoptic area [12]. Endogenous Ang-(1-7) in the rat hypothalamus is present in concentrations comparable to Ang I and Ang II [25]. Microiontophoretic application of Ang-(1-7) into the PVN augments the excitability of the neurons in this region [7, 8, 38], an effect that can be selectively blocked by the specific MasR antagonist A-779 [8]. Therefore, considering the key contribution of hypothalamic regions in the control of sympathetic activity [3, 45], a functional role for Ang-(1-7) in specific hypothalamic nuclei in autonomic cardiovascular regulation is very likely. Surprisingly, however, this functional aspect was evaluated only in few studies. In anesthetized rats, Gomes da Silva et al. [54] found that microinjection of Ang-(1-7) into the PVN increased renal sympathetic activity (RSNA) and this effect was mediated by MasR stimulation. This finding was confirmed by the fact that microinjection of the MasR antagonist A-779 into the PVN reduced RSNA, and this reduction was similar in magnitude to the effect observed with muscimol, a powerful neuronal inhibitor [54]. Blockade of MasR in the PVN also prevents the increases in arterial pressure associated to intermittent hypoxia [30]. Evidence also suggests that endogenous Ang-(1-7) in the PVN facilitates the cardiac sympathetic afferent reflex and the sympathetic outflow in renovascular hypertension [122]. Altogether, these studies indicate a sympathoexcitatory action for Ang-(1-7) in the PVN on autonomic cardiovascular control, at least in some specific conditions.

ACE2 overexpression in the subfornical organ, on the other hand, inhibits both pressor and drinking responses resulting from icv administration of Ang-II and this effect is attributed to a decrease in Ang II levels associated to an increase in Ang-(1-7) levels, leading to the activation of MasR [40]. Subfornical organ is a brain region known for its involvement in the regulation of autonomic cardiovascular function. Circulating peptide access to hypothalamic regions is likely mediated via subfornical organ [113]. In this regard, adenovirus-induced overexpression of ACE2 specifically in the PVN attenuates the hypertensive response evoked by peripheral infusion of Ang II [118]. In addition, MasR expression was increased while AT_1R expression was decreased at the PVN in the presence of ACE2 overexpression [118]. Since ACE2 efficiently hydrolyzes AngII to Ang-(1-7), a possible counteracting effect of ACE2/Ang-(1-7)/MasR on Ang II/AT_1R stimulatory effects at the PVN on autonomic cardiovascular control needs to be considered. Evidence also suggests that Ang-(1–7) may play a protective role in the anterior hypothalamic area in hypertensive conditions. In sinoaortic denervated rats, but not in normotensive rats, injection of Ang-(1-7) into the anterior hypothalamic area decreases blood pressure, an effect blocked by A-779 and thus mediated via MasR [63]. In addition, injections of Ang-(1-7) into the anterior hypothalamic area of SHRs decreases blood pressure, an effect mediated by nitric oxide (NO) generation [23]. In fact, Ang-(1-7) induced an increase in NOS activity and neuronal NOS expression in hypothalami from SHR [23].

Other Angiotensin-(1-7) Effects in the Brain

Stroke is one of the leading causes of death and impaired quality of life as a result of neurological deficit. It has been shown that central administration of Ang-(1-7) reduces brain damage and improves neurological outcome in ischemic stroke elicited by endothelin1-induced middle cerebral artery occlusion (MCAO) [88]. Similar protective effect of Ang-(1-7) has been documented in a rat model of permanent MCAO [66, 67]. Supporting these findings, ICV administration of the ACE2 activator diminazene aceturate under similar conditions to those used for Ang-(1-7) effectively decreased the intracerebral infarct and behavioral deficits resulting from endothelin-1-induced MCAO [88]. These protective actions of Ang-(1-7) and diminazene aceturate were abolished by coadministration of the MasR blocker, suggesting that is a MasR-mediated effect and supporting the involvement of Ang-(1-7) in the protective effects of the ACE2 activator [88, 121]. In another model of stroke as in stroke-prone SHR, which is an established animal model of hypertension-induced hemorrhagic stroke, central administration of Ang-(1-7) increases lifespan and improves the neurological status of these rats as well as decreases microglial numbers in the striatum, implying attenuation of cerebral inflammation [102].

The beneficial actions of Ang-(1-7) in ischemic stroke were due to anti-inflammatory mechanisms. Thus, reductions in inducible nitric oxide synthase (NOS) gene expression, in the pro-inflammatory interleukin (IL)-1b and IL-6, and in microglial activation were elicited by Ang-(1-7) [101, 121]. In addition, Ang-(1-7) decreased the levels of oxidative stress and suppressed NF-kB activity, a transcriptional regulator involved in inflammation, which was accompanied by a reduction of pro-inflammatory cytokines and cyclooxynease-2 (COX-2) in the peri-infarct regions [66]. The beneficial effects of Ang-(1-7) on ischemic stroke were reversed by the MasR antagonist but not by the AT2 receptor antagonist, suggesting the involvement of MasR. Neuronal overexpression of ACE2 protects the brain from ischemia-induced damage [27], and this effect was greater in older animals [144]. Blockade of the MasR pathway in the brain partially abolished the beneficial effects of ACE2 overexpression, suggesting that Ang-(1-7) production that results from ACE2 overexpression may exert this protective effect [143].

Ang-(1-7) promotes brain angiogenesis. Infusion of Ang-(1-7) for 4 weeks promotes endothelial cell proliferation and increases brain capillary density in rats with permanent MCAO, which was accompanied by endothelial nitric oxide synthase (eNOS) activation and upregulation of NO. Furthermore, Ang-(1-7)-induced brain angiogenesis attenuates the reduction of regional cerebral blood flow during subsequent ischemia and leads to the improvement in stroke outcome [67]. Supporting the beneficial effects of the depressor arm of the RAS, it has been shown that the expression of ACE2, Ang-(1-7), and MasR was upregulated after acute cerebral ischemic stroke in rats [85].

It has been shown that Ang-(1-7) exerts a protective role in blood–brain barrier damage. In cerebral ischemia reperfusion injury- and hypoxia-induced blood–brain barrier damage, Ang-(1-7) promotes the expression of zonula occludens-1 and claudin-5, which are proteins associated with tight junction in cerebral endothelial cells of the blood–brain barrier [130].

In an attempt to elucidate the cellular target for the protective effect elicited by Ang-(1-7) in the brain, we investigated the different cellular types protected by Ang-(1-7) by transmission electron microscopy in the model of brain damage induced by Shiga toxin 2 (Stx2)-producing enterohemorrhagic *Escherichia coli*. Stx2 induced neurodegeneration and axon demyelination. Ang-(1-7) prevented neuronal damage and hampered the Stx2-induced demyelination through the stimulation of MasR [53].

Evidence suggests that Ang-(1-7) could have a role in the central motor control because it is able to affect the dopamine and GABA release in the dorsal striatum [120]. This assumption is supported by recent results of Costa-Besada and colleagues [28] showing that MasR immunostaining is present in dopaminergic neurons and glial cells in the substantia nigra of rats, monkeys, and humans. In addition, these authors reported that the Ang-(1-7)/MasR axis in the dopaminergic system is downregulated with aging and this could facilitate the aging-related vulnerability to neurodegeneration [28].

Several recent studies demonstrated that ACE2/Ang-(1-7)/MasR interacts centrally with different neurotransmitters, including gamma-aminobutyric acid (GABA), dopamine, and norepinephrine (NE) [51, 120, 129]. Therefore, besides autonomic cardiovascular control, central Ang-(1-7) might be involved in several brain mechanisms, from hormone release to cognitive behavior. It has been reported that Ang-(1–7) is involved in learning and memory processes that occur in central limbic regions, such as the hippocampus [61, 78] and amygdala [2, 119]. Ang-(1-7) has been shown to enhance long-term potentiation in the CA1 region of the hippocampus [61] and in the lateral amygdala of rats [2]. Long-term potentiation and long-term depression are two forms of activity-dependent synaptic plasticity involved in learning and memory. MasR are highly expressed in the hippocampus [87], a critical brain structure for memory processing. In addition, Ang-(1-7)/MasR axis has been shown to facilitate object recognition memory performance [78]. In diabetic rats, Ang(1-7) treatment upregulates the expression of glial fibrillary acidic protein (GFAP) and glial cell line-derived neurotrophic factor (GDNF), promoting neuron survival in the hippocampus, an effect blocked by treatment with A-779 [141]. In addition, Ang-(1-7) improves the cognitive deficits in rats subjected to permanent bilateral occlusion of the common carotid arteries [134]. Finally, the improvement of cognitive aspects in humans taking captopril has been attributed to the higher availability of Ang-(1-7) [126].

Central Ang-(1-7)/MasR axis is strongly associated with emotional stress and anxiety modulation [42]. Ang-(1-7) psychotropic activity was one of the first biological actions reported for this heptapeptide [64]. Mice lacking the MasR presented increased anxiety behavior [127]. It is worth to mention that the later finding was published even before the association between Mas protein and Ang-(1-7) [111]. Central administration of Ang-(1-7) attenuates the stress-induced tachycardia [80] and reduces anxiety [68]. More recently, it was shown that Ang-(1-7) ICV injection attenuated anxious-like behavior in two different transgenic rat lines [4, 68, 69]. In addition, it was shown, for the first time, that Ang-(1-7) was effective in attenuating depression phenotype [4, 68]. Further, it is interesting to observe that hypertensive patients who suffer from depression exhibit mood improvement after treatment with ACE inhibitor captopril [146], which in turn was shown to increase Ang-(1-7) levels [21, 74]. Recently, it was shown that anxiety-like behavior of hypertensive rats was rescued in enalapril-treated animals after MasR blockade, unveiling an anti-depressive role for endogenous Ang-(1-7) [4]. Similar findings were found in mice by increasing ACE2 activity [129].

Evidence suggests that the amygdala may be a site involved in these ansiolitic effects [14, 95, 129]. MasR are robustly expressed in GABAergic neurons in the basolateral amygdala (BLA) and ACE2 overexpression increases the spontaneous postsynaptic inhibitory currents in this region [129]. Apart from modulating GABAergic actions, a direct functional interaction between MasR and the AT_1R in the amygdala is also a possibility. Halbach and colleagues found that the increase in field potentials in the BLA induced by Ang II was reversed in MasR knockout mice, where a decrease in field potentials was observed [60]. Another area implicated in the integration of emotional behavior and sympathetic drive to the periphery is the periaqueductal gray (PAG). Xing et al. [135] showed that Ang-(1-7) interaction with MasR inhibits neuronal activity of dorsolateral PAG by NO-dependent signaling pathway via a MasR activation [135], an effect that was impaired in chronic heart failure animals [136].

Neuromodulatory Actions of Ang-(1-7)

Ang-(1-7) possesses two main characteristics of a "neuromodulator," a diffuse site of action and the ability to influence the release and the respective effects of certain neurotransmitters [75]. NO has been strongly associated as a signaling molecule that mediates Ang-(1-7) effects in neurons. NO in the CNS acts as a sympatho-inhibitory molecule in meditating sympathetic outflow [96]. Ang-(1-7) is present in neurons containing NO synthase [18] and central administration of Ang-(1-7) stimulates NO release [142]. In the brain, Ang-(1-7) increases NO levels via activation of the MasR and the Ang II AT2R [39, 52]. Ang-(1-7) increases neuronal NO synthase-derived NO levels, increasing neuronal potassium current in catecholaminergic neurons [140]. In the PAG, Ang-(1–7) plays an inhibitory role via a NO-dependent signaling pathway, and spontaneous firing activity of PAG neurons is largely reduced in the presence of Ang-(1-7) [135].

Consistent evidences suggest that Ang-(1-7) can modulate GABA actions in the brain. GABA is a key neurotransmitter in key regions controlling the sympathetic tone [59, 96]. Ang-(1-7) increases extracellular GABA concentration in the rat striatum and this effect is blocked by the NO synthase inhibitor L-NAME, suggesting the participation of NO for the effect of Ang-(1-7) on GABA release [120]. MasR mRNA is present in a high percentage of GABAergic neurons in the BLA and expression of MasR mRNA is upregulated by ACE2 overexpression [129]. In addition, ACE2 overexpression increases the frequency of spontaneous inhibitory postsynaptic currents in the BLA, an effect that is eliminated by central infusion of the MasR antagonist. These findings strongly suggest that ACE2 may influence GABA neurotransmission within the BLA via MasR activation [129]. Interestingly, the BLA is a site of action for Ang-(1-7) in attenuating acute emotional stress responses [95]. Curiously, transgenic rats that overexpress Ang-(1-7) are less responsive to diazepam (a $GABA_A$ receptor-modulating drug) on anxiety behavior modulation when compared to controls [89].

Additional evidence suggests that Ang-(1-7) modulates the release of substance P in the hypothalamus [34], glutamate and taurine in the caudal ventrolateral medulla [128], and dopamine in the hypothalamus [97] and in the basal ganglia [28, 120].

Concerning NE release, Ang-(1-7) exerts a sympathoinhibitory action on sympathetic neurons from hypothalami of normotensive and SHR rats [50]. In neurons from hypothalamus and brainstem of normotensive and SHR, Ang-(1-7) leads to a decrease in NE levels in the synaptic cleft. This effect results from a decrease in neurotransmitter release and synthesis and an increase in its neuronal uptake. Thus, Ang-(1-7) through MasR elicits a decrease in NE release in the hypothalamus of normotensive and hypertensive rats in a bradykinin/NO-dependent manner through the cGMP/PKG pathway, which in turn, maintains low NE outflow [51, 52]. Ang-(1-7) also influences NE biosynthesis. The peptide through an AT_2R-mediated mechanism downregulates tyrosine hydroxylase, the rate-limiting enzyme in catecholamines biosynthesis, by increasing its degradation through stimulation of the ubiquitin-proteasome system, reducing in consequence NE biosynthesis in neurons from the hypothalamus of normotensive and SHR rats [83].

NE transporter (NET) regulates the clearance of NE from the synaptic cleft. Changes in the activity of NET should have a significant impact on the concentration and duration of NE present in the synaptic cleft; thus, NET is essential for a fine-tuned control of sympathetic activity. Ang-(1-7) induces a long-term stimulatory effect on NE neuronal uptake by increasing NET transcription and expression [84]. Altogether, these results showed that Ang-(1-7) elicits a negative neuromodulatory role on NE neurotransmission, thus contributing to the modulation of NE homeostasis and maintaining appropriate synaptic NE levels during hypertensive conditions.

References

1. Agarwal D, Welsch MA, Keller JN, Francis J. Chronic exercise modulates RAS components and improves balance between pro-and anti-inflammatory cytokines in the brain of SHR. Basic Res Cardiol. 2011;106:1069–85.
2. Albrecht D. Angiotensin-(1-7)-induced plasticity changes in the lateral amygdala are mediated by COX-2 and NO. Learn Mem. 2007;14:177–84.
3. Allen AM. Inhibition of the hypothalamic paraventricular nucleus in spontaneously hypertensive rats dramatically reduces sympathetic vasomotor tone. Hypertension. 2002;39:275–80.
4. Almeida-Santos AF, Kangussu LM, Moreira FA, Santos RA, Aguiar DC, Campagnole-Santos MJ. Anxiolytic-and antidepressant-like effects of angiotensin-(1–7) in hypertensive transgenic (mRen2) 27 rats. Clin Sci. 2016;130:1247–55.
5. Alzamora AC, Santos RA, Campagnole-Santos MJ. Hypotensive effect of ANG II and ANG-(1–7) at the caudal ventrolateral medulla involves different mechanisms. Am J Physiol-Regul Integr Comp Physiol. 2002;283:R1187–95.
6. Alzamora AC, Santos RA, Campagnole-Santos MJ. Baroreflex modulation by angiotensins at the rat rostral and caudal ventrolateral medulla. Am J Physiol-Regul Integr Comp Physiol. 2006;290:R1027–34.
7. Ambühl P, Felix D, Imboden H, Khosla MC, Ferrario CM. Effects of angiotensin analogues and angiotensin receptor antagonists on paraventricular neurones. Regul Pept. 1992;38:111–20.
8. Ambühl P, Felix D, Khosla MC. [7-D-ALA]-angiotensin-(1–7): selective antagonism of angiotensin-(1–7) in the rat paraventricular nucleus. Brain Res Bull. 1994;35:289–91.
9. Arnold AC, Sakima A, Ganten D, Ferrario CM, Diz DI. Modulation of reflex function by endogenous angiotensins in older transgenic rats with low glial angiotensinogen. Hypertension. 2008;51:1326–31.
10. Arnold AC, Sakima A, Kasper SO, Vinsant S, Garcia-Espinosa MA, Diz DI. The brain renin-angiotensin system and cardiovascular responses to stress: insights from transgenic rats with low brain angiotensinogen. J Appl Physiol. 2012;113:1929–36.
11. Barnes KL, Knowles WD, Ferrario CM. Angiotensin II and angiotensin (1–7) excite neurons in the canine medulla in vitro. Brain Res Bull. 1990;24:275–80.
12. Becker LK, Etelvino GM, Walther T, Santos RAS, Campagnole-Santos MJ. Immunofluorescence localization of the receptor Mas in cardiovascular-related areas of the rat brain. AJP Heart Circ Physiol. 2007;293:H1416–24.
13. Becker LK, Santos RA, Campagnole-Santos MJ. Cardiovascular effects of angiotensin II and angiotensin-(1–7) at the RVLM of trained normotensive rats. Brain Res. 2005;1040:121–8.
14. Bild W, Ciobica A. Angiotensin-(1–7) central administration induces anxiolytic-like effects in elevated plus maze and decreased oxidative stress in the amygdala. J Affect Disord. 2013;145:165–71.
15. Block CH, Santos RAS, Brosnihan KB, Ferrario CM. Immunocytochemical localization of angiotensin-(1–7) in the rat forebrain. Peptides. 1988;9:1395–401.

16. Britto RR, Santos RA, Fagundes-Moura CR, Khosla MC, Campagnole-Santos MJ. Role of angiotensin-(1-7) in the modulation of the baroreflex in renovascular hypertensive rats. Hypertension. 1997;30:549–56.

17. Brosnihan KB, Schiavone MT, Sprunger AE, Chappell MC, Rizzo M, Ferrario CM. In vivo release of angiotensin II from the rat hypothalamus. Hypertension. 1988;11:I158.

18. Calka J, Block CH. Angiotensin-(1–7) and nitric oxide synthase in the hypothalamo-neurohypophysial system. Brain Res Bull. 1993;30:677–85.

19. Campagnole-Santos MJ, Diz DI, Santos RA, Khosla MC, Brosnihan KB, Ferrario CM. Cardiovascular effects of angiotensin-(1-7) injected into the dorsal medulla of rats. Am J Physiol-Heart Circ Physiol. 1989;257:H324–9.

20. Campagnole-Santos MJ, Heringer SB, Batista EN, Khosla MC, Santos RA. Differential baroreceptor reflex modulation by centrally infused angiotensin peptides. Am J Physiol-Regul Integr Comp Physiol. 1992;263:R89–94.

21. Campbell DJ, Kladis A, Duncan A-M. Effects of converting enzyme inhibitors on angiotensin and bradykinin peptides. Hypertension. 1994;23:439–49.

22. Cangussu LM, de Castro UGM, do Pilar Machado R, Silva ME, Ferreira PM, RAS d S, Campagnole-Santos MJ, Alzamora AC. Angiotensin-(1-7) antagonist, A-779, microinjection into the caudal ventrolateral medulla of renovascular hypertensive rats restores baroreflex bradycardia. Peptides. 2009;30:1921–7.

23. Cerrato BD, Frasch AP, Nakagawa P, Longo-Carbajosa N, Peña C, Höcht C, Gironacci MM. Angiotensin-(1–7) upregulates central nitric oxide synthase in spontaneously hypertensive rats. Brain Res. 2012;1453:1–7.

24. Chan YS, Wong TM. Relationship of rostral ventrolateral medullary neurons and angiotensin in the central control of blood pressure. Neurosignals. 1995;4:133–41.

25. Chappell MC, Brosnihan KB, Diz DI, Ferrario CM. Identification of angiotensin-(1-7) in rat brain. Evidence for differential processing of angiotensin peptides J Biol Chem. 1989;264:16518–23.

26. Chaves GZ, Caligiorne SM, Santos RA, Khosla MC, Campagnole-Santos MJ. Modulation of the baroreflex control of heart rate by angiotensin-(1–7) at the nucleus tractus solitarii of normotensive and spontaneously hypertensive rats. J Hypertens. 2000;18:1841–8.

27. Chen J, Zhao Y, Chen S, Wang J, Xiao X, Ma X, Penchikala M, Xia H, Lazartigues E, Zhao B. Neuronal over-expression of ACE2 protects brain from ischemia-induced damage. Neuropharmacology. 2014;79:550–8.

28. Costa-Besada MA, Valenzuela R, Garrido-Gil P, Villar-Cheda B, Parga JA, Lanciego JL, Labandeira-Garcia JL. Paracrine and intracrine angiotensin 1-7/Mas receptor axis in the substantia nigra of rodents, monkeys, and humans. Mol Neurobiol. 2018;55(7):5847–67.

29. Couto AS, Baltatu O, Santos RA, Ganten D, Bader M, Campagnole-Santos MJ. Differential effects of angiotensin II and angiotensin-(1-7) at the nucleus tractus solitarii of transgenic rats with low brain angiotensinogen. J Hypertens. 2002;20:919–25.

30. da Silva AQG, Fontes MAP, Kanagy NL. Chronic infusion of angiotensin receptor antagonists in the hypothalamic paraventricular nucleus prevents hypertension in a rat model of sleep apnea. Brain Res. 2011;1368:231–8.

31. de Moura MM, dos Santos RAS, Campagnole-Santos MJ, Todiras M, Bader M, Alenina N, Haibara AS. Altered cardiovascular reflexes responses in conscious Angiotensin-(1-7) receptor Mas-knockout mice. Peptides. 2010;31:1934–9.

32. Díez-Freire C, Vázquez J, Correa de Adjounian MF, Ferrari MF, Yuan L, Silver X, Torres R, Raizada MK. ACE2 gene transfer attenuates hypertension-linked pathophysiological changes in the SHR. Physiol Genomics. 2006;27:12–9.

33. Diz DI, Garcia-Espinosa MA, Gallagher PE, Ganten D, Ferrario CM, Averill DB. Angiotensin-(1−7) and Baroreflex function in nucleus Tractus Solitarii of (mRen2) 27 transgenic rats. J Cardiovasc Pharmacol. 2008;51:542.

34. Diz DI, Pirro NT. Differential actions of angiotensin II and angiotensin-(1-7) on transmitter release. Hypertension. 1992;19:II41.

35. Donoghue M, Hsieh F, Baronas E, Godbout K, Gosselin M, Stagliano N, Donovan M, Woolf B, Robison K, Jeyaseelan R. A novel angiotensin-converting enzyme–related carboxypeptidase (ACE2) converts angiotensin I to angiotensin 1-9. Circ Res. 2000;87:e1–9.
36. Doobay MF, Talman LS, Obr TD, Tian X, Davisson RL, Lazartigues E. Differential expression of neuronal ACE2 in transgenic mice with overexpression of the brain renin-angiotensin system. Am J Physiol-Regul Integr Comp Physiol. 2007;292:R373–81.
37. Du D, Chen J, Liu M, Zhu M, Jing H, Fang J, Shen L, Zhu D, Yu J, Wang J. The effects of angiotensin II and angiotensin-(1–7) in the rostral ventrolateral medulla of rats on stress-induced hypertension. PLoS One. 2013;8:e70976.
38. Felix D, Khosla MC, Barnes KL, Imboden H, Montani B, Ferrario CM. Neurophysiological responses to angiotensin-(1-7). Hypertension. 1991;17:1111–4.
39. Feng Y, Xia H, Cai Y, Halabi CM, Becker LK, Santos RA, Speth RC, Sigmund CD, Lazartigues E. Brain-selective overexpression of human angiotensin-converting enzyme type 2 attenuates neurogenic hypertension. Circ Res. 2010;106:373–82.
40. Feng Y, Yue X, Xia H, Bindom SM, Hickman PJ, Filipeanu CM, Wu G, Lazartigues E. Angiotensin-converting enzyme 2 overexpression in the Subfornical organ prevents the angiotensin II-mediated Pressor and drinking responses and is associated with angiotensin II type 1 receptor Downregulation. Circ Res. 2008;102:729–36.
41. Fontes M, Baltatu O, Caligiorne S, Campagnole-Santos M, Ganten D, Bader M, Santos R. Angiotensin peptides acting at rostral ventrolateral medulla contribute to hypertension of TGR (mREN2) 27 rats. Physiol Genomics. 2000;2:137–42.
42. Fontes MAP, Martins Lima A, dos Santos RAS. Brain angiotensin-(1–7)/Mas axis: a new target to reduce the cardiovascular risk to emotional stress. Neuropeptides. 2016;56:9–17.
43. Fontes MAP, Pinge MM, Naves V, Campagnole-Santos MJ, Lopes OU, Khosla MC, Santos RAS. Cardiovascular effects produced by microinjection of angiotensins and angiotensin antagonists into the ventrolateral medulla of freely moving rats. Brain Res. 1997;750:305–10.
44. Fontes MAP, Silva LCS, Campagnole-Santos MJ, Khosla MC, Guertzenstein PG, Santos RAS. Evidence that angiotensin-(1–7) plays a role in the central control of blood pressure at the ventro-lateral medulla acting through specific receptors. Brain Res. 1994;665:175–80.
45. Fontes MAP, Tagawa T, Polson JW, Cavanagh S-J, Dampney RAL. Descending pathways mediating cardiovascular response from dorsomedial hypothalamic nucleus. Am J Physiol-Heart Circ Physiol. 2001;280:H2891–901.
46. Freund M, Walther T, von Bohlen und Halbach O. Immunohistochemical localization of the angiotensin-(1–7) receptor Mas in the murine forebrain. Cell Tissue Res. 2012;348:29–35.
47. Ganten D, Speck G. The brain renin-angiotensin system: a model for the synthesis of peptides in the brain. Biochem Pharmacol. 1978;27:2379–89.
48. Gao J, Marc Y, Iturrioz X, Leroux V, Balavoine F, Llorens-Cortes C. A new strategy for treating hypertension by blocking the activity of the brain renin–angiotensin system with aminopeptidase a inhibitors. Clin Sci. 2014;127:135–48.
49. Garcia-Espinosa MA, Shaltout HA, Gallagher PE, Chappell MC, Diz DI. In vivo expression of angiotensin-(1-7) lowers blood pressure and improves baroreflex function in transgenic (mRen2) 27 rats. J Cardiovasc Pharmacol. 2012;60:150.
50. Gironacci MM, Carbajosa NAL, Goldstein J, Cerrato BD. Neuromodulatory role of angiotensin-(1–7) in the central nervous system. Clin Sci. 2013;125:57–65.
51. Gironacci MM, Valera MS, Yujnovsky I, Pena C. Angiotensin-(1–7) inhibitory mechanism of norepinephrine release in hypertensive rats. Hypertension. 2004;44:783–7.
52. Gironacci MM, Vatta M, Rodriguez-Fermepín M, Fernández BE, Peña C. Angiotensin-(1–7) reduces norepinephrine release through a nitric oxide mechanism in rat hypothalamus. Hypertension. 2000;35:1248–52.
53. Goldstein J, Carden TR, Perez MJ, Taira CA, Höcht C, Gironacci MM. Angiotensin-(1–7) protects from brain damage induced by Shiga toxin 2-producing enterohemorrhagic Escherichia coli. Am J Physiol-Regul Integr Comp Physiol. 2016;311:R1173–85.
54. Gomes da Silva AQ, Sousa dos Santos RA, Peliky Fontes MA. Blockade of endogenous angiotensin-(1–7) in the hypothalamic Paraventricular nucleus reduces renal sympathetic tone. Hypertension. 2005;46:341–8.

55. Gouveia TLF, Frangiotti MIB, de Brito JMV, de Castro Neto EF, Sakata MM, Febba AC, Casarini DE, Amado D, Cavalheiro EA, Almeida SS. The levels of renin–angiotensin related components are modified in the hippocampus of rats submitted to pilocarpine model of epilepsy. Neurochem Int. 2012;61:54–62.

56. Grobe JL, Xu D, Sigmund CD. An intracellular renin-angiotensin system in neurons: fact, hypothesis, or fantasy. Physiology. 2008;23:187–93.

57. Guimaraes PS, Oliveira MF, Braga JF, Nadu AP, Schreihofer A, Santos RA, Campagnole-Santos MJ. Increasing angiotensin-(1–7) levels in the brain attenuates metabolic syndrome–related risks in fructose-fed rats. Hypertension. 2014;63:1078–85.

58. Guimaraes PS, Santiago NM, Xavier CH, Velloso EP, Fontes MA, Santos RA, Campagnole-Santos MJ. Chronic infusion of angiotensin-(1-7) into the lateral ventricle of the brain attenuates hypertension in DOCA-salt rats. Am J Physiol-Heart Circ Physiol. 2012;303:H393–400.

59. Guyenet PG. The sympathetic control of blood pressure. Nat Rev Neurosci. 2006;7:335–46.

60. Halbach OVBU, Walther T, Bader M, Albrecht D. Interaction between Mas and the angiotensin AT1 receptor in the amygdala. J Neurophysiol. 2000;83:2012–21.

61. Hellner K, Walther T, Schubert M, Albrecht D. Angiotensin-(1–7) enhances LTP in the hippocampus through the G-protein-coupled receptor Mas. Mol Cell Neurosci. 2005;29:427–35.

62. Heringer-Walther S, Batista ÉN, Walther T, Khosla MC, Santos RA, Campagnole-Santos MJ. Baroreflex improvement in SHR after ACE inhibition involves angiotensin-(1-7). Hypertension. 2001;37:1309–14.

63. Höcht C, Gironacci MM, Mayer MA, Schuman M, Bertera FM, Taira CA. Involvement of angiotensin-(1–7) in the hypothalamic hypotensive effect of captopril in sinoaortic denervated rats. Regul Pept. 2008;146:58–66.

64. Hoły Z, Braszko J, Kupryszewski G, Witczuk B, Wiśniewski K. Angiotensin II--derived peptides devoid of phenylalanine in position 8 have full psychotropic activity of the parent hormone. J Physiol Pharmacol Off J Pol Physiol Soc. 1992;43:183–92.

65. Isa K, Arnold AC, Westwood BM, Chappell MC, Diz DI. Angiotensin-converting enzyme inhibition, but not AT1 receptor blockade, in the solitary tract nucleus improves baroreflex sensitivity in anesthetized transgenic hypertensive (mRen2) 27 rats. Hypertens Res. 2011;34:1257–62.

66. Jiang T, Gao L, Guo J, Lu J, Wang Y, Zhang Y. Suppressing inflammation by inhibiting the NF-κB pathway contributes to the neuroprotective effect of angiotensin-(1-7) in rats with permanent cerebral ischaemia. Br J Pharmacol. 2012;167:1520–32.

67. Jiang T, Yu J-T, Zhu X-C, Zhang Q-Q, Tan M-S, Cao L, Wang H-F, Lu J, Gao Q, Zhang Y-D. Angiotensin-(1–7) induces cerebral ischaemic tolerance by promoting brain angiogenesis in a Mas/eNOS-dependent pathway. Br J Pharmacol. 2014;171:4222–32.

68. Kangussu LM, Almeida-Santos AF, Bader M, Alenina N, Fontes MAP, Santos RA, Aguiar DC, Campagnole-Santos MJ. Angiotensin-(1-7) attenuates the anxiety and depression-like behaviors in transgenic rats with low brain angiotensinogen. Behav Brain Res. 2013;257:25–30.

69. Kangussu LM, Almeida-Santos AF, Moreira FA, Fontes MA, Santos RA, Aguiar DC, Campagnole-Santos MJ. Reduced anxiety-like behavior in transgenic rats with chronically overproduction of angiotensin-(1–7): role of the Mas receptor. Behav Brain Res. 2017;331:193–8.

70. Kangussu LM, Guimaraes PS, Nadu AP, Melo MB, Santos RA, Campagnole-Santos MJ. Activation of angiotensin-(1–7)/Mas axis in the brain lowers blood pressure and attenuates cardiac remodeling in hypertensive transgenic (mRen2) 27 rats. Neuropharmacology. 2015;97:58–66.

71. Kar S, Gao L, Belatti DA, Curry PL, Zucker IH. Central angiotensin (1-7) enhances baroreflex gain in conscious rabbits with heart failure. Hypertension. 2011;58:627–34.

72. Kar S, Gao L, Zucker IH. Exercise training normalizes ACE and ACE2 in the brain of rabbits with pacing-induced heart failure. J Appl Physiol. 2010;108:923–32.

73. Karamyan VT, Speth RC. Enzymatic pathways of the brain renin–angiotensin system: unsolved problems and continuing challenges. Regul Pept. 2007;143:15–27.

74. Kohara K, Brosnihan KB, Ferrario CM. Angiotensin(1-7) in the spontaneously hypertensive rat. Peptides. 1993;14(5):883–91.

75. Kow LM, Pfaff DW. Neuromodulatory actions of peptides. Annu Rev Pharmacol Toxicol. 1988;28:163–88.
76. Krob HA, Vinsant SL, Ferrario CM, Friedman DP. Angiotensin-(1–7) immunoreactivity in the hypothalamus of the (mRen-2d)27 transgenic rat. Brain Res. 1998;798:36–45.
77. Lautner RQ, Villela DC, Fraga-Silva RA, Silva NC, Verano-Braga T, Costa-Fraga F, Jankowski J, Jankowski V, De Sousa FB, Alzamora AC. Discovery and characterization of alamandine, a novel component of the renin-angiotensin system. Circ Res. 2013;112:1104–11.
78. Lazaroni TLN, Raslan ACS, Fontes WRP, de Oliveira ML, Bader M, Alenina N, Moraes MFD, dos Santos RA, Pereira GS. Angiotensin-(1–7)/Mas axis integrity is required for the expression of object recognition memory. Neurobiol Learn Mem. 2012;97:113–23.
79. Li P, Sun H-J, Cui B-P, Zhou Y-B, Han Y. Angiotensin-(1–7) in the rostral ventrolateral medulla modulates enhanced cardiac sympathetic afferent reflex and sympathetic activation in renovascular hypertensive rats. Hypertension. 2013;61:820–7.
80. Lima AM, Xavier CH, Ferreira AJ, Raizada MK, Wallukat G, Velloso EPP, dos Santos RAS, Fontes MAP. Activation of angiotensin-converting enzyme 2/angiotensin-(1–7)/Mas axis attenuates the cardiac reactivity to acute emotional stress. Am J Physiol-Heart Circ Physiol. 2013;305:H1057–67.
81. Lima DX, Campagnole-Santos MJ, Fontes MA, Khosla MC, Santos RA. Haemorrhage increases the pressor effect of angiotensin-(1–7) but not of angiotensin II at the rat rostral ventrolateral medulla. J Hypertens. 1999;17:1145–52.
82. Lima DX, Fontes MAP, Oliveira RC, Campagnole-Santos MJ, Khosla MC, Santos RA. Pressor action of angiotensin I at the ventrolateral medulla: effect of selective angiotensin blockade. Immunopharmacology. 1996;33:305–7.
83. Lopez Verrilli MA, Pirola CJ, Pascual MM, Dominici FP, Turyn D, Gironacci MM. Angiotensin-(1–7) through AT2 receptors mediates tyrosine hydroxylase degradation via the ubiquitin–proteasome pathway. J Neurochem. 2009;109:326–35.
84. Lopez Verrilli MA, Rodriguez Fermepín M, Longo Carbajosa N, Landa S, Cerrato BD, García S, Fernandez BE, Gironacci MM. Angiotensin-(1-7) through Mas receptor up-regulates neuronal norepinephrine transporter via Akt and Erk1/2-dependent pathways: Ang-(1-7) on norepinephrine neuronal uptake. J Neurochem. 2012;120:46–55.
85. Lu J, Jiang T, Wu L, Gao L, Wang Y, Zhou F, Zhang S, Zhang Y. The expression of angiotensin-converting enzyme 2–angiotensin-(1–7)–Mas receptor axis are upregulated after acute cerebral ischemic stroke in rats. Neuropeptides. 2013;47:289–95.
86. Mahon JM, Allen M, Herbert J, Fitzsimons JT. The association of thirst, sodium appetite and vasopressin release with c-fos expression in the forebrain of the rat after intracerebroventricular injection of angiotensin II, angiotensin-(1–7) or carbachol. Neuroscience. 1995;69:199–208.
87. Martin KA, Grant SGN, Hockfield S. The mas proto-oncogene is developmentally regulated in the rat central nervous system. Dev Brain Res. 1992;68:75–82.
88. Mecca AP, Regenhardt RW, O'Connor TE, Joseph JP, Raizada MK, Katovich MJ, Sumners C. Cerebroprotection by angiotensin-(1–7) in endothelin-1-induced ischaemic stroke. Exp Physiol. 2011;96:1084–96.
89. Moura Santos D, Ribeiro Marins F, Limborço-Filho M, de Oliveira ML, Hamamoto D, Xavier CH, Moreira FA, Santos RAS, Campagnole-Santos MJ, Peliky Fontes MA. Chronic overexpression of angiotensin-(1-7) in rats reduces cardiac reactivity to acute stress and dampens anxious behavior. Stress. 2017;20:189–96.
90. Muratani H, Averill DB, Ferrario CM. Effect of angiotensin II in ventrolateral medulla of spontaneously hypertensive rats. Am J Physiol-Regul Integr Comp Physiol. 1991;260:R977–84.
91. Nakagaki T, Hirooka Y, Ito K, Kishi T, Hoka S, Sunagawa K. Role of angiotensin-(1-7) in rostral ventrolateral medulla in blood pressure regulation via sympathetic nerve activity in Wistar-Kyoto and spontaneous hypertensive rats. Clin Exp Hypertens. 2011;33:223–30.
92. Nautiyal M, Shaltout HA, de Lima DC, do Nascimento K, Chappell MC, Diz DI. Central angiotensin-(1–7) improves vagal function independent of blood pressure in hypertensive (mRen2) 27 rats. Hypertension. 2012;60:1257–65.

93. Oliveira DR, Santos RA, Santos GF, Khosla MC, Campagnole-Santos MJ. Changes in the baroreflex control of heart rate produced by central infusion of selective angiotensin antagonists in hypertensive rats. Hypertension. 1996;27:1284–90.

94. Oliveira RC, Campagnole-Santos MJ, Santos RA. The pressor effect of angiotensin-(1-7) in the rat rostral ventrolateral medulla involves multiple peripheral mechanisms. Clinics. 2013;68:245–52.

95. Oscar CG, de Figueirêdo Müller-Ribeiro FC, de Castro LG, Lima AM, Campagnole-Santos MJ, Santos RAS, Xavier CH, Fontes MAP. Angiotensin-(1–7) in the basolateral amygdala attenuates the cardiovascular response evoked by acute emotional stress. Brain Res. 2015;1594:183–9.

96. Patel KP. Role of paraventricular nucleus in mediating sympathetic outflow in heart failure. Heart Fail Rev. 2000;5:73–86.

97. Pawlak R, Napiorkowska-Pawlak D, Takada Y, Urano T, Nagai N, Ihara H, Takada A. The differential effect of angiotensin II and angiotensin 1-7 on norepinephrine, epinephrine, and dopamine concentrations in rat hypothalamus: the involvement of angiotensin receptors. Brain Res Bull. 2001;54:689–94.

98. Pereira MG, Souza LL, Becari C, Duarte DA, Camacho FR, Oliveira JAC, Gomes MD, Oliveira EB, Salgado MCO, Garcia-Cairasco N. Angiotensin II–independent angiotensin-(1–7) formation in rat HippocampusNovelty and significance. Hypertension. 2013;62:879–85.

99. Potts PD, Horiuchi J, Coleman MJ, Dampney RAL. The cardiovascular effects of angiotensin-(1–7) in the rostral and caudal ventrolateral medulla of the rabbit. Brain Res. 2000;877:58–64.

100. Qadri F, Wolf A, Waldmann T, Rascher W, Unger T. Sensitivity of hypothalamic paraventricular nucleus to C- and N-terminal angiotensin fragments: vasopressin release and drinking. J Neuroendocrinol. 1998;10:275–81.

101. Regenhardt RW, Desland F, Mecca AP, Pioquinto DJ, Afzal A, Mocco J, Sumners C. Anti-inflammatory effects of angiotensin-(1-7) in ischemic stroke. Neuropharmacology. 2013;71:154–63.

102. Regenhardt RW, Mecca AP, Desland F, Ritucci-Chinni PF, Ludin JA, Greenstein D, Banuelos C, Bizon JL, Reinhard MK, Sumners C. Centrally administered angiotensin-(1–7) increases the survival of stroke-prone spontaneously hypertensive rats. Exp Physiol. 2014;99:442–53.

103. Rice GI, Jones AL, Grant PJ, Carter AM, Turner AJ, Hooper NM. Circulating activities of angiotensin-converting enzyme, its homolog, angiotensin-converting enzyme 2, and neprilysin in a family study. Hypertension. 2006;48:914–20.

104. Rice GI, Thomas DA, Grant PJ, Turner AJ, Hooper NM. Evaluation of angiotensin-converting enzyme (ACE), its homologue ACE2 and neprilysin in angiotensin peptide metabolism. Biochem J. 2004;383:45–51.

105. Rowe BP, Saylor DL, Speth RC, Absher DR. Angiotensin-(1–7) binding at angiotensin II receptors in the rat brain. Regul Pept. 1995;56:139–46.

106. Sakima A, Averill DB, Gallagher PE, Kasper SO, Tommasi EN, Ferrario CM, Diz DI. Impaired heart rate Baroreflex in older rats. Hypertension. 2005;46:333–40.

107. Sakima A, Averill DB, Kasper SO, Jackson L, Ganten D, Ferrario CM, Gallagher PE, Diz DI. Baroreceptor reflex regulation in anesthetized transgenic rats with low glia-derived angiotensinogen. Am J Physiol-Heart Circ Physiol. 2007;292:H1412–9.

108. Santos R, Campagnole-Santos M, Baracho NC, Fontes M, Silva LC, Neves LA, Oliveira DR, Caligiorne SM, Rodrigues AR, Gropen C. Characterization of a new angiotensin antagonist selective for angiotensin-(1–7): evidence that the actions of angiotensin-(1–7) are mediated by specific angiotensin receptors. Brain Res Bull. 1994;35:293–8.

109. Santos RA, Brosnihan KB, Chappell MC, Pesquero J, Chernicky CL, Greene LJ, Ferrario CM. Converting enzyme activity and angiotensin metabolism in the dog brainstem. Hypertension. 1988;11:I153.

110. Santos RA, Campagnole-Santos MJ, Andrade SP. Angiotensin-(1–7): an update. Regul Pept. 2000;91:45–62.

111. Santos RAS, e Silva ACS, Maric C, Silva DMR, Machado RP, de Buhr I, Heringer-Walther S, Pinheiro SVB, Lopes MT, Bader M, Mendes EP, Lemos VS, Campagnole-Santos MJ, Schultheiss H-P, Speth R, Walther T. Angiotensin-(1-7) is an endogenous ligand for the G protein-coupled receptor Mas. Proc Natl Acad Sci. 2003;100:8258–63.

112. Schiavone MT, Santos RA, Brosnihan KB, Khosla MC, Ferrario CM. Release of vasopressin from the rat hypothalamo-neurohypophysial system by angiotensin-(1-7) heptapeptide. Proc Natl Acad Sci. 1988;85:4095–8.

113. Shekhar A. Angiotensin type 1 receptor antagonists—a novel approach to augmenting post-traumatic stress disorder and phobia therapies? Biol Psychiatry. 2014;75:836–7.

114. Silva DMR, Vianna HR, Cortes SF, Campagnole-Santos MJ, Santos RAS, Lemos VS. Evidence for a new angiotensin-(1–7) receptor subtype in the aorta of Sprague–Dawley rats. Peptides. 2007;28:702–7.

115. Silva LCS, Fontes MAP, Campagnole-Santos MJ, Khosla MC, Campos RR, Guertzenstein PG, Santos RAS. Cardiovascular effects produced by micro-injection of angiotensin-(1–7) on vasopressor and vasodepressor sites of the ventrolateral medulla. Brain Res. 1993;613:321–5.

116. Silva-Barcellos NM, Caligiorne S, dos Santos RA, Frézard F. Site-specific microinjection of liposomes into the brain for local infusion of a short-lived peptide. J Control Release. 2004;95:301–7.

117. Silva-Barcellos NM, Frézard F, Caligiorne S, Santos RA. Long-lasting cardiovascular effects of liposome-entrapped angiotensin-(1-7) at the rostral ventrolateral medulla. Hypertension. 2001;38:1266–71.

118. Sriramula S, Cardinale JP, Lazartigues E, Francis J. ACE2 overexpression in the paraventricular nucleus attenuates angiotensin II-induced hypertension. Cardiovasc Res. 2011;92:401–8.

119. Staschewski J, Kulisch C, Albrecht D. Different isoforms of nitric oxide synthase are involved in angiotensin-(1–7)-mediated plasticity changes in the amygdala in a gender-dependent manner. Neuroendocrinology. 2011;94:191–9.

120. Stragier B, Hristova I, Sarre S, Ebinger G, Michotte Y. In vivo characterization of the angiotensin-(1–7)-induced dopamine and γ-aminobutyric acid release in the striatum of the rat. Eur J Neurosci. 2005;22:658–64.

121. Sumners C, Horiuchi M, Widdop RE, McCarthy C, Unger T, Steckelings UM. Protective arms of the renin–angiotensin-system in neurological disease. Clin Exp Pharmacol Physiol. 2013;40:580–8.

122. Sun H-J, Li P, Chen W-W, Xiong X-Q, Han Y. Angiotensin II and angiotensin-(1-7) in Paraventricular nucleus modulate cardiac sympathetic afferent reflex in Renovascular hypertensive rats. PLoS One. 2012;7:e52557.

123. Tipnis SR, Hooper NM, Hyde R, Karran E, Christie G, Turner AJ. A human homolog of angiotensin-converting enzyme cloning and functional expression as a captopril-insensitive carboxypeptidase. J Biol Chem. 2000;275:33238–43.

124. Tsuchihashi T, Kagiyama S, Matsumura K, Abe I, Fujishima M. Effects of chronic oral treatment with imidapril and TCV-116 on the responsiveness to angiotensin II in ventrolateral medulla of SHR. J Hypertens. 1999;17:917–22.

125. Vickers C, Hales P, Kaushik V, Dick L, Gavin J, Tang J, Godbout K, Parsons T, Baronas E, Hsieh F. Hydrolysis of biological peptides by human angiotensin-converting enzyme-related carboxypeptidase. J Biol Chem. 2002;277:14838–43.

126. von Bohlen und Halbach O, Albrecht D. The CNS renin-angiotensin system. Cell Tissue Res. 2006;326:599–616.

127. Walther T, Balschun D, Voigt J-P, Fink H, Zuschratter W, Birchmeier C, Ganten D, Bader M. Sustained long term potentiation and anxiety in mice lacking the Mas Protooncogene. J Biol Chem. 1998;273:11867–73.

128. Wang J, Peng Y-J, Zhu D-N. Amino acids modulate the hypotensive effect of angiotensin-(1-7) at the caudal ventrolateral medulla in rats. Regul Pept. 2005;129:1–7.

129. Wang L, de Kloet AD, Pati D, Hiller H, Smith JA, Pioquinto DJ, Ludin JA, Oh SP, Katovich MJ, Frazier CJ. Increasing brain angiotensin converting enzyme 2 activity decreases

anxiety-like behavior in male mice by activating central Mas receptors. Neuropharmacology. 2016;105:114–23.

130. Wu J, Zhao D, Wu S, Wang D. Ang-(1–7) exerts protective role in blood–brain barrier damage by the balance of TIMP-1/MMP-9. Eur J Pharmacol. 2015;748:30–6.

131. Xia H, Lazartigues E. Angiotensin-converting enzyme 2 in the brain: properties and future directions. J Neurochem. 2008;107:1482–94.

132. Xia H, Sriramula S, Chhabra KH, Lazartigues E. Brain angiotensin-converting enzyme type 2 shedding contributes to the development of neurogenic Hypertension Novelty and significance. Circ Res. 2013;113:1087–96.

133. Xiao L, Gao L, Lazartigues E, Zucker IH. Brain-selective overexpression of angiotensin-converting enzyme 2 attenuates sympathetic nerve activity and enhances baroreflex function in chronic heart failure. Hypertension. 2011;58:1057–65.

134. Xie W, Zhu D, Ji L, Tian M, Xu C, Shi J. Angiotensin-(1-7) improves cognitive function in rats with chronic cerebral hypoperfusion. Brain Res. 2014;1573:44–53.

135. Xing J, Kong J, Lu J, Li J. Angiotensin-(1-7) inhibits neuronal activity of dorsolateral periaqueductal gray via a nitric oxide pathway. Neurosci Lett. 2012;522:156–61.

136. Xing J, Lu J, Li J. Role of angiotensin-(1–7) and Mas-R-nNOS pathways in amplified neuronal activity of dorsolateral periaqueductal gray after chronic heart failure. Neurosci Lett. 2014;563:6–11.

137. Xue B, Zhang Z, Beltz TG, Guo F, Hay M, Johnson AK. Estrogen regulation of the brain renin-angiotensin system in protection against angiotensin II-induced sensitization of hypertension. Am J Physiol-Heart Circ Physiol. 2014;307:H191–8.

138. Xue B, Zhang Z, Johnson RF, Guo F, Hay M, Johnson AK. Central endogenous angiotensin-(1–7) protects against aldosterone/NaCl-induced hypertension in female rats. Am J Physiol-Heart Circ Physiol. 2013;305:H699–705.

139. Yamazato M, Yamazato Y, Sun C, Diez-Freire C, Raizada MK. Overexpression of angiotensin-converting enzyme 2 in the rostral ventrolateral medulla causes long-term decrease in blood pressure in the spontaneously hypertensive rats. Hypertension. 2007;49:926–31.

140. Yang R-F, Yin J-X, Li Y-L, Zimmerman MC, Schultz HD. Angiotensin-(1–7) increases neuronal potassium current via a nitric oxide-dependent mechanism. Am J Physiol-Cell Physiol. 2011;300:C58–64.

141. Zhang D, Xiao Q, Luo H, Zhao K. Effects of angiotensin-(1-7) on hippocampal expressions of GFAP and GDNF and cognitive function in rats with diabetes mellitus. Nan Fang Yi Ke Da Xue Xue Bao. 2015;35:646–51.

142. Zhang Y, Lu J, Shi J, Lin X, Dong J, Zhang S, Liu Y, Tong Q. Central administration of angiotensin-(1–7) stimulates nitric oxide release and upregulates the endothelial nitric oxide synthase expression following focal cerebral ischemia/reperfusion in rats. Neuropeptides. 2008;42:593–600.

143. Zheng J, Li G, Chen S, Bihl J, Buck J, Zhu Y, Xia H, Lazartigues E, Chen Y, Olson JE. Activation of the ACE2/Ang-(1–7)/Mas pathway reduces oxygen–glucose deprivation-induced tissue swelling, ROS production, and cell death in mouse brain with angiotensin II overproduction. Neuroscience. 2014a;273:39–51.

144. Zheng J-L, Li G-Z, Chen S-Z, Wang J-J, Olson JE, Xia H-J, Lazartigues E, Zhu Y-L, Chen Y-F. Angiotensin converting enzyme 2/Ang-(1–7)/Mas Axis protects brain from ischemic injury with a tendency of age-dependence. CNS Neurosci Ther. 2014b;20:452–9.

145. Zhou L-M, Shi Z, Gao J, Han Y, Yuan N, Gao X-Y, Zhu G-Q. Angiotensin-(1–7) and angiotension II in the rostral ventrolateral medulla modulate the cardiac sympathetic afferent reflex and sympathetic activity in rats. Pflüg Arch-Eur J Physiol. 2010;459:681–8.

146. Zubenko GS, Nixon RA. Mood-elevating effect of captopril in depressed patients. Am J Psychiatry. 1984;1:41.

Heart – Coronary Vessels and Cardiomyocytes

Anderson José Ferreira, Carlos Henrique Castro, and Robson Augusto Souza Santos

Introduction

The renin-angiotensin system (RAS) plays a key role in several target organs, such as heart and blood vessels, exerting a powerful control in the maintenance of homeostasis [1–4]. In addition to the angiotensin-converting enzyme (ACE)/angiotensin (Ang) II/AT$_1$ receptor axis, the RAS has a counter-regulatory axis composed by ACE2, Ang-(1-7), and Mas receptor. Ang-(1-7) is a biologically active component of the RAS which binds to Mas, inducing many beneficial actions, such as vasodilatation, antifibrosis, antihypertrophic, and antiproliferative effects [5–13]. This peptide is produced mainly through the action of ACE2, which has approximately 400-fold less affinity to Ang I than to Ang II [14–16]; therefore, Ang II is the major substrate for Ang-(1-7) synthesis. In fact, the conversion of Ang II to Ang-(1-7) by ACE2 is important to regulate the RAS activity since Ang-(1-7) induces opposite effects to those elicited by Ang II [7–14]. Additionally, ACE2 can form Ang-(1–7) less efficiently through hydrolysis of Ang I to Ang-(1-9) with subsequent Ang-(1-7) formation [14].

A. J. Ferreira (✉)
Department of Morphology, Biological Sciences Institute, Federal University of Minas Gerais, Belo Horizonte, MG, Brazil
e-mail: anderson@icb.ufmg.br

C. H. Castro
Department of Physiological Sciences, Biological Sciences Institute, Federal University of Goiás, Goiânia, GO, Brazil

R. A. S. Santos
National Institute of Science and Technology in Nano Biopharmaceutics – Department of Physiology and Biophysics, Biological Sciences Institute, Federal University of Minas Gerais, Belo Horizonte, MG, Brazil

Department of Physiology and Pharmacology, Biological Sciences Institute, Federal University of Minas Gerais, Belo Horizonte, MG, Brazil

© Springer Nature Switzerland AG 2019
R. A. S. Santos (ed.), *Angiotensin-(1-7)*,
https://doi.org/10.1007/978-3-030-22696-1_5

The relevance of the RAS is highlighted by the success obtained in therapeutic strategies based on the pharmacological inhibition of this system in cardiovascular diseases [17–22]. Blockade of the RAS with ACE inhibitors (ACEi) or AT_1 receptor antagonists (ARBs) improves the outcomes of patients with hypertension, acute myocardial infarction and chronic systolic heart failure [23–25]. Importantly, it has been shown that administration of ACEi and ARBs causes substantial increases in plasma Ang-(1-7) levels, leading to the assumption that part of their clinical effects might be mediated by this heptapeptide [26–28]. Indeed, some effects of ACEi and ARBs can be blocked or attenuated by A-779, a Mas antagonist, confirming the role of Ang-(1-7) in the actions of these compounds [29]. The beneficial effects of Ang-(1-7), as well as its likely participation in the effects of the ACEi and ARBs, represent evidences for the potential target of the ACE2/Ang-(1-7)/Mas axis as a therapeutic strategy

Cardiomyocytes and Coronary Vessels

The heart is one of the most important target for the actions of the ACE2/Ang-(1-7)/Mas axis. Since the discovery of Ang-(1-7) in the late 1980s [30, 31], several studies have demonstrated important effects of this peptide in hearts. The presence of Ang-(1-7) and its receptor Mas in cardiomyocytes [32, 33], sinoatrial node [34], and coronary vessels [35, 36] and the ability of the heart to produce Ang-(1-7) [37, 38] are evidence of the role of this peptide in cardiac tissues. Additionally, ACE2 is expressed in myofibroblasts, cardiomyocytes [39–41], as well as in coronary vessels [42, 43]. Classical pharmacotherapeutic agents used to treat heart failure, including ACEi, ARBs, and aldosterone receptor blockers, increase ACE2 activity and/or expression, indicating its importance in the cardiac diseases establishment and progression [44–46].

It has been observed that Ang-(1-7) modulates the inotropism of hearts. Overexpression of Ang-(1-7) in the heart using an engineered fusion protein leads to an increased left ventricular contractile function [11]. In contrast, Mas-deficient mice present an impairment of heart function [33]. However, in rabbit right papillary muscles, Ang-(1-7), through its binding to Mas, induces a negative inotropic effect modulated by the endocardial endothelium and nitric oxide, independently of AT1 or AT2 receptors activation [47].

Although further elucidations regarding the signaling pathways in cardiomyocytes involving Mas activation are necessary, some mechanisms have been proposed. Transgenic rats harboring an Ang-(1-7)-producing fusion protein in hearts show an increased Ca^{2+} transient amplitude, faster Ca^{2+} uptake, and increased expression of SERCA2 [11]. In keeping with these results, cardiomyocytes from Mas-deficient mice present slower $[Ca^{2+}]_i$ transients accompanied by a lower Ca^{2+} ATPase expression in the sarcoplasmic reticulum [33, 48]. Strikingly, acute exposure to Ang-(1-7) has no demonstrable effects on Ca^{2+} transients [48] or on left ventricular myocyte L-type Ca^{2+} current responses [49]. Although acute Ang-(1-7) treatment failed to alter Ca^{2+} handling in ventricular myocytes of rats, these findings suggest an important role of Ang-(1-7)/Mas in the long-term maintenance of the Ca^{2+} homeostasis in the heart.

In addition to the changes in calcium-handling proteins, Ang-(1-7) plays its effects in the heart by stimulating the nitric oxide (NO) production. Indeed, Ang-(1-7) via Mas increases the synthesis of NO through a mechanism involving the activation of the endothelial NO synthase (eNOS) and AKT [48]. These effects were abolished by A-779 and are absent in cardiomyocytes from Mas-deficient mice [48]. In spontaneously hypertensive rats, Ang-(1-7) infusion upregulates cardiac NOS expression and activity through an AT2- and bradykinin-dependent mechanism [50].

Also, Gomes et al. [51] found that the treatment of isolated cardiomyocytes of rats with Ang-(1-7) efficiently prevents the Ang II-induced hypertrophy by modulating the calcineurin/NFAT signaling cascade. These effects were blocked by NOS inhibition and by guanylyl cyclase inhibitors, indicating that these effects are mediated by the NO/cGMP pathway. Also, Ang-(1-7) inhibits serum-stimulated mitogen-activated protein kinase (MAPK) activation in cardiac myocytes [52] and prevents the Ang II-mediated phosphorylation of ERK1/2 and Rho kinase in hearts in a dose-dependent manner [53].

Ang-(1-7)/ACE2/Mas axis is also important in the maintenance of the heart structure. Genetic deletion of Mas receptor leads to higher levels of collagen types I and III and fibronectin in both right and left ventricles from adult mice [33, 54]. Interestingly, neonatal mice presented a similar pattern of ECM protein expression as observed in adult mice [54], indicating that the structural disturbances seen in hearts of adult mice are not due to chronic adaptative alterations. Mas activation is also involved in the development of the gestational cardiac remodeling. Pharmacological blockade or genetic deletion of Mas attenuated the pregnancy-induced myocyte hypertrophy and increased the collagen type III deposition in left ventricles from pregnant normotensive rats [35]. Additionally, exercise training increases Ang-(1-7) levels and upregulates Mas receptor in hypertrophied rat hearts [55], suggesting that ACE2/Ang-(1-7)/Mas axis plays a role in physiological cardiac hypertrophy.

The vasodilatory action of Ang-(1-7) has been reported in several vascular beds, including coronary arteries [5, 6, 13, 56–63]. In fact, early studies have reported the endothelium as the major site for generation [64] and metabolism [26] of Ang-(1-7). In addition to Ang-(1-7), endothelial cells also express ACE2 and Mas [65, 66].

Initially, the coronary vasodilator effect of Ang-(1-7) was observed in coronary vessels of dogs [56, 67] and pigs [57, 68]. This effect was abolished by NOS inhibitor and attenuated by bradykinin B_2 receptor antagonist. Accordingly, Ang-(1-7) potentiates the vasodilator effect of bradykinin in coronary vessels of dogs [56] and rat [69]. van Esch et al. [70] demonstrated that Ang-(1-7) blocked the AT1-induced vasoconstriction but did not affect the coronary circulation when applied alone at nanomolar or micromolar concentrations in isolated rat hearts. However, Ang-(1-7) or the Mas agonist, AVE 0991, at picomolar concentrations, induced potent coronary vasodilation in healthy hearts through Mas activation and the NO-guanylate cyclase pathway [36, 43]. These effects were absent in pressure overload-induced hypertrophic hearts, but the pre-treatment with an AT1 receptor antagonist restored the Ang-(1-7)-induced coronary vasodilation in these hearts [43, 71]. Ang-(1-7) also evokes coronary vasodilation in isolated perfused mice hearts, through a mechanism involving interaction of its receptor Mas with AT1 and AT2 receptors leading to release of prostaglandins and nitric oxide [72]. In addition, isolated perfused hearts from Mas-deficient mice presented a higher coronary perfusion pressure [33], which confirms the role of Mas in the control of the coronary circulation. Altogether, these data show that Ang-(1-7) produces a complex coronary effect, involving NO release and interactions among Mas and AT1, AT2, and B2 receptors.

References

1. Hall JE, Guyton AC, Mizelle HL. Role of the renin-angiotensin system in control of sodium excretion and arterial pressure. Acta Physiol Scand Suppl [Internet]. 1990/01/01. 1990;591:48–62. Available from: http://www.ncbi.nlm.nih.gov/entrez/query.fcgi?cmd=Retrieve&db=PubMed&dopt=Citation&list_uids=2220409.
2. Guyton AC. Kidneys and fluids in pressure regulation. Small volume but large pressure changes. Hypertension [Internet]. 1992/01/01. 1992;19(1 Suppl):I2–8. Available from: http://www.ncbi.nlm.nih.gov/entrez/query.fcgi?cmd=Retrieve&db=PubMed&dopt=Citation&list_uids=1730451.
3. Santos RASS, Campagnole-Santos MJ, Andrade SP. Angiotensin-(1-7): an update. Regul Pept [Internet]. 2000;91(1–3):45–62. Available from: http://www.ncbi.nlm.nih.gov/entrez/query.fcgi?cmd=Retrieve&db=PubMed&dopt=Citation&list_uids=10967201.
4. Marshall RP. The pulmonary renin-angiotensin system. Curr Pharm Des [Internet]. 2003/02/07. 2003;9(9):715–22. Available from: http://www.ncbi.nlm.nih.gov/entrez/query.fcgi?cmd=Retrieve&db=PubMed&dopt=Citation&list_uids=12570789.
5. le Tran Y, Forster C. Angiotensin-(1-7) and the rat aorta: modulation by the endothelium. J Cardiovasc Pharmacol [Internet]. 1997;30(5):676–82. Available from: http://www.ncbi.nlm.nih.gov/entrez/query.fcgi?cmd=Retrieve&db=PubMed&dopt=Citation&list_uids=9388051.
6. Santos RAS, Simoes e Silva AC, Maric C, DMR S, Machado RP, de Buhr I, et al. Angiotensin-(1-7) is an endogenous ligand for the G protein-coupled receptor Mas. Proc Natl Acad Sci U S A [Internet]. 2003;100(14):8258–63. Available from: http://www.ncbi.nlm.nih.gov/entrez/query.fcgi?cmd=Retrieve&db=PubMed&dopt=Citation&list_uids=12829792.
7. Santos RAS, Ferreira AJ, Nadu AP, Braga ANG, de Almeida AP, Campagnole-Santos MJ, et al. Expression of an angiotensin-(1-7)-producing fusion protein produces cardioprotective effects in rats. Physiol Genomics. 2004;17(3):292–9.
8. Grobe JL, Mecca AP, Lingis M, Shenoy V, Bolton TA, Machado JM, et al. Prevention of angiotensin II-induced cardiac remodeling by angiotensin-(1-7). Am J Physiol Heart Circ Physiol [Internet]. 2007;292(2):H736–42. Available from: http://www.ncbi.nlm.nih.gov/entrez/query.fcgi?cmd=Retrieve&db=PubMed&dopt=Citation&list_uids=17098828.
9. Mercure C, Yogi A, Callera GE, Aranha AB, Bader M, Ferreira AJ, et al. Angiotensin(1-7) Blunts Hypertensive Cardiac Remodeling by a Direct Effect on the Heart. Circ Res [Internet]. 2008/10/11. 2008 Nov 21 [cited 2013 Aug 13];103(11). Available from: http://www.ncbi.nlm.nih.gov/entrez/query.fcgi?cmd=Retrieve&db=PubMed&dopt=Citation&list_uids=18845809.
10. Nadu AP, Ferreira AJ, Reudelhuber TL, Bader M, Santos RA. Reduced isoproterenol-induced renin-angiotensin changes and extracellular matrix deposition in hearts of TGR(A1-7)3292 rats. J Am Soc Hypertens [Internet]. 2008/09/01. 2008;2(5):341–8. Available from: http://www.ncbi.nlm.nih.gov/entrez/query.fcgi?cmd=Retrieve&db=PubMed&dopt=Citation&list_uids=20409916.
11. Ferreira AJ, Castro CH, Guatimosim S, Almeida PWM, Gomes ERM, Dias-Peixoto MF, et al. Attenuation of isoproterenol-induced cardiac fibrosis in transgenic rats harboring an angiotensin-(1-7)-producing fusion protein in the heart. Ther Adv Cardiovasc Dis [Internet]. 2010 Apr [cited 2013 Aug 13];4(2):83–96. Available from: http://www.ncbi.nlm.nih.gov/entrez/query.fcgi?cmd=Retrieve&db=PubMed&dopt=Citation&list_uids=20051448.

12. Santiago NM, Guimaraes PS, Sirvente RA, Oliveira LA, Irigoyen MC, Santos RA, et al. Lifetime overproduction of circulating Angiotensin-(1-7) attenuates deoxycorticosterone acetate-salt hypertension-induced cardiac dysfunction and remodeling. Hypertension [Internet]. 2010/03/10. 2010;55(4):889–96. Available from: http://www.ncbi.nlm.nih.gov/entrez/query.fcgi?cmd=Retrieve&db=PubMed&dopt=Citation&list_uids=20212262.

13. Savergnini SQ, Beiman M, Lautner RQ, de Paula-Carvalho V, Allahdadi K, Pessoa DC, et al. Vascular relaxation, antihypertensive effect, and cardioprotection of a novel peptide agonist of the MAS receptor. Hypertension [Internet]. 2010;56(1):112–20. Available from: http://www.ncbi.nlm.nih.gov/entrez/query.fcgi?cmd=Retrieve&db=PubMed&dopt=Citation&list_uids=20479330

14. Vickers C, Hales P, Kaushik V, Dick L, Gavin J, Tang J, et al. Hydrolysis of biological peptides by human angiotensin-converting enzyme-related carboxypeptidase. J Biol Chem [Internet]. 2002;277(17):14838–43. Available from: http://www.ncbi.nlm.nih.gov/entrez/query.fcgi?cmd=Retrieve&db=PubMed&dopt=Citation&list_uids=11815627

15. Tipnis SR, Hooper NM, Hyde R, Karran E, Christie G, Turner AJ. A human homolog of angiotensin-converting enzyme. Cloning and functional expression as a captopril-insensitive carboxypeptidase. J Biol Chem [Internet]. 2000;275(43):33238–43. Available from: http://www.ncbi.nlm.nih.gov/entrez/query.fcgi?cmd=Retrieve&db=PubMed&dopt=Citation&list_uids=10924499.

16. Donoghue M, Hsieh F, Baronas E, Godbout K, Gosselin M, Stagliano N, et al. A novel angiotensin-converting enzyme-related carboxypeptidase (ACE2) converts angiotensin I to angiotensin 1-9. Circ Res [Internet]. 2000;87(5):E1–9. Available from: http://www.ncbi.nlm.nih.gov/entrez/query.fcgi?cmd=Retrieve&db=PubMed&dopt=Citation&list_uids=10969042.

17. Ferrario CM. The renin-angiotensin system: importance in physiology and pathology. J Cardiovasc Pharmacol [Internet]. 1990/01/01. 1990;15(Suppl 3):S1–5. Available from: http://www.ncbi.nlm.nih.gov/entrez/query.fcgi?cmd=Retrieve&db=PubMed&dopt=Citation&list_uids=1691411.

18. Nicholls MG, Richards AM, Agarwal M. The importance of the renin-angiotensin system in cardiovascular disease. J Hum Hypertens [Internet]. 1998/07/09. 1998;12(5):295–9. Available from: http://www.ncbi.nlm.nih.gov/entrez/query.fcgi?cmd=Retrieve&db=PubMed&dopt=Citation&list_uids=9655650.

19. Marshall RP, McAnulty RJ, Laurent GJ. Angiotensin II is mitogenic for human lung fibroblasts via activation of the type 1 receptor. Am J Respir Crit Care Med. 2000;161:1999–2004.

20. Fleming I, Kohlstedt K, Busse R. The tissue renin-angiotensin system and intracellular signalling. Curr Opin Nephrol Hypertens [Internet]. 2005/12/13. 2006;15(1):8–13. Available from: http://www.ncbi.nlm.nih.gov/entrez/query.fcgi?cmd=Retrieve&db=PubMed&dopt=Citation&list_uids=16340660.

21. Cargill RI, Lipworth BJ. Lisinopril attenuates acute hypoxic pulmonary vasoconstriction in humans. Chest [Internet]. 1996;109(2):424–9. Available from: http://www.ncbi.nlm.nih.gov/pubmed/8620717.

22. Orfanos SE, Armaganidis A, Glynos C, Psevdi E, Kaltsas P, Sarafidou P, et al. Pulmonary capillary endothelium-bound angiotensin-converting enzyme activity in acute lung injury. Circulation [Internet]. 2000;102(16):2011–8. Available from: http://www.ncbi.nlm.nih.gov/pubmed/11034953.

23. Schiffrin EL. Vascular and cardiac benefits of angiotensin receptor blockers. Am J Med [Internet]. 2002/10/29. 2002;113(5):409–18. Available from: http://www.ncbi.nlm.nih.gov/entrez/query.fcgi?cmd=Retrieve&db=PubMed&dopt=Citation&list_uids=12401536.

24. Ma TK, Kam KK, Yan BP, Lam YY. Renin-angiotensin-aldosterone system blockade for cardiovascular diseases: current status. Br J Pharmacol [Internet]. 2010/07/02. 2010;160(6):1273–92. Available from: http://www.ncbi.nlm.nih.gov/entrez/query.fcgi?cmd=Retrieve&db=PubMed&dopt=Citation&list_uids=20590619.

25. Vijayaraghavan K, Deedwania P. Renin-angiotensin-aldosterone blockade for cardiovascular disease prevention. Cardiol Clin [Internet]. 2011/01/25. 2011;29(1):137–56. Available from: http://www.ncbi.nlm.nih.gov/entrez/query.fcgi?cmd=Retrieve&db=PubMed&dopt=Citation&list_uids=21257105.

26. Chappell MC, Pirro NT, Sykes A, Ferrario CM. Metabolism of angiotensin-(1-7) by angiotensin-converting enzyme. Hypertension [Internet]. 1998;31(1 Pt 2):362–7. Available from: http://www.ncbi.nlm.nih.gov/entrez/query.fcgi?cmd=Retrieve&db=PubMed&dopt=Citation&list_uids=9453329.

27. Iyer SN, Ferrario CM, Chappell MC. Angiotensin-(1-7) contributes to the antihypertensive effects of blockade of the renin-angiotensin system. Hypertension [Internet]. 1998;31(1 Pt 2):356–61. Available from: http://www.ncbi.nlm.nih.gov/entrez/query.fcgi?cmd=Retrieve&db=PubMed&dopt=Citation&list_uids=9453328.

28. Davie AP, McMurray JJ. Effect of angiotensin-(1-7) and bradykinin in patients with heart failure treated with an ACE inhibitor. Hypertension [Internet]. 1999;34(3):457–60. Available from: http://www.ncbi.nlm.nih.gov/entrez/query.fcgi?cmd=Retrieve&db=PubMed&dopt=Citation&list_uids=10489393.

29. Britto RR, Santos RA, Fagundes-Moura CR, Khosla MC, Campagnole-Santos MJ. Role of angiotensin-(1-7) in the modulation of the baroreflex in renovascular hypertensive rats. Hypertension [Internet]. 1997/10/10. 1997;30(3 Pt 2):549–56. Available from: http://www.ncbi.nlm.nih.gov/entrez/query.fcgi?cmd=Retrieve&db=PubMed&dopt=Citation&list_uids=9322980.

30. Santos RAS, Brosnihan KB, Chappell MC, Pesquero J, Chernicky CLCL, Greene LLJ, et al. Converting enzyme activity and angiotensin metabolism in the dog brainstem. Hypertension [Internet]. 1988;11(2 Pt 2):I153–7. Available from: http://www.ncbi.nlm.nih.gov/entrez/query.fcgi?cmd=Retrieve&db=PubMed&dopt=Citation&list_uids=2831145.

31. Schiavone MT, Santos RA, Brosnihan KB, Khosla MC, Ferrario CM. Release of vasopressin from the rat hypothalamo-neurohypophysial system by angiotensin-(1-7) heptapeptide. Proc Natl Acad Sci U S A [Internet]. 1988;85(11):4095–8. Available from: http://www.ncbi.nlm.nih.gov/entrez/query.fcgi?cmd=Retrieve&db=PubMed&dopt=Citation&list_uids=3375255.

32. Averill DB, Ishiyama Y, Chappell MC, Ferrario CM. Cardiac angiotensin-(1-7) in ischemic cardiomyopathy. Circulation [Internet]. 2003;108(17):2141–6. Available from: http://www.ncbi.nlm.nih.gov/entrez/query.fcgi?cmd=Retrieve&db=PubMed&dopt=Citation&list_uids=14517166.

33. Santos RAS, Castro CH, Gava E, Pinheiro SVB, Almeida AP, Paula RD De, et al. Impairment of in vitro and in vivo heart function in angiotensin-(1-7) receptor MAS knockout mice. Hypertension [Internet]. 2006 May [cited 2013 Aug 8];47(5):996–1002. Available from: http://www.ncbi.nlm.nih.gov/pubmed/16567589.

34. Ferreira AJ, Moraes PL, Foureaux G, Andrade AB, Santos RA, Almeida AP. The Angiotensin-(1-7)/mas receptor axis is expressed in sinoatrial node cells of rats. J Histochem Cytochem [Internet]. 2011/05/25. 2011;59(8):761–8. Available from: http://www.ncbi.nlm.nih.gov/entrez/query.fcgi?cmd=Retrieve&db=PubMed&dopt=Citation&list_uids=21606202.

35. Carmos-Silva C, de Almeida JFQ, Macedo LM, Melo MBB, Pedrino GR, Santos FFCA, et al. Mas receptor contributes to pregnancy-induced cardiac remodeling. Clin Sci (Lond) [Internet]. 2016 Sep 13 [cited 2017 Feb 17]. Available from: http://www.ncbi.nlm.nih.gov/pubmed/27624141.

36. Moraes PL, Kangussu LM, Castro CH, Santos RA, Ferreira AJ, Almeida AP. Vasodilator effect of Angiotensin- (1-7) on vascular coronary bed of rats: Role of Mas, ACE and ACE2. Protein Pept Lett. 2017;24:869.

37. Trask AJ, Averill DB, Ganten D, Chappell MC, Ferrario CM. Primary role of angiotensin-converting enzyme-2 in cardiac production of angiotensin-(1-7) in transgenic Ren-2 hypertensive rats. Am J Physiol Hear Circ Physiol [Internet]. 2007/02/20. 2007;292(6):H3019-24. Available from: http://www.ncbi.nlm.nih.gov/entrez/query.fcgi?cmd=Retrieve&db=PubMed&dopt=Citation&list_uids=17308000.

38. Neves LA, Almeida AP, Khosla MC, Santos RA. Metabolism of angiotensin I in isolated rat hearts. Effect of angiotensin converting enzyme inhibitors. Biochem Pharmacol [Internet]. 1995 Oct 26 [cited 2018 Mar 24];50(9):1451–9. Available from: http://www.ncbi.nlm.nih.gov/pubmed/7503796.

39. Guy JL, Lambert DW, Turner AJ, Porter KE. Functional angiotensin-converting enzyme 2 is expressed in human cardiac myofibroblasts. Exp Physiol [Internet]. 2008/01/29. 2008;93(5):579–88. Available from: http://www.ncbi.nlm.nih.gov/entrez/query.fcgi?cmd=Re trieve&db=PubMed&dopt=Citation&list_uids=18223028.

40. Gallagher PE, Ferrario CM, Tallant EA. Regulation of ACE2 in cardiac myocytes and fibroblasts. Am J Physiol Hear Circ Physiol [Internet]. 2008/10/14. 2008;295(6):H2373-9. Available from: http://www.ncbi.nlm.nih.gov/entrez/query.fcgi?cmd=Retrieve&db=PubMed &dopt=Citation&list_uids=18849338.

41. Ferreira AJ, Shenoy V, Qi Y, Fraga-Silva RA, Santos RAS, Katovich MJ, et al. Angiotensin-converting enzyme 2 activation protects against hypertension-induced cardiac fibrosis involving extracellular signal-regulated kinases. Exp Physiol [Internet]. 2011 Mar [cited 2013 Aug 13];96(3):287–94. Available from: http://www.ncbi.nlm.nih.gov/entrez/query.fcgi?cmd=Retri eve&db=PubMed&dopt=Citation&list_uids=21148624.

42. Oudit GY, Crackower MA, Backx PH, Penninger JM. The role of ACE2 in cardiovascular physiology. Trends Cardiovasc Med [Internet]. 2003;13(3):93–101. Available from: http:// www.ncbi.nlm.nih.gov/entrez/query.fcgi?cmd=Retrieve&db=PubMed&dopt=Citation&l ist_uids=12691672.

43. Souza APS, Sobrinho DBS, Almeida JFQ, Alves GMM, Macedo LM, Porto JE, et al. Angiotensin II type 1 receptor blockade restores angiotensin-(1-7)-induced coronary vaso-dilation in hypertrophic rat hearts. Clin Sci (Lond) [Internet]. 2013 Nov 1 [cited 2013 Aug 13];125(9):449–59. Available from: http://www.ncbi.nlm.nih.gov/pubmed/23718715.

44. Ferrario CM, Jessup J, Chappell MC, Averill DB, Brosnihan KB, Tallant EA, et al. Effect of angiotensin-converting enzyme inhibition and angiotensin II receptor blockers on cardiac angiotensin-converting enzyme 2. Circulation [Internet]. 2005;111(20):2605–10. Available from: http://www.ncbi.nlm.nih.gov/entrez/query.fcgi?cmd=Retrieve&db=PubMed&dopt=Cit ation&list_uids=15897343.

45. Keidar S, Gamliel-Lazarovich A, Kaplan M, Pavlotzky E, Hamoud S, Hayek T, et al. Mineralocorticoid receptor blocker increases angiotensin-converting enzyme 2 activity in congestive heart failure patients. Circ Res [Internet]. 2005/09/24. 2005;97(9):946–53. Available from: http://www.ncbi.nlm.nih.gov/entrez/query.fcgi?cmd=Retrieve&db=PubMed&dopt=Cit ation&list_uids=16179584.

46. Kaiqiang J, Minakawa M, Fukui K, Suzuki Y, Fukuda I. Olmesartan improves left ventricular function in pressure-overload hypertrophied rat heart by blocking angiotensin II receptor with synergic effects of upregulation of angiotensin converting enzyme 2. Ther Adv Cardiovasc Dis [Internet]. 2009;3(2):103–11. Available from: http://www.ncbi.nlm.nih.gov/entrez/query.fcgi? cmd=Retrieve&db=PubMed&dopt=Citation&list_uids=19171689.

47. Castro-Chaves P, Pintalhao M, Fontes-Carvalho R, Cerqueira R, Leite-Moreira AF. Acute modulation of myocardial function by angiotensin 1-7. Peptides [Internet]. 2009/06/16. 2009;30(9):1714–9. Available from: http://www.ncbi.nlm.nih.gov/entrez/query.fcgi?cmd=Re trieve&db=PubMed&dopt=Citation&list_uids=19524627.

48. Dias-Peixoto MF, Santos RAS, Gomes ERM, Alves MNM, Almeida PWM, Greco L, et al. Molecular mechanisms involved in the angiotensin-(1-7)/Mas signaling pathway in cardiomyocytes. Hypertension [Internet]. 2008 Sep [cited 2013 Aug 13];52(3):542–8. Available from: http://www.ncbi.nlm.nih.gov/entrez/query.fcgi?cmd=Retrieve&db=PubMed&dopt=Citation &list_uids=18695148.

49. Zhou P, Cheng CP, Li T, Ferrario CM, Cheng H-J. Modulation of cardiac L-type Ca 2+ current by angiotensin-(1-7): Normal versus heart failure HHS Public Access. Ther Adv Cardiovasc Dis. 2015;9(6):342–53.

50. Costa MA, Verrilli MAL, Gomez KA, Nakagawa P, Peña C, Arranz C, et al. Angiotensin- (1-7) upregulates cardiac nitric oxide synthase in spontaneously hypertensive rats. 2010;2:1205–11.

51. Gomes ERM, Lara AA, Almeida PWM, Guimaraes D, Resende RR, Campagnole-Santos MJ, et al. Angiotensin-(1-7) prevents cardiomyocyte pathological remodeling through a nitric oxide/guanosine 3′,5′-cyclic monophosphate-dependent pathway. Hypertension [Internet].

2009/12/10. 2010 Jan [cited 2013 Aug 13];55(1):153–60. Available from: http://www.ncbi.nlm.nih.gov/pubmed/19996065.

52. Tallant EA, Clark MA. Molecular mechanisms of inhibition of vascular growth by angiotensin-(1-7). Hypertension [Internet]. 2003;42(4):574–9. Available from: http://www.ncbi.nlm.nih.gov/entrez/query.fcgi?cmd=Retrieve&db=PubMed&dopt=Citation&list_uids=12953014.

53. Giani JF, Gironacci MM, Munoz MC, Turyn D, Dominici FP. Angiotensin-(1-7) has a dual role on growth-promoting signalling pathways in rat heart in vivo by stimulating STAT3 and STAT5a/b phosphorylation and inhibiting angiotensin II-stimulated ERK1/2 and Rho kinase activity. Exp Physiol [Internet]. 2008/05/02. 2008;93(5):570–8. Available from: http://www.ncbi.nlm.nih.gov/entrez/query.fcgi?cmd=Retrieve&db=PubMed&dopt=Citation&list_uids=18448663.

54. Gava E, de Castro CH, Ferreira AJ, Colleta H, Melo MB, Alenina N, et al. Angiotensin-(1-7) receptor Mas is an essential modulator of extracellular matrix protein expression in the heart. Regul Pept Elsevier BV. 2012;175(1–3):30–42.

55. Filho AG, Ferreira AJ, Santos SHS, Neves SRS, Silva Camargos ER, Becker LK, et al. Selective increase of angiotensin(1-7) and its receptor in hearts of spontaneously hypertensive rats subjected to physical training. Exp Physiol. 2008;93:589–98.

56. Brosnihan KB, Li P, Ferrario CM. Angiotensin-(1-7) dilates canine coronary arteries through kinins and nitric oxide. Hypertension [Internet]. 1996;27(3 Pt 2):523–8. Available from: http://www.ncbi.nlm.nih.gov/entrez/query.fcgi?cmd=Retrieve&db=PubMed&dopt=Citation&list_uids=8613197.

57. Porsti I, Bara AT, Busse R, Hecker M, Porsti I, Bara AT, et al. Release of nitric oxide by angiotensin-(1-7) from porcine coronary endothelium: implications for a novel angiotensin receptor. Br J Pharmacol [Internet]. 1994;111(3):652–4. Available from: http://www.ncbi.nlm.nih.gov/entrez/query.fcgi?cmd=Retrieve&db=PubMed&dopt=Citation&list_uids=8019744.

58. Feterik K, Smith L, Katusic ZS. Angiotensin-(1-7) causes endothelium-dependent relaxation in canine middle cerebral artery. Brain Res [Internet]. 2000;873(1):75–82. Available from: http://www.ncbi.nlm.nih.gov/entrez/query.fcgi?cmd=Retrieve&db=PubMed&dopt=Citation&list_uids=10915812.

59. Meng W, Busija DW. Comparative effects of angiotensin-(1-7) and angiotensin II on piglet pial arterioles. Stroke [Internet]. 1993/12/01. 1993;24(12):2041–4; discussion 2045. Available from: http://www.ncbi.nlm.nih.gov/entrez/query.fcgi?cmd=Retrieve&db=PubMed&dopt=Citation&list_uids=8248986.

60. Osei SY, Ahima RS, Minkes RK, Weaver JP, Khosla MC, Kadowitz PJ. Differential responses to angiotensin-(1-7) in the feline mesenteric and hindquarters vascular beds. Eur J Pharmacol [Internet]. 1993;234(1):35–42. Available from: http://www.ncbi.nlm.nih.gov/entrez/query.fcgi?cmd=Retrieve&db=PubMed&dopt=Citation&list_uids=7682513.

61. Ren Y, Garvin JL, Carretero OA. Vasodilator action of angiotensin-(1-7) on isolated rabbit afferent arterioles. Hypertension [Internet]. 2002;39(3):799–802. Available from: http://www.ncbi.nlm.nih.gov/entrez/query.fcgi?cmd=Retrieve&db=PubMed&dopt=Citation&list_uids=11897767.

62. Oliveira MA, Fortes ZB, Santos RA, Kosla MC, De Carvalho MH. Synergistic effect of angiotensin-(1-7) on bradykinin arteriolar dilation in vivo. Peptides [Internet] 1999;20(10):1195–1201. . Available from: http://www.ncbi.nlm.nih.gov/entrez/query.fcgi?cmd=Retrieve&db=PubMed&dopt=Citation&list_uids=10573291.

63. Fernandes L, Fortes ZB, Nigro D, Tostes RC, Santos RA, Catelli De Carvalho MH. Potentiation of bradykinin by angiotensin-(1-7) on arterioles of spontaneously hypertensive rats studied in vivo. Hypertension [Internet]. 2001;37(2 Part 2):703–9. Available from: http://www.ncbi.nlm.nih.gov/entrez/query.fcgi?cmd=Retrieve&db=PubMed&dopt=Citation&list_uids=11230360.

64. Santos RA, Brosnihan KB, Jacobsen DW, DiCorleto PE, Ferrario CM. Production of angiotensin-(1-7) by human vascular endothelium. Hypertension [Internet]. 1992;19(2 Suppl):II56–61. Available from: http://www.ncbi.nlm.nih.gov/entrez/query.fcgi?cmd=Retrieve&db=PubMed&dopt=Citation&list_uids=1310484.

65. Burrell LM, Johnston CI, Tikellis C, Cooper ME. ACE2, a new regulator of the renin-angiotensin system. Trends Endocrinol Metab [Internet]. 2004/04/28. 2004;15(4):166–9. Available from: http://www.ncbi.nlm.nih.gov/entrez/query.fcgi?cmd=Retrieve&db=PubMed&dopt=Citation&list_uids=15109615.

66. Sampaio WO, Souza dos Santos RA, Faria-Silva R, da Mata Machado LT, Schiffrin EL, Touyz RM. Angiotensin-(1-7) through receptor Mas mediates endothelial nitric oxide synthase activation via Akt-dependent pathways. Hypertension [Internet]. 2007 Jan [cited 2013 Aug 13];49(1):185–92. Available from: http://www.ncbi.nlm.nih.gov/pubmed/17116756.

67. Brosnihan KB, Li P, Tallant EA, Ferrario CM. Angiotensin-(1-7): a novel vasodilator of the coronary circulation. Biol Res [Internet]. 1998;31(3):227–34. Available from: http://www.ncbi.nlm.nih.gov/entrez/query.fcgi?cmd=Retrieve&db=PubMed&dopt=Citation&list_uids=9830510.

68. Gorelik G, Carbini LA, Scicli AG. Angiotensin 1-7 induces bradykinin-mediated relaxation in porcine coronary artery. J Pharmacol Exp Ther [Internet]. 1998;286(1):403–10. Available from: http://www.ncbi.nlm.nih.gov/entrez/query.fcgi?cmd=Retrieve&db=PubMed&dopt=Citation&list_uids=9655885.

69. Almeida AP, Frabregas BC, Madureira MM, Santos RJA, Campagnole-Santos MJ, Santos RJA. Angiotensin-(1-7) potentiates the coronary vasodilatatory effect of bradykinin in the isolated rat heart. Braz J Med Biol Res [Internet]. 2000;33(6):709–13. Available from: http://www.ncbi.nlm.nih.gov/entrez/query.fcgi?cmd=Retrieve&db=PubMed&dopt=Citation&list_uids=10829099.

70. van Esch JH, Oosterveer CR, Batenburg WW, van Veghel R, Jan Danser AH. Effects of angiotensin II and its metabolites in the rat coronary vascular bed: is angiotensin III the preferred ligand of the angiotensin AT2 receptor? Eur J Pharmacol [Internet]. 2008/05/31. 2008;588(2–3):286–93. Available from: http://www.ncbi.nlm.nih.gov/entrez/query.fcgi?cmd=Retrieve&db=PubMed&dopt=Citation&list_uids=18511032.

71. Nunes ADC, Souza APS, Macedo LM, Alves PH, Pedrino GR, Colugnati DB, et al. Influence of antihypertensive drugs on aortic and coronary effects of Ang-(1-7) in pressure-overloaded rats. Brazilian J Med Biol Res. 2017;50(4):1–8.

72. Castro CH, Santos RA, Ferreira AJ, Bader M, Alenina N, Almeida AP. Evidence for a functional interaction of the angiotensin-(1-7) receptor Mas with AT1 and AT2 receptors in the mouse heart. Hypertension [Internet]. 2005;46(4):937–42. Available from: http://www.ncbi.nlm.nih.gov/entrez/query.fcgi?cmd=Retrieve&db=PubMed&dopt=Citation&list_uids=16157793.

Angiotensin-(1-7) and the Heart

Carlos M. Ferrario, Che Ping Cheng, and Jasmina Varagic

Introduction

Thirty years ago, the cardiology field achieved a major step forward with the landmark publication of the beneficial effects of angiotensin converting enzyme (ACE) inhibition on congestive heart failure (HF) [1, 2]. Those studies showed for the first time that the addition of enalapril to conventional therapy was associated with reduction in death and improved symptoms in patients with severe congestive HF [1]. The same year saw the publication of our first report regarding the biological actions of angiotensin-(1-7) [Ang-(1-7)] [3], which is now recognized as the effector hormone that within the renin angiotensin aldosterone system (RAAS) counterbalances the pro-hypertensive and growth-inducing actions of angiotensin II (Ang II) [4, 5]. For readers less familiar with the topic at hand, our laboratory then linked the antihypertensive actions to ACE inhibition to increased circulating Ang-(1-7) levels as vascular endothelial ACE degraded Ang-(1-7) into angiotensin-(1-5) [Ang-(1-5)] [6]. This finding led to explore the possibility that Ang-(1-7) antihypertensive and cardio-renal protective actions in part explained ACE inhibitors' mechanism of action [7]. Further studies in rodents [8–11] and hypertensive patients [7, 12–14] revealed a role of Ang-(1-7)

C. M. Ferrario (✉)
Department of Surgery, Winston Salem, NC, USA
e-mail: cferrari@wakehealth.edu

C. P. Cheng
Cardiovascular Center, Winston Salem, NC, USA
e-mail: ccheng@wakehealth.edu

J. Varagic
Department of Surgery, Winston Salem, NC, USA

Wake Forest School of Medicine, Winston Salem, NC, USA
e-mail: jvaragic@wakehealth.edu

© Springer Nature Switzerland AG 2019
R. A. S. Santos (ed.), *Angiotensin-(1-7)*,
https://doi.org/10.1007/978-3-030-22696-1_6

in contributing to the antihypertensive and cardio-renal protective functions by ACE inhibition. On the other hand, it has been reported that blockade of ACE with ramipril prevented the adverse cardiac effects of high Ang-(1-7) doses in rats with sub-total nephrectomy [15].

It was a natural corollary to the learning of Ang-(1-7) biological actions that attention was paid to its potential role as a modulator of cardiac function. Earlier studies documented an uncoupling of Ang-(1-7) concentrations in canine's coronary sinus blood during the acute phase of myocardial infarction both before and following administration of the ACE inhibitor benazeprilat [16]. A more complete study showed selective increased expression of immunoreactive Ang-(1-7) confined to cardiac myocytes of the penumbra region of the ischemic myocardium surrounding the rat's left ventricle [17]. Although local Ang-(1-7) generation in the intact heart was demonstrated by Santos et al. [16] and Campbell et al. [18], the question of whether local cardiac tissue production of Ang-(1-7) is the effector pathway opposing adverse myocardial remodeling in patients remains to be fully answered. Variances in methodology, experimental conditions, and species differences might explain discordant effects of ACE inhibition on Ang-(1-7) across the heart circulation [19–22].

This chapter summarizes what we know about the role of Ang-(1-7) as a modulator of heart function. Review of the literature is based on search of the PubMed database, inclusion of the original studies performed by our laboratory, and studies examined were restricted to English language reports.

Angiotensin-(1-7) and Heart Function

This book reviews the role that Ang-(1-7) plays in counterbalancing Ang II mechanisms of action in normal and disease stages [5, 23–27]. A commented list of key articles linking Ang-(1–7) actions with heart function in health and disease is presented in Table 1. Several studies document Ang-(1-7) actions in reversing myocyte hypertrophy [28], exhibiting antiproliferative actions [29–31], counteracting inflammatory signals, and expression of radical oxygen species (ROS) [32, 33], facilitating nitric oxide (NO) release [34, 35] and inhibiting vascular atherosclerosis [36]. Signal transduction pathways in which Ang-(1-7) acts as a ligand are complex, engaging not only the Mas receptor [37, 38] but also the binding of the heptapeptide to AT_2-receptors (AT_2-R) and bradykinin B2 receptors [39, 40]. Decarboxylation of Ang-(1-7) or cleavage of phenylalanine [8] in angiotensin A [41] yields a peptide mirroring Ang-(1-7) mechanisms of action through binding to the Mas-related G protein couple receptor D (MrgD) [42–44].

Loot et al.'s original studies [45] provided the first comprehensive functional document of Ang-(1-7) acting as a cardioprotective agent in myocardial infarction and HF. In this study [45], the investigators assessed the effect of an 8-week intravenous infusion of Ang-(1-7) in left ventricular end-diastolic pressure and dP/dt in the progression to HF induced by coronary artery ligation. The improvement in left ventricular function was accompanied by maintenance of baseline coronary

Table 1 Ang-(1-7) and Heart Function

Year	Reference	Summary of findings
2014	Alghamri et al. [48]	Myocardial aminopeptidase degrades Ang-(1-7) into Ang-(2-7).
2009	Al-Maghrebi et al. [82]	In streptozotocin-treated SHR, Ang-(1-7) restores left ventricular function from 40 min of global ischemia in isolated perfused hearts. Treatment with the Ang-(1-7) receptor antagonist A779 recovers cardiac function.
2000	Almeida et al. [138]	Another study showing potentiation of coronary vasodilator effects of bradykinin by Ang-(1-7).
2003	Averill et al. [17]	First demonstration of increased Ang-(1-7) immunoreactivity limited to cardiac myocytes and ventricular tissue surrounding myocardial infarction in Lewis rats.
1996	Brosnihan et al. [34]	Ang-(1-7) induces coronary artery vasodilation through stimulation of kinin and nitric oxide.
2004	Campbell et al. [18]	Ang-(1-7) levels in coronary sinus and arterial blood of HF patients do not change following acute administration of an ACE inhibitor.
2005	Castro et al. [139]	Ang-(1-7) produces complex effects in isolated, perfused mouse hearts leading to the release of prostaglandins and nitric oxide through interaction with types 1 and 2 Ang II receptors.
2006	Castro et al. [23]	Genetic deletion or pharmacological blockade of Mas receptors in mice worsens recovery following ischemia/reperfusion.
2009	Castro et al. [140]	In rabbit, right-papillary muscle Ang-(1-7) through its binding to Mas receptor induces a negative inotropic effect modulated by the endocardial endothelium and NO; this response is independent of AT_1 or AT_2 receptors activation.
2016	Chang et al. [141]	Ang-(1-7) downregulates hypoxia pro-apoptotic signaling cascade by decreasing protein levels of hypoxia-inducible factor 1α (HIF-1α) and insulin-like growth factor binding protein-3 (IGFBP3). Ang-(1-7) activates the IGF1R/PI3K/Akt signaling pathways. Silencing the Mas receptor or exposure to the Ang-(1-7)-antagonist A779 abrogated these effects.
2010	Costa et al. [142]	Further evidence of Ang-(1-7) ability to stimulate cardiac NOS in ventricles from SHR through an AT_2 receptor mechanism.
2013	Cunha et al. [51]	Oral treatment with AVE 0991 reduces blood pressure and cardiac remodeling in 2 K-1C hypertensive rats.
1999	Davie and McMurray [94]	Report no effects of Ang-(1-7) on forearm blood flow in HF patients.
2013	De Almeida et al. [143]	These interesting studies document enhanced Ca^{++} intracellular transients in cardiac myocytes during co-administration of aldosterone and Ang-(1-7). The cross-talk may be related to enhanced NO release by Ang-(1-7).
2015	De Almeida et al. [52]	Ang-(1-7) protective signaling against DOCA-induced diastolic dysfunction occurs independently of BP attenuation and is mediated by the activation of pathways involved in Ca^{++} handling, hypertrophy, and survival.
2004	De Mello [118]	Antiarrhythmic Ang (1-7) actions may be explained by activation of the sodium pump which hyperpolarizes the cardiac myocytes and re-establishes impulse conduction during ischemia/reperfusion.
2009	De Mello [119]	Ang (1-7) has effects on cardiac myocytes cell volume that is counteracted by ouabain.

(continued)

Table 1 (continued)

Year	Reference	Summary of findings
2014	De Mello [144]	Ang-(1-7) prevents impairment of cell communication and impulse propagation in cardiac myocytes obtained from Sprague Dawley rats.
2015	De Mello [145]	Ang-(1-7) increases the inward calcium current in cardiomyocytes from WKY rats.
2007	De Mello et al. [146]	First direct demonstration of Ang-(1-7) cardioprotective actions through activation of the sodium pump, hyperpolarization of the cell membrane, and increased conduction velocity in cardiomyopathic hamsters.
2017	De Moraes et al. [128]	Further confirmation of Ang-(1-7) vasodilator actions in the coronary circulation.
2008	Dias-Peixoto et al. [147]	Extending the seminal studies of De Mello [118], these authors showed that chronic Mas-deficiency leads to impaired Ca^{++} handling in cardiomyocytes.
2012	Dias-Peixoto et al. [129]	Broad examination of Mas receptor expression by Western blots demonstrates downregulation of the protein in isoproterenol-induced HF or 21 days post-myocardial infarction.
2016	Diniz et al. [130]	Experimental hyperthyroidism that causes cardiac hypertrophy in Wistar rats is associated with increased cardiac Ang-(1-7), ACE2, and Mas receptor levels.
2012	Dong et al. [84]	In a rat model of diabetic cardiomyopathy, 4 weeks after ACE2 gene transfer, increased cardiac ACE2 expression is accompanied by enhanced matrix metalloproteinase-2 (MMP-2) activity, reduced myocardial fibrosis, improved left ventricular (LV) ejection fractions, and decreased LV volumes. Authors propose that ACE2 overexpression may enhance collagen degradation by MMP-2.
2008	Ebermann et al. [85]	AVE0991, a nonpeptide angiotensin-(1-7) receptor agonist, rescues cardiac function in rats with streptozotocin-induced diabetes mellitus.
2005	Ferrario et al. [148]	Chronic administration of lisinopril or losartan is associated with increases in circulating Ang-(1-7) and cardiac ACE2 gene transcripts.
2007	Ferreira et al. [73]	AVE-0991, reported as a selective Ang-(1-7) agonist, improves contractile variables in isolated hearts and diminishes the worsening of cardiac function following coronary artery occlusion. Similar results were reported by the same authors in another publication in which isoproterenol is used to induce cardiac injury [55].
2010	Ferreira et al. [54]	Transgenic (TG) rats expressing an Ang-(1-7)-producing fusion protein showed less deposition of cardiac collagen and fibronectin in response to isoproterenol-induced HF.
2011	Ferreira et al. [149]	An ACE2 activator has cardioprotective actions in SHR associated with reduction in ERK phosphorylation and increase in Ang-(1-7).
2008	Filho et al. [132]	In SHR exposed to swimming training, Ang-(1-7) levels and Mas receptor mRNA are increased in their hypertrophic left ventricles.
2008	Gallagher et al. [150]	ANG II reduced ACE2 activity and ACE2 mRNA in rat neonatal cardiac myocytes and cardiac fibroblasts, effects blocked by losartan. Ang-(1-7) counteracted the downregulation of ACE2 in these rodent cells.
2012	Gava et al. [151]	Investigated the impact of genetic deletion of the Mas receptor or AT_2 receptors on the expression of specific extracellular matrix (ECM) proteins in atria, right ventricles, and atrioventricular (AV) valves of neonatal and adult mice. Findings demonstrate Mas receptor involvement in the expression of ECM proteins within both the ventricular myocardium and AV valve.

Table 1 (continued)

Year	Reference	Summary of findings
2007–2010	Giani et al. [133, 152–154]	A series of studies from these investigators in Argentina elucidate the signaling molecules mediating Ang-(1-7) actions. Ang-(1-7) stimulates STAT3 and STAT5a/b phosphorylation and stimulates the phosphorylation of JACK2, IRS-1 and Akt pathways in the rat heart.
2010	Gomes et al. [155]	Transgenic rat with increased levels of circulating Ang-(1-7) [TGR[A1-7]3292] are protected from Ang II-induced pathological remodeling of ventricular cardiomyocytes. Cardiomyocytes from TGR(A1-7)3292 rats infused with Ang II presented increased expression levels of neuronal NO synthase and cGMP.
2004	Goulter et al. [156]	ACE and ACE2 mRNAs upregulated in human ventricular myocardium from donors with idiopathic dilated cardiomyopathy and ischemic cardiomyopathy.
2006	Grobe et al. [157]	In DOCA-salt-induced hypertensive rats, Ang-(1-7) shows antifibrotic actions that are independent of blood pressure or cardiac hypertrophy.
2007	Grobe et al. [74]	Prevention of cardiac remodeling and interstitial fibrosis by Ang-(1-7) infusion independent of blood pressure.
2012	Guimaraes et al. [158]	Assessed the effect of exercise training on collagen deposition and RAS components in the heart of FVB/N mice lacking Mas receptor. The data confirm a role of Ang-(1-7)/Mas axis in mitigating Ang II-mediated cardiac remodeling.
2017	Guo et al. [56]	Sirtuin-3-mediated deacetylation of FoxO3a, which triggers SOD2 expression, participates in the intracellular mechanisms underlying Ang-(1-7) actions.
2015	Hao et al. [159]	Combination of Ang-(1-7) and the ACE inhibitor perindopril counteracts left ventricular remodeling in diabetic rats.
2015	Hao et al. [86]	Treatment with ANG-(1-7) prevents diabetes-induced right ventricular fibrosis and dysfunction.
2010	He et al. [160]	AVE0991 prevents Ang II-inducing myocardial hypertrophy in a dose-dependent fashion, a process that may be associated with the inhibition of TGF-beta1/Smad2 signaling in neonatal cardiomyocytes.
2017	Hisatake et al. [95]	High serum ACE2 concentration associated with lower serum Ang-(1-7) levels are present in patients with a diagnosis of acute HF.
2004	Ishiyama et al. [161]	First demonstration that Ang II exerts a negative feedback on ACE2 through AT_1 receptors.
2005	Iwata et al. [162]	In adult rat cardiac fibroblasts, Ang-(1-7) inhibits collagen synthesis and opposes the effects of Ang II on endothelin-1 and leukemia inhibitory factor mRNAs.
2018	Jesus et al. [163]	Alamandine via MrgD receptors induces AMPK/NO signaling to counteract Ang II-induced cardiac hypertrophy in C57BL/6 mice.
2016	Joviano Santos et al. [121]	Explored anti-arrhythmic Ang-(1-7) actions in a variety of experimental setups.
2007	Kozlovski et al. [164]	Ang-(1-7) in isolated guinea pig heart induces coronary vasodilation mediated by endogenous bradykinin and nitric oxide release through endothelial B(2) receptors.
1990	Kumagai et al. [165]	Coronary artery vasoconstriction is reported from delivery of high doses of Ang-(1-7) (10^{-5}) to isolated heart preparations from Syrian hamsters with and without cardiomyopathy.

(continued)

Table 1 (continued)

Year	Reference	Summary of findings
2009	Lei et al. [166]	Ang (1-7) and enalaprilat improve left ventricular function of isolated rat heart perfused by burn serum.
2017	Lei et al. [87]	Exposing H9c2 cells to high glucose enhances leptin and phosphorylated (p)-MAPK pathway expressions. Levels of leptin and p-p38 MAPK/p-extracellular signal-regulated protein kinase 1/2 (ERK1/2), but not p-c-Jun N-terminal kinase, were significantly suppressed by treatment of the cells with Ang-(1-7).
2009	Li et al. [167]	Demonstrates beneficial effect of Ang-(1-7) injection in mice with 5/6 nephrectomy.
2015	Liang et al. [168]	Pressure overload from aortic stenosis in SD rats is accompanied by increases in plasma ACE, ACE2, and Ang-(1-7). These changes occurred in association with increased myocardial ACE2 mRNA. Telmisartan reverses these changes in the ACE2/Ang-(1-7) axis together with increased cardiac Mas mRNA and protein.
2008	Liao et al. [169]	In isolated rat hearts, Ang-(1-7) reverses oxidative stress induced by ischemic–reperfusion injury.
2010	Lin et al. [170]	Ang-(1-7) upregulates ACE2 expression in human cardiac fibroblasts extending original studies of Ishiyama et al. [161] and Gallagher et al. [150, 171] showing that Ang II inhibits ACE2 expression.
2010	Lin et al. [170]	Human cardiac fibroblasts respond to Ang II by increasing ACE2 expression through AT_1 receptors coupled to activation of ERK-MAP signaling pathways. Ang-(1-7) augments ACE2 expression through a Mas receptor mechanism, suggesting that the heptapeptide acts as a positive feedback loop on ACE2.
2010	Liu et al. [172]	Ang-(1-7) reverses atrial fibrillation and associated atrial fibrosis induced in canines by chronic rapid pacing.
2011	Liu et al. [173]	In chronically paced canine hearts, atrial ionic currents and action potential duration correlated with atrial mRNA expression of I(TO) Kv4.3 and I(CaL)α1C subunits. Changes induced by the arrhythmogenic pacing stimuli were reversed by irbesartan and Ang-(1-7) but not enalapril administration.
2012	Liu et al. [174]	A rat model of Adriamycin-induced dilated cardiomyopathy shows that Ang-(1-7) attenuates left ventricular dysfunction and myocardial apoptosis by downregulating caspase-3 and Bax and upregulating anti-apoptotic protein B-cell lymphoma-extra-large (Bcl-xL) expression.
2002	Loot et al. [45]	In this classic study of Ang-(1-7) cardioprotection mechanisms, the authors show for the first time a beneficial effect of Ang-(1-7) on HF progression due to occlusion of a coronary artery in Sprague Dawley rats.
2015	Luo et al. [175]	Ang-(1-7) counteracts impairment of cardiac Ca^{++} in a rat model of HF due to coronary artery occlusion.
2011	Marques et al. [75]	An oral formulation developed by including Ang-(1-7) in hydroxypropyl beta-cyclodextrin (HPbetaCD) in normal, infarcted, and isoproterenol-treated rats improves cardiac function.
2013	Martins et al. [176]	Ang-(1-7) attenuates air-jet-induced emotional stress tachycardia in Sprague Dawley rats.
2012	McCollum et al. [177]	Cardiac fibroblasts from neonatal rat hearts reveal reduced $^{(3)}$H-thymidine-leucine and $^{(3)}$H-thymidine-proline incorporation by Ang-(1-7) and Ang II- or ET-1-stimulated increase in phospho-ERK1 and -ERK2. Authors suggest that Ang-(1-7) upregulated DUSP1 to reduce MAP kinase activity and synthesis of mitogenic prostaglandins.

Table 1 (continued)

Year	Reference	Summary of findings
2012	McCollum et al. [58]	ANG-(1-7) upregulates DUSP-1 to reduce ANG II-stimulated ERK activation.
2005	Mendes et al. [59]	Cardiac content of Ang II is reduced by chronic Ang-(1-7) infusion.
2014	Meng et al. [178]	A779, an Ang-(1-7) antagonist, exacerbates cardiac damage and inflammatory response due to chronic Ang II infusion in rats.
2008	Mercure et al. [60]	Targeted Ang II or Ang-(1-7) overproduction in the heart of transgenic mice does not alter myocardial contractility. However, an eight-fold cardiac Ang-(1-7) increase is associated with reduced cardiac hypertrophy and fibrosis compared to non-transgenic mice. Cardiac Ang(1-7) selectively modulates some of the downstream signaling effectors of cardiac remodeling.
2014	Mori et al. [88]	Ang-(1-7) treatment ameliorates myocardial hypertrophy and fibrosis with normalization of diastolic dysfunction in db/db mice.
1997	Neves et al. [122]	These early studies in isolated-perfused hearts show that Ang-(1-7) facilitates reperfusion arrhythmias.
2014	Niu et al. [179]	Ang-(1-7) inhibits cell proliferation and AVP-stimulated collagen production in isolated cardiac fibroblasts from neonatal rats by inactivating Mas receptor-calcineurin-NF-κB signaling pathway.
2017	Pachauri et al. [180]	Ang-(1-7) improves ischemic preconditioning in rat's heart.
2008	Pan et al. [181]	Transcript expression of matrix metalloproteinase MMP-9 and TIMP-2 is downregulated by Ang(1-7) in human cardio fibroblasts.
2008	Pan et al. [181]	Report that Ang-(1-7)-induced decreased ratios of MMPs to TIMPs mRNAs in human cardiac fibroblasts attenuate Ang II-induced cardiac remodeling.
2015	Papinska et al. [89]	Short-term Ang-(1-7) infusion in db/db mice improves heart function as well as bone marrow and blood levels of endothelial and mesenchymal stem cells.
2016	Papinska et al. [90]	Ang-(1-7) improves heart function and reduces oxidative stress in db/db mice.
2012	Patel et al. [97]	The Ang II receptor antagonist – Irbesartan – and Ang-(1-7) prevented cardiac hypertrophy and improved cardiac remodeling in pressure-overloaded ACE2-null mice by suppressing NADPH oxidase and normalizing pathological signaling pathways.
2015	Patel et al. [182]	Treatment of wild-type male C57BL/6 mice with recombining ACE2 (rhACE2) prevents Ang II-induced cardiac hypertrophy, diastolic dysfunction while A779 prevented these beneficial effects and precipitated systolic dysfunction. rhACE2 effectively antagonized Ang II-mediated myocardial fibrosis, whereas myocardial oxidative stress and matrix metalloproteinase 2 activity were further increased by Ang-(1-7) inhibition even in the presence of rhACE2. Authors concluded that blocking Ang-(1–7) action prevents the therapeutic effects of rhACE2 in the setting of elevated Ang II.
2011	Peltonen et al. [183]	Decreased expression of Mas-receptor in human aortic valve disease.
1994	Porsti et al. [184]	Ang-(1-7) elicited a concentration-dependent dilator response ($ED_{50}>$ or $= 2$ μM) in porcine coronary artery rings markedly attenuated by the nitric oxide synthase inhibitor, NG-nitro-L-arginine, and abolished after removal of the endothelium.

(continued)

Table 1 (continued)

Year	Reference	Summary of findings
2011	Qi et al. [185]	Lentivirus-mediated overexpression of Ang-(1-7) prevents myocardial infarct-induced cardiac dysfunction and increased cardiac ACE2 and bradykinin B2 mRNAs in Sprague Dawley rats. Parallel investigation in neonatal cardiac myocytes showed a beneficial effect of Ang-(1-7) on inflammatory cytokines.
2014	Raffai et al. [186]	Bradykinin responses potentiated by Ang-(1-7) and captopril not affected by the BK1 antagonist SSR240612 and remain augmented in the presence of either N-nitro-L-arginine methyl ester hydrochloride plus indomethacin or TRAM-34 plus UCL-1684.
1999	Roks et al. [110]	ACE activity in human plasma and human atrial tissue inhibited by Ang-(1-7) with an IC_{50} of 3.0 and 4.0 micromol/L. Ang-(1-7) blocks Ang II-mediated vasoconstriction in human internal mammary arteries.
2010	Santiago et al. [63]	TG rats overexpressing an angiotensin Ang-(1-7)- fusion protein leading to increased circulating levels of the heptapeptide are protected against cardiac dysfunction and fibrosis and also present an attenuated increase in blood pressure after DOCA-salt-induced hypertension.
1990	Santos et al. [16]	In this first evaluation of Ang-(1-7) mechanism of action, Ang-(1-7) levels across the coronary circulation of the dog do not change in response to acute myocardial infarction or delivery of an ACE inhibitor.
2006	Santos et al. [187]	Extensive functional and molecular variables of genetic deletion of the Mas receptor in mice demonstrate a critical role of Ang-(1-7) in heart function.
2010	Shah et al. [188]	Ang-(1-7) increased ANP secretion at high atrial pacing via the Mas/PI3K/Akt pathway and the activation of $Na^{(+)}/H^{(+)}$ exchanger-1 and CaMKII.
2013	Souza et al. [189]	Documents a critical role of the Mas receptor in mediating coronary artery vasodilator activity through NO-related AT_2 mechanisms.
2005	Tallant et al. [28]	Transfection of cultured myocytes with an antisense oligonucleotide to the Mas receptor blocked the Ang-(1-7)-mediated inhibition of serum-stimulated MAPK activation.
2016	Tanno et al. [65]	The AT_1 receptor antagonist olmesartan reveals a role of ACE2/Ang-(1-7)/mas axis and Nox4 expression in mediating cardiac hypertrophy in transgenic mice overexpressing renin in the liver.
2017	Teixeira et al. [190]	Ang-(1-7) binds to AT_1 receptors without activating Gq, but triggering beta-arrestins 1 and 2 recruitment and activation. Authors conclude that Ang-(1-7) cardioprotective actions are mediated in part by acting as an endogenous beta-arrestin biased agonist of type 1 Ang II receptors.
2010	Trask et al. [191]	Chronic inhibition of ACE2 worsens cardiac remodeling and fibrosis in transgenic rats expressing the ren-2 gene ([mRen2]27).
2018	Tyrankiewicz et al. [101]	Changes in ACE/ACE2 balance from a panel of nine angiotensins in plasma, heart, and aorta of Tgalphaq*44 mice reveal upregulation of the ACE2/Ang-(1-7) pathway during early-stage HF. End-stage HF associates with downregulation of ACE2/Ang-(1-7).
2010	Varagic et al. [67]	In SHR, deleterious effects of a high salt diet on cardiac function and structure are associated with reduced cardiac Ang-(1-7) content and ACE2 mRNA.
2011	Velkoska et al. [192]	In a rat model of renal impairment induced by subtotal nephrectomy, Ang-(1-7) increases cardiac ACE activity and reduces ACE2 activity. Authors suggest that in the presence of significant compromise renal function, Ang-(1-7) displays deleterious effects.

Table 1 (continued)

Year	Reference	Summary of findings
2010	Wang et al. [79]	Studies in rats and mice deficient in Mas receptor or overexpressing Ang-(1-7) exclusively in the heart provide evidence that the beneficial effects may be related to circulating Ang-(1-7) and its stimulation of cardiac progenitor cells.
2005	Wang et al. [68]	Cardiac hypertrophy and fibrosis attenuated by chronic administration of Ang-(1-7) to rats with suprarenal coarctation of the aorta.
2014	Wang et al. [193]	Ang-(1-7) attenuates the increased diastolic intracellular Ca^{++} during reperfusion, restores the decreased peak Ca^{++} transients during ischemia, and reverses the decreased amplitude of Ca^{++} transient throughout the ischemia/reperfusion (I/R) periods in isolated rat ventricular myocytes. Ang-(1-7) suppresses the reactive oxygen species production in I/R, especially during the ischemic phase.
2016	Wang et al. [194]	The intermediate-conductance Ca^{++}-activated K^+ channel (KCa3.1) is a critical target of the ACE2/Ang-(1-7)/mas axis through inhibiting the ERK1/2 pathway.
2016	Wang et al. [195]	In a comparison of the cardioprotective actions of Ang II receptor antagonists in rats with hypertension induced by aortic constriction, the ACE2/Ang-(1-7)/mas axis appears to be engaged in the antihypertensive mechanisms associated with chronic administration of olmesartan, candesartan, and losartan.
2017	Wang et al. [78]	Blockade of Ang II receptors with two orally active antagonists confirms cardioprotective actions of the ACE2/Ang-(1-7)/mas axis in Sprague Dawley rats post-myocardial infarction.
2017	Wang et al. [135]	Comparison of responses to obesity-induced hypertension in male and female mice demonstrates a role for Mas agonists to provide cardioprotective activity.
2010	Wang, Y et al.	In mice, Ang-(1-7) cardioprotection after myocardial infarction may be secondary to stimulation of cardiac progenitor cells.
2011	Watts et al. [196]	Ang-(1-7) protects right ventricular function following pulmonary embolism.
2012	Zeng et al. [80]	The nonpeptide Ang-(1-7) analog (AVE 0991) improves cardiac function in rats post-myocardial infarction.
2017	Zhang et al. [102]	Ang-(1-7) increases $[Ca^{++}]_iT$ and produces positive inotropic and lusitropic effects in the LV and myocytes of isoproterenol-induced HF. These effects are mediated by the Mas receptor and involve activation of NO/BK pathways.
2015	Zhao et al. [197]	A model of hypoxia/reperfusion documents Ang-(1-7) effects on preventing mitochondrial dysfunction and induction of Akt phosphorylation.
2015	Zhao et al. [198]	Ang-(1-7) prevents acute electrical remodeling in dogs with acute atrial tachycardia via the PI3K/Akt/NO signaling pathway.
2015	Zhao et al. [81]	Cardiac angiogenesis stimulated by Ang-(1-7) occurs via increased expression of cardiac VEGF-D and MMP-9.
2016	Zhao et al. [199]	The study extends the antiarrhythmic Ang-(1-7) actions in canines by demonstrating that the heptapeptide may alleviate atrial structural and electrical remodeling in part by atrial natriuretic peptide secretion.
2015	Zhou et al. [103]	In rat myocytes collected from isoproterenol-induced HF, Ang-(1-7) increases iCa^{++},L responses.
2003	Zisman et al. [21]	Failing human hearts demonstrate increased Ang-(1-7)-forming activity from both Ang I and Ang II.

(continued)

Table 1 (continued)

Year	Reference	Summary of findings
2003	Zisman et al. [22]	Ang-(1-7) is formed in the heart of transplantation recipients. Values of Ang-(1-7) in human coronary sinus correlate with cardiac Ang II levels.
2011	Zong et al. [104]	Protective actions of telmisartan and losartan in Adriamycin-induced HF in rats linked to circulating Ang-(1-7).

flow and restoration of bradykinin-induced maximal coronary flow vasodilation. Loot's et al. [45] original demonstration was in keeping with studies reporting Ang-(1-7) release in the coronary sinus blood draining the ischemic heart of canines [16] and the presence of robust Ang-(1-7)-specific immunostaining of heart tissue surrounding the scar region following myocardial infarction [17]. Over the next 15 years, Ang-(1-7) cardioprotective actions have been confirmed and expanded (Table 1) through the additional observations of a critical role of angiotensin converting enzyme 2 (ACE2) [46] as a component of the counterbalancing axis composed of Ang-(1-7) and the Mas receptor axis.

Studies in intact animal models of hypertension [4, 8, 10, 47–69], myocardial infarction [17, 45, 48, 70–81], and experimentally induced diabetes [19, 82–87] or genetically induced obesity [88–91] report improved cardiac function and reversal of cardiac adverse remodeling induced through augmenting the expression or activity of Ang-(1-7).

Failure of the cardiac pump to maintain adequate tissue perfusion pressure is a complex process engaging activation of neuro-hormonal mechanisms via enhanced release of norepinephrine by sympathetic nerve endings and tissue Ang II levels. Table 1 includes an extensive number of preclinical studies supporting a role for Ang-(1-7) as counteracting HF mechanisms [45, 73, 90, 92–104]. A recent study showed high Ang-(1-7) circulating levels in hypertensive patients with HF and preserved ejection fraction [105]. Reported Ang-(1-7) actions in inducing vasodilator responses in part mediated by potentiating the actions of bradykinin, NO, and prostaglandins may explain the benefit of the heptapeptide in halting heart failure progression [106, 107]. Extending early original studies by Benter et al. [49], Brosnihan et al. [35], and Iyers et al. [11], Paul et al. [108] confirmed an interaction between Ang-(1-7) and bradykinin in mediating the vasodilator actions of the heptapeptide. As reviewed by Lee et al. [106] and Schindler et al. [109], confirmation of the beneficial effects of Ang-(1-7) in preclinical studies to HF patients remains limited. While studies confirm Ang-(1-7) vasodilator actions in human forearm and internal mammary artery [110, 111], there is a dearth of direct investigative approaches in which Ang-(1-7) is evaluated as preventing or counteracting HF evolution in human subjects. Both Zisman et al. [21] and Campbell et al. [18] report increased Ang-(1-7) in HF patients receiving ACE inhibitors, while a preliminary study presented in abstract form suggests increased myocardial Mas receptor expression during the remodeling stage [112]. The development of orally active compounds with Ang-(1-7) activity included in hydroxy-propyl-beta-cyclodextrin (HPB-CD) [113] provides a potential avenue for the assessment of Ang-(1-7) cardioprotection in randomized clinical trials. Shenoy et al. [114, 115] bioencapsulation of ACE2 is another strategy that could be used in HF patients.

Progression of hypertensive stages associated with adverse cardiac remodeling augments the risk for developing atrial fibrillation [116, 117]. Since our original observation showing that Ang-(1-7) enhanced the electrogenic sodium pump [118], antiarrhythmic Ang-(1-7) actions have been extended and confirmed in additional studies [118–122].

The initial observation that Ang-(1-7) mechanism of action could not be explained by the coupling of the ligand to known AT_1 or AT_2 receptors [123] led Santos and collaborators in Brazil to explore the nature of the receptor responsible for Ang-(1-7) actions. This effort led to the identification of the orphaned Mas oncogene receptor as the protein accounting for Ang-(1-7) cellular actions [37, 124]. The direct involvement of the Mas receptor in mediating cardiac myocyte growth was then documented by Tallant et al. [28]. To date, numerous studies confirm the role of the Mas receptor in mediating Ang-(1-7) cardiovascular actions (Table 1 and references [64, 112, 125–135]). Altogether activation of the Mas receptor includes NO production by phosphoinositide 3-kinase (PI3K)-dependent Akt phosphorylation of NO synthase [136] and inhibition of mitogen-activated protein kinase (MAPK) phosphorylation [137] leading to release of arachidonic acid as well as prostacyclin-mediated production of cyclic adenosine monophosphate (cAMP) and of cAMP-dependent protein kinase activation [27].

Newer studies document that Ang-(1-7) restores myocyte L-type calcium current ($I_{Ca,L}$) fluxes in experimentally induced HF through a Mas receptor signaling mechanism involving activation of NO/bradykinin pathways. Ang-(1-7)-provoked increases in [Ca^{++}] transients induce positive inotropic and lusitropic effects in left ventricular function and myocyte harvested from isolated failing hearts contractility in HF [103].

Summary

After almost two decades of skepticism as to whether Ang-(1-7) was anything other than an inert metabolite of Ang II [18, 94], the existence of a counterbalancing arm opposing Ang II mechanisms of actions is no longer denied. The modulatory antifibrotic and anti-hypertropic actions of Ang-(1-7) are well documented in multiple experimental conditions and species. Synthesis and expression of Ang-(1-7) in human hearts is documented, but clinical translation of these findings to therapies counteracting the adverse cardiac remodeling associated with heart disease remains to be fully achieved. The development of therapeutic orally active formulations involving prolonging Ang-(1-7) half-life and oral absorption and comparative efforts based on ACE2 should facilitate the design of randomized clinical trials to test the effectiveness of augmenting the activity of the ACE2/Ang-(1-7)/mas axis in the evolution of atrial fibrillation, cardiac recovery post-myocardial infarction, and HF progression.

Acknowledgments Research conducted on this topic by the authors of this review and their collaborators was supported by the National Heart, Lung, and Blood Institute of the National Institutes of Health (NHLBI of the NIH) grant 2P01HL051952.

Literature Cited

1. Group CTS. Effects of enalapril on mortality in severe congestive heart failure. Results of the Cooperative North Scandinavian Enalapril Survival Study (CONSENSUS). N Engl J Med. 1987;316(23):1429–35.
2. Michel D. Consensus with CONSENSUS (Cooperative North Scandinavian Enalapril Survival Study). Fortschr Med. 1988;106(5):81–2.
3. Schiavone MT, Santos RA, Brosnihan KB, Khosla MC, Ferrario CM. Release of vasopressin from the rat hypothalamo-neurohypophysial system by angiotensin-(1-7) heptapeptide. Proc Natl Acad Sci U S A. 1988;85(11):4095–8.
4. Ferrario CM. Angiotension-(1-7) and antihypertensive mechanisms. J Nephrol. 1998;11(6):278–83.
5. Ferrario CM, Chappell MC, Tallant EA, Brosnihan KB, Diz DI. Counterregulatory actions of angiotensin-(1-7). Hypertension. 1997;30(3 Pt 2):535–41.
6. Chappell MC, Pirro NT, Sykes A, Ferrario CM. Metabolism of angiotensin-(1-7) by angiotensin-converting enzyme. Hypertension. 1998;31(1 Pt 2):362–7.
7. Ferrario CM, Jaiswal N, Yamamoto K, Diz DI, Schiavone MT. Hypertensive mechanisms and converting enzyme inhibitors. Clin Cardiol. 1991;14(8 Suppl 4):IV56–62; discussion IV83–90.
8. Ferrario CM, Averill DB, Brosnihan KB, Chappell MC, Iskandar SS, Dean RH, et al. Vasopeptidase inhibition and Ang-(1-7) in the spontaneously hypertensive rat. Kidney Int. 2002;62(4):1349–57.
9. Iyer SN, Chappell MC, Averill DB, Diz DI, Ferrario CM. Vasodepressor actions of angiotensin-(1-7) unmasked during combined treatment with lisinopril and losartan. Hypertension. 1998;31(2):699–705.
10. Iyer SN, Ferrario CM, Chappell MC. Angiotensin-(1-7) contributes to the antihypertensive effects of blockade of the renin-angiotensin system. Hypertension. 1998;31(1 Pt 2):356–61.
11. Iyer SN, Yamada K, Diz DI, Ferrario CM, Chappell MC. Evidence that prostaglandins mediate the antihypertensive actions of angiotensin-(1-7) during chronic blockade of the renin-angiotensin system. J Cardiovasc Pharmacol. 2000;36(1):109–17.
12. Ferrario CM, Martell N, Yunis C, Flack JM, Chappell MC, Brosnihan KB, et al. Characterization of angiotensin-(1-7) in the urine of normal and essential hypertensive subjects. Am J Hypertens. 1998;11(2):137–46.
13. Ferrario CM, Smith RD, Brosnihan B, Chappell MC, Campese VM, Vesterqvist O, et al. Effects of omapatrilat on the renin-angiotensin system in salt-sensitive hypertension. Am J Hypertens. 2002;15(6):557–64.
14. Luque M, Martin P, Martell N, Fernandez C, Brosnihan KB, Ferrario CM. Effects of captopril related to increased levels of prostacyclin and angiotensin-(1-7) in essential hypertension. J Hypertens. 1996;14(6):799–805.
15. Burrell LM, Gayed D, Griggs K, Patel SK, Velkoska E. Adverse cardiac effects of exogenous angiotensin 1-7 in rats with subtotal nephrectomy are prevented by ACE inhibition. PLoS One. 2017;12(2):e0171975.
16. Santos RA, Brum JM, Brosnihan KB, Ferrario CM. The renin-angiotensin system during acute myocardial ischemia in dogs. Hypertension. 1990;15(2 Suppl):I121–7.
17. Averill DB, Ishiyama Y, Chappell MC, Ferrario CM. Cardiac angiotensin-(1-7) in ischemic cardiomyopathy. Circulation. 2003;108(17):2141–6.

18. Campbell DJ, Zeitz CJ, Esler MD, Horowitz JD. Evidence against a major role for angiotensin converting enzyme-related carboxypeptidase (ACE2) in angiotensin peptide metabolism in the human coronary circulation. J Hypertens. 2004;22(10):1971–6.
19. Mahmood A, Jackman HL, Teplitz L, Igic R. Metabolism of angiotensin I in the coronary circulation of normal and diabetic rats. Peptides. 2002;23(6):1171–5.
20. Neves LA, Almeida AP, Khosla MC, Santos RA. Metabolism of angiotensin I in isolated rat hearts. Effect of angiotensin converting enzyme inhibitors. Biochem Pharmacol. 1995;50(9):1451–9.
21. Zisman LS, Keller RS, Weaver B, Lin Q, Speth R, Bristow MR, et al. Increased angiotensin-(1-7)-forming activity in failing human heart ventricles: evidence for upregulation of the angiotensin-converting enzyme homologue ACE2. Circulation. 2003;108(14):1707–12.
22. Zisman LS, Meixell GE, Bristow MR, Canver CC. Angiotensin-(1-7) formation in the intact human heart: in vivo dependence on angiotensin II as substrate. Circulation. 2003;108(14):1679–81.
23. Castro CH, Santos RA, Ferreira AJ, Bader M, Alenina N, Almeida AP. Effects of genetic deletion of angiotensin-(1-7) receptor Mas on cardiac function during ischemia/reperfusion in the isolated perfused mouse heart. Life Sci. 2006;80(3):264–8.
24. Chappel MC, Ferrario CM. ACE and ACE2: their role to balance the expression of angiotensin II and angiotensin-(1-7). Kidney Int. 2006;70(1):8–10.
25. Ferrario CM. Angiotensin-converting enzyme 2 and angiotensin-(1-7): an evolving story in cardiovascular regulation. Hypertension. 2006;47(3):515–21.
26. Ferrario CM. ACE2: more of Ang-(1-7) or less Ang II? Curr Opin Nephrol Hypertens. 2011;20(1):1–6.
27. Ferrario CM, Ahmad S, Joyner J, Varagic J. Advances in the renin angiotensin system focus on angiotensin-converting enzyme 2 and angiotensin-(1-7). Adv Pharmacol. 2010;59:197–233.
28. Tallant EA, Ferrario CM, Gallagher PE. Angiotensin-(1-7) inhibits growth of cardiac myocytes through activation of the mas receptor. Am J Physiol Heart Circ Physiol. 2005;289(4):H1560–6.
29. Pei N, Mao Y, Wan P, Chen X, Li A, Chen H, et al. Angiotensin II type 2 receptor promotes apoptosis and inhibits angiogenesis in bladder cancer. J Exp Clin Cancer Res. 2017;36(1):77.
30. Pei N, Wan R, Chen X, Li A, Zhang Y, Li J, et al. Angiotensin-(1-7) decreases cell growth and angiogenesis of human nasopharyngeal carcinoma xenografts. Mol Cancer Ther. 2016;15(1):37–47.
31. Tallant EA, Diz DI, Ferrario CM. State-of-the-art lecture. Antiproliferative actions of angiotensin-(1-7) in vascular smooth muscle. Hypertension. 1999;34(4 Pt 2):950–7.
32. Chappell MC, Al Zayadneh EM. Angiotensin-(1-7) and the regulation of anti-fibrotic signaling pathways. J Cell Signal. 2017;2(1).
33. Rodrigues Prestes TR, Rocha NP, Miranda AS, Teixeira AL, Simoes ESAC. The anti-inflammatory potential of ACE2/Angiotensin-(1-7)/Mas receptor axis: evidence from basic and clinical research. Curr Drug Targets. 2017;18(11):1301–13.
34. Brosnihan KB, Li P, Ferrario CM. Angiotensin-(1-7) dilates canine coronary arteries through kinins and nitric oxide. Hypertension. 1996;27(3 Pt 2):523–8.
35. Brosnihan KB, Li P, Tallant EA, Ferrario CM. Angiotensin-(1-7): a novel vasodilator of the coronary circulation. Biol Res. 1998;31(3):227–34.
36. Yang JM, Dong M, Meng X, Zhao YX, Yang XY, Liu XL, et al. Angiotensin-(1-7) dose-dependently inhibits atherosclerotic lesion formation and enhances plaque stability by targeting vascular cells. Arterioscler Thromb Vasc Biol. 2013;33(8):1978–85.
37. Santos RA, Simoes e Silva AC, Maric C, Silva DM, Machado RP, de Buhr I, et al. Angiotensin-(1-7) is an endogenous ligand for the G protein-coupled receptor Mas. Proc Natl Acad Sci U S A. 2003;100(14):8258–63.
38. Santos RAS, Sampaio WO, Alzamora AC, Motta-Santos D, Alenina N, Bader M, et al. The ACE2/Angiotensin-(1-7)/MAS Axis of the renin-angiotensin system: focus on Angiotensin-(1-7). Physiol Rev. 2018;98(1):505–53.

39. Bruce E, Shenoy V, Rathinasabapathy A, Espejo A, Horowitz A, Oswalt A, et al. Selective activation of angiotensin AT2 receptors attenuates progression of pulmonary hypertension and inhibits cardiopulmonary fibrosis. Br J Pharmacol. 2015;172(9):2219–31.

40. Villela D, Leonhardt J, Patel N, Joseph J, Kirsch S, Hallberg A, et al. Angiotensin type 2 receptor (AT2R) and receptor Mas: a complex liaison. Clin Sci (Lond). 2015;128(4):227–34.

41. Jankowski V, Vanholder R, van der Giet M, Tolle M, Karadogan S, Gobom J, et al. Mass-spectrometric identification of a novel angiotensin peptide in human plasma. Arterioscler Thromb Vasc Biol. 2007;27(2):297–302.

42. Lautner RQ, Villela DC, Fraga-Silva RA, Silva N, Verano-Braga T, Costa-Fraga F, et al. Discovery and characterization of alamandine: a novel component of the renin-angiotensin system. Circ Res. 2013;112(8):1104–11.

43. Qaradakhi T, Apostolopoulos V, Zulli A. Angiotensin (1-7) and Alamandine: similarities and differences. Pharmacol Res. 2016;111:820–6.

44. Villela DC, Passos-Silva DG, Santos RA. Alamandine: a new member of the angiotensin family. Curr Opin Nephrol Hypertens. 2014;23(2):130–4.

45. Loot AE, Roks AJ, Henning RH, Tio RA, Suurmeijer AJ, Boomsma F, et al. Angiotensin-(1-7) attenuates the development of heart failure after myocardial infarction in rats. Circulation. 2002;105(13):1548–50.

46. Crackower MA, Sarao R, Oudit GY, Yagil C, Kozieradzki I, Scanga SE, et al. Angiotensin-converting enzyme 2 is an essential regulator of heart function. Nature. 2002;417(6891):822–8.

47. Agrawal V, Gupta JK, Qureshi SS, Vishwakarma VK. Role of cardiac renin angiotensin system in ischemia reperfusion injury and preconditioning of heart. Indian Heart J. 2016;68(6):856–61.

48. Alghamri MS, Morris M, Meszaros JG, Elased KM, Grobe N. Novel role of aminopeptidase-A in angiotensin-(1-7) metabolism post myocardial infarction. Am J Physiol Heart Circ Physiol. 2014;306(7):H1032–40.

49. Benter IF, Ferrario CM, Morris M, Diz DI. Antihypertensive actions of angiotensin-(1-7) in spontaneously hypertensive rats. Am J Phys. 1995;269(1 Pt 2):H313–9.

50. Benter IF, Yousif MH, Anim JT, Cojocel C, Diz DI. Angiotensin-(1-7) prevents development of severe hypertension and end-organ damage in spontaneously hypertensive rats treated with L-NAME. Am J Physiol Heart Circ Physiol. 2006;290(2):H684–91.

51. Cunha TM, Lima WG, Silva ME, Souza Santos RA, Campagnole-Santos MJ, Alzamora AC. The nonpeptide ANG-(1-7) mimic AVE 0991 attenuates cardiac remodeling and improves baroreflex sensitivity in renovascular hypertensive rats. Life Sci. 2013;92(4–5):266–75.

52. de Almeida PW, Melo MB, Lima Rde F, Gavioli M, Santiago NM, Greco L, et al. Beneficial effects of angiotensin-(1-7) against deoxycorticosterone acetate-induced diastolic dysfunction occur independently of changes in blood pressure. Hypertension. 2015;66(2):389–95.

53. Eatman D, Wang M, Socci RR, Thierry-Palmer M, Emmett N, Bayorh MA. Gender differences in the attenuation of salt-induced hypertension by angiotensin (1-7). Peptides. 2001;22(6):927–33.

54. Ferreira AJ, Castro CH, Guatimosim S, Almeida PW, Gomes ER, Dias-Peixoto MF, et al. Attenuation of isoproterenol-induced cardiac fibrosis in transgenic rats harboring an angiotensin-(1-7)-producing fusion protein in the heart. Ther Adv Cardiovasc Dis. 2010;4(2):83–96.

55. Ferreira AJ, Oliveira TL, Castro MC, Almeida AP, Castro CH, Caliari MV, et al. Isoproterenol-induced impairment of heart function and remodeling are attenuated by the nonpeptide angiotensin-(1-7) analogue AVE 0991. Life Sci. 2007;81(11):916–23.

56. Guo L, Yin A, Zhang Q, Zhong T, O'Rourke ST, Sun C. Angiotensin-(1-7) attenuates angiotensin II-induced cardiac hypertrophy via a Sirt3-dependent mechanism. Am J Physiol Heart Circ Physiol. 2017;312(5):H980–H91.

57. Iyer SN, Averill DB, Chappell MC, Yamada K, Allred AJ, Ferrario CM. Contribution of angiotensin-(1-7) to blood pressure regulation in salt-depleted hypertensive rats. Hypertension. 2000;36(3):417–22.

58. McCollum LT, Gallagher PE, Ann TE. Angiotensin-(1-7) attenuates angiotensin II-induced cardiac remodeling associated with upregulation of dual-specificity phosphatase 1. Am J Physiol Heart Circ Physiol. 2012;302(3):H801–10.
59. Mendes AC, Ferreira AJ, Pinheiro SV, Santos RA. Chronic infusion of angiotensin-(1-7) reduces heart angiotensin II levels in rats. Regul Pept. 2005;125(1–3):29–34.
60. Mercure C, Yogi A, Callera GE, Aranha AB, Bader M, Ferreira AJ, et al. Angiotensin(1-7) blunts hypertensive cardiac remodeling by a direct effect on the heart. Circ Res. 2008;103(11):1319–26.
61. Pei Z, Meng R, Li G, Yan G, Xu C, Zhuang Z, et al. Angiotensin-(1-7) ameliorates myocardial remodeling and interstitial fibrosis in spontaneous hypertension: role of MMPs/TIMPs. Toxicol Lett. 2010;199(2):173–81.
62. Pendergrass KD, Pirro NT, Westwood BM, Ferrario CM, Brosnihan KB, Chappell MC. Sex differences in circulating and renal angiotensins of hypertensive mRen(2). Lewis but not normotensive Lewis rats. Am J Physiol Heart Circ Physiol. 2008;295(1):H10–20.
63. Santiago NM, Guimaraes PS, Sirvente RA, Oliveira LA, Irigoyen MC, Santos RA, et al. Lifetime overproduction of circulating Angiotensin-(1-7) attenuates deoxycorticosterone acetate-salt hypertension-induced cardiac dysfunction and remodeling. Hypertension. 2010;55(4):889–96.
64. Tan Z, Wu J, Ma H. Regulation of angiotensin-converting enzyme 2 and mas receptor by Ang-(1-7) in heart and kidney of spontaneously hypertensive rats. J Renin-Angiotensin-Aldosterone Syst. 2011;12(4):413–9.
65. Tanno T, Tomita H, Narita I, Kinjo T, Nishizaki K, Ichikawa H, et al. Olmesartan inhibits cardiac hypertrophy in mice overexpressing renin independently of blood pressure: its beneficial effects on ACE2/Ang(1-7)/Mas Axis and NADPH oxidase expression. J Cardiovasc Pharmacol. 2016;67(6):503–9.
66. Trask AJ, Averill DB, Ganten D, Chappell MC, Ferrario CM. Primary role of angiotensin-converting enzyme-2 in cardiac production of angiotensin-(1-7) in transgenic Ren-2 hypertensive rats. Am J Physiol Heart Circ Physiol. 2007;292(6):H3019–24.
67. Varagic J, Ahmad S, Brosnihan KB, Groban L, Chappell MC, Tallant EA, et al. Decreased cardiac Ang-(1-7) is associated with salt-induced cardiac remodeling and dysfunction. Ther Adv Cardiovasc Dis. 2010;4(1):17–25.
68. Wang LJ, He JG, Ma H, Cai YM, Liao XX, Zeng WT, et al. Chronic administration of angiotensin-(1-7) attenuates pressure-overload left ventricular hypertrophy and fibrosis in rats. Di Yi Jun Yi Da Xue Xue Bao. 2005;25(5):481–7.
69. Yamamoto K, Ohishi M, Katsuya T, Ito N, Ikushima M, Kaibe M, et al. Deletion of angiotensin-converting enzyme 2 accelerates pressure overload-induced cardiac dysfunction by increasing local angiotensin II. Hypertension. 2006;47(4):718–26.
70. Effect of a stable angiotensin-(1-7) analogue on progenitor cell recruitment and cardiovascular function post myocardial infarction. J Am Heart Assoc. 2015;4(4).
71. Bove CM, Gilson WD, Scott CD, Epstein FH, Yang Z, Dimaria JM, et al. The angiotensin II type 2 receptor and improved adjacent region function post-MI. J Cardiovasc Magn Reson. 2005;7(2):459–64.
72. Burchill LJ, Velkoska E, Dean RG, Griggs K, Patel SK, Burrell LM. Combination renin-angiotensin system blockade and angiotensin-converting enzyme 2 in experimental myocardial infarction: implications for future therapeutic directions. Clin Sci (Lond). 2012;123(11):649–58.
73. Ferreira AJ, Jacoby BA, Araujo CA, Macedo FA, Silva GA, Almeida AP, et al. The nonpeptide angiotensin-(1-7) receptor Mas agonist AVE-0991 attenuates heart failure induced by myocardial infarction. Am J Physiol Heart Circ Physiol. 2007;292(2):H1113–9.
74. Grobe JL, Mecca AP, Lingis M, Shenoy V, Bolton TA, Machado JM, et al. Prevention of angiotensin II-induced cardiac remodeling by angiotensin-(1-7). Am J Physiol Heart Circ Physiol. 2007;292(2):H736–42.

75. Marques FD, Ferreira AJ, Sinisterra RD, Jacoby BA, Sousa FB, Caliari MV, et al. An oral formulation of angiotensin-(1-7) produces cardioprotective effects in infarcted and isoproterenol-treated rats. Hypertension. 2011;57(3):477–83.
76. McMurray J, Davie AP. Angiotensin-(1-7) attenuates the development of heart failure after myocardial infarction in rats. Circulation. 2002;106(20):e147; author reply e47.
77. Seva Pessoa B, Becher PM, Van Veghel R, De Vries R, Tempel D, Sneep S, et al. Effect of a stable Angiotensin-(1-7) analogue on progenitor cell recruitment and cardiovascular function post myocardial infarction. J Am Heart Assoc. 2015;5:4(s).
78. Wang J, He W, Guo L, Zhang Y, Li H, Han S, et al. The ACE2-Ang (1-7)-Mas receptor axis attenuates cardiac remodeling and fibrosis in post-myocardial infarction. Mol Med Rep. 2017;16(2):1973–81.
79. Wang Y, Qian C, Roks AJ, Westermann D, Schumacher SM, Escher F, et al. Circulating rather than cardiac angiotensin-(1-7) stimulates cardioprotection after myocardial infarction. Circ Heart Fail. 2010;3(2):286–93.
80. Zeng WT, Chen WY, Leng XY, Tang LL, Sun XT, Li CL, et al. Impairment of cardiac function and remodeling induced by myocardial infarction in rats are attenuated by the nonpeptide angiotensin-(1-7) analog AVE 0991. Cardiovasc Ther. 2012;30(3):152–61.
81. Zhao W, Zhao T, Chen Y, Sun Y. Angiotensin 1-7 promotes cardiac angiogenesis following infarction. Curr Vasc Pharmacol. 2015;13(1):37–42.
82. Al-Maghrebi M, Benter IF, Diz DI. Endogenous angiotensin-(1-7) reduces cardiac ischemia-induced dysfunction in diabetic hypertensive rats. Pharmacol Res. 2009;59(4):263–8.
83. Coutinho DC, Monnerat-Cahli G, Ferreira AJ, Medei E. Activation of angiotensin-converting enzyme 2 improves cardiac electrical changes in ventricular repolarization in streptozotocin-induced hyperglycaemic rats. Europace. 2014;16(11):1689–96.
84. Dong B, Yu QT, Dai HY, Gao YY, Zhou ZL, Zhang L, et al. Angiotensin-converting enzyme-2 overexpression improves left ventricular remodeling and function in a rat model of diabetic cardiomyopathy. J Am Coll Cardiol. 2012;59(8):739–47.
85. Ebermann L, Spillmann F, Sidiropoulos M, Escher F, Heringer-Walther S, Schultheiss HP, et al. The angiotensin-(1-7) receptor agonist AVE0991 is cardioprotective in diabetic rats. Eur J Pharmacol. 2008;590(1–3):276–80.
86. Hao PP, Yang JM, Zhang MX, Zhang K, Chen YG, Zhang C, et al. Angiotensin-(1-7) treatment mitigates right ventricular fibrosis as a distinctive feature of diabetic cardiomyopathy. Am J Physiol Heart Circ Physiol. 2015;308(9):H1007–19.
87. Lei Y, Xu Q, Zeng B, Zhang W, Zhen Y, Zhai Y, et al. Angiotensin-(1-7) protects cardiomyocytes against high glucose-induced injuries through inhibiting reactive oxygen species-activated leptin-p38 mitogen-activated protein kinase/extracellular signal-regulated protein kinase 1/2 pathways, but not the leptin-c-Jun N-terminal kinase pathway in vitro. J Diabetes Investig. 2017;8(4):434–45.
88. Mori J, Patel VB, Abo Alrob O, Basu R, Altamimi T, Desaulniers J, et al. Angiotensin 1-7 ameliorates diabetic cardiomyopathy and diastolic dysfunction in db/db mice by reducing lipotoxicity and inflammation. Circ Heart Fail. 2014;7(2):327–39.
89. Papinska AM, Mordwinkin NM, Meeks CJ, Jadhav SS, Rodgers KE. Angiotensin-(1-7) administration benefits cardiac, renal and progenitor cell function in db/db mice. Br J Pharmacol. 2015.
90. Papinska AM, Soto M, Meeks CJ, Rodgers KE. Long-term administration of angiotensin (1-7) prevents heart and lung dysfunction in a mouse model of type 2 diabetes (db/db) by reducing oxidative stress, inflammation and pathological remodeling. Pharmacol Res. 2016;107:372–80.
91. Patel VB, Basu R, Oudit GY. ACE2/Ang 1-7 axis: a critical regulator of epicardial adipose tissue inflammation and cardiac dysfunction in obesity. Adipocytes. 2016;5(3):306–11.
92. Chamsi-Pasha MA, Shao Z, Tang WH. Angiotensin-converting enzyme 2 as a therapeutic target for heart failure. Curr Heart Fail Rep. 2014;11(1):58–63.
93. Cole-Jeffrey CT, Pepine CJ, Katovich MJ, Grant MB, Raizada MK, Hazra S. Beneficial effects of Angiotensin-(1-7) on CD34+ cells from patients with heart failure. J Cardiovasc Pharmacol. 2018;71(3):155–9.

94. Davie AP, McMurray JJ. Effect of angiotensin-(1-7) and bradykinin in patients with heart failure treated with an ACE inhibitor. Hypertension. 1999;34(3):457–60.
95. Hisatake S, Kiuchi S, Kabuki T, Oka T, Dobashi S, Ikeda T. Serum angiotensin-converting enzyme 2 concentration and angiotensin-(1-7) concentration in patients with acute heart failure patients requiring emergency hospitalization. Heart Vessel. 2017;32(3):303–8.
96. Iwata M, Cowling RT, Yeo SJ, Greenberg B. Targeting the ACE2-Ang-(1-7) pathway in cardiac fibroblasts to treat cardiac remodeling and heart failure. J Mol Cell Cardiol. 2011;51(4):542–7.
97. Patel VB, Bodiga S, Fan D, Das SK, Wang Z, Wang W, et al. Cardioprotective effects mediated by angiotensin II type 1 receptor blockade and enhancing angiotensin 1-7 in experimental heart failure in angiotensin-converting enzyme 2-null mice. Hypertension. 2012;59(6):1195–203.
98. Patel VB, Lezutekong JN, Chen X, Oudit GY. Recombinant human ACE2 and the Angiotensin 1-7 Axis as potential new therapies for heart failure. Can J Cardiol. 2017;33(7):943–6.
99. Patel VB, Zhong JC, Grant MB, Oudit GY. Role of the ACE2/Angiotensin 1-7 Axis of the renin-angiotensin system in heart failure. Circ Res. 2016;118(8):1313–26.
100. Sukumaran V, Veeraveedu PT, Gurusamy N, Lakshmanan AP, Yamaguchi K, Ma M, et al. Olmesartan attenuates the development of heart failure after experimental autoimmune myocarditis in rats through the modulation of ANG 1-7 mas receptor. Mol Cell Endocrinol. 2012;351(2):208–19.
101. Tyrankiewicz U, Olkowicz M, Skorka T, Jablonska M, Orzylowska A, Bar A, et al. Activation pattern of ACE2/Ang-(1-7) and ACE/Ang II pathway in course of heart failure assessed by multiparametric MRI in vivo in Tgalphaq*44 mice. J Appl Physiol (1985). 2018;124(1):52–65.
102. Zhang X, Cheng HJ, Zhou P, Kitzman DW, Ferrario CM, Li WM, et al. Cellular basis of angiotensin-(1-7)-induced augmentation of left ventricular functional performance in heart failure. Int J Cardiol. 2017;236:405–12.
103. Zhou P, Cheng CP, Li T, Ferrario CM, Cheng HJ. Modulation of cardiac L-type Ca2+ current by angiotensin-(1-7): normal versus heart failure. Ther Adv Cardiovasc Dis. 2015;9(6):342–53.
104. Zong WN, Yang XH, Chen XM, Huang HJ, Zheng HJ, Qin XY, et al. Regulation of angiotensin-(1–7) and angiotensin II type 1 receptor by telmisartan and losartan in adriamycin-induced rat heart failure. Acta Pharmacol Sin. 2011;32(11):1345–50.
105. Yu J, Wu Y, Zhang Y, Zhang L, Ma Q, Luo X. Role of ACE2-Ang (1-7)-Mas receptor axis in heart failure with preserved ejection fraction with hypertension. Zhong Nan Da Xue Xue Bao Yi Xue Ban. 2018;43(7):738–46.
106. Lee VC, Lloyd EN, Dearden HC, Wong K. A systematic review to investigate whether Angiotensin-(1-7) is a promising therapeutic target in human heart failure. Int J Pept. 2013;2013:260346.
107. Stewart MH, Lavie CJ, Ventura HO. Future pharmacological therapy in hypertension. Curr Opin Cardiol. 2018l;33(4):408–15.
108. Paula RD, Lima CV, Britto RR, Campagnole-Santos MJ, Khosla MC, Santos RA. Potentiation of the hypotensive effect of bradykinin by angiotensin-(1-7)-related peptides. Peptides. 1999;20(4):493–500.
109. Schindler C, Bramlage P, Kirch W, Ferrario CM. Role of the vasodilator peptide angiotensin-(1-7) in cardiovascular drug therapy. Vasc Health Risk Manag. 2007;3(1):125–37.
110. Roks AJ, van Geel PP, Pinto YM, Buikema H, Henning RH, de Zeeuw D, et al. Angiotensin-(1-7) is a modulator of the human renin-angiotensin system. Hypertension. 1999;34(2):296–301.
111. Ueda S, Masumori-Maemoto S, Ashino K, Nagahara T, Gotoh E, Umemura S, et al. Angiotensin-(1-7) attenuates vasoconstriction evoked by angiotensin II but not by noradrenaline in man. Hypertension. 2000;35(4):998–1001.
112. Batlle Perales M, Perez-Villa F, Lazaro A, et al. The Ang(1-7) mas receptor expression is increased in myocardial tissue from heart failure patients that are in a highly active remodelling stage. Eur J Heart Fail. 2009:8.

113. Bertagnolli M, Casali KR, De Sousa FB, Rigatto K, Becker L, Santos SH, et al. An orally active angiotensin-(1-7) inclusion compound and exercise training produce similar cardiovascular effects in spontaneously hypertensive rats. Peptides. 2014;51:65–73.

114. Shenoy V, Kwon KC, Rathinasabapathy A, Lin S, Jin G, Song C, et al. Oral delivery of Angiotensin-converting enzyme 2 and Angiotensin-(1-7) bioencapsulated in plant cells attenuates pulmonary hypertension. Hypertension. 2014;64(6):1248–59.

115. Shenoy V, Qi Y, Katovich MJ, Raizada MK. ACE2, a promising therapeutic target for pulmonary hypertension. Curr Opin Pharmacol. 2011;11(2):150–5.

116. Ogunsua AA, Shaikh AY, Ahmed M, McManus DD. Atrial fibrillation and hypertension: mechanistic, epidemiologic, and treatment parallels. Methodist Debakey Cardiovasc J. 2015;11(4):228–34.

117. Verdecchia P, Angeli F, Reboldi G. Hypertension and atrial fibrillation: doubts and certainties from basic and clinical studies. Circ Res. 2018;122(2):352–68.

118. De Mello WC. Angiotensin (1-7) re-establishes impulse conduction in cardiac muscle during ischaemia-reperfusion. The role of the sodium pump. J Renin Angiotensin Aldosterone Syst. 2004;5(4):203–8.

119. De Mello WC. Cell swelling, impulse conduction, and cardiac arrhythmias in the failing heart. Opposite effects of angiotensin II and angiotensin (1-7) on cell volume regulation. Mol Cell Biochem. 2009;330(1–2):211–7.

120. De Mello WC. Angiotensin (1-7) reduces the cell volume of swollen cardiac cells and decreases the swelling-dependent chloride current. Implications for cardiac arrhythmias and myocardial ischemia. Peptides. 2010;31(12):2322–4.

121. Joviano-Santos JV, Santos-Miranda A, Joca HC, Cruz JS, Ferreira AJ. New insights into the elucidation of angiotensin-(1-7) in vivo antiarrhythmic effects and its related cellular mechanisms. Exp Physiol. 2016;101(12):1506–16.

122. Neves LA, Almeida AP, Khosla MC, Campagnole-Santos MJ, Santos RA. Effect of angiotensin-(1-7) on reperfusion arrhythmias in isolated rat hearts. Braz J Med Biol Res. 1997;30(6):801–9.

123. Tallant EA, Lu X, Weiss RB, Chappell MC, Ferrario CM. Bovine aortic endothelial cells contain an angiotensin-(1-7) receptor. Hypertension. 1997;29(1 Pt 2):388–93.

124. Lemos VS, Silva DM, Walther T, Alenina N, Bader M, Santos RA. The endothelium-dependent vasodilator effect of the nonpeptide Ang(1–7) mimic AVE 0991 is abolished in the aorta of mas-knockout mice. J Cardiovasc Pharmacol. 2005;46(3):274–9.

125. Abwainy A, Babiker F, Akhtar S, Benter IF. Endogenous angiotensin-(1-7)/Mas receptor/NO pathway mediates the cardioprotective effects of pacing postconditioning. Am J Physiol Heart Circ Physiol. 2016;310(1):H104–12.

126. Carmos-Silva C, Almeida JF, Macedo LM, Melo MB, Pedrino GR. Santos FF, et al. Clin Sci (Lond): Mas receptor contributes to pregnancy-induced cardiac remodeling; 2016.

127. Chen Q, Yang Y, Huang Y, Pan C, Liu L, Qiu H. Angiotensin-(1-7) attenuates lung fibrosis by way of Mas receptor in acute lung injury. J Surg Res. 2013;185(2):740–7.

128. de Moraes PL, Kangussu LM, Castro CH, Almeida AP, Santos RAS, Ferreira AJ. Vasodilator Effect of Angiotensin-(1-7) on Vascular Coronary Bed of Rats: Role of Mas, ACE and ACE2. Protein Pept Lett. 2017;24(9):869–75.

129. Dias-Peixoto MF, Ferreira AJ, Almeida PW, Braga VB, Coutinho DC, Melo DS, et al. The cardiac expression of Mas receptor is responsive to different physiological and pathological stimuli. Peptides. 2012;35(2):196–201.

130. Diniz GP, Senger N, Carneiro-Ramos MS, Santos RA, Barreto-Chaves ML. Cardiac ACE2/angiotensin 1–7/Mas receptor axis is activated in thyroid hormone-induced cardiac hypertrophy. Ther Adv Cardiovasc Dis. 2016;10(4):192–202.

131. Ferreira AJ, Moraes PL, Foureaux G, Andrade AB, Santos RA, Almeida AP. The angiotensin-(1-7)/Mas receptor axis is expressed in sinoatrial node cells of rats. J Histochem Cytochem. 2011;59(8):761–8.

132. Filho AG, Ferreira AJ, Santos SH, Neves SR, Silva Camargos ER, Becker LK, et al. Selective increase of angiotensin(1–7) and its receptor in hearts of spontaneously hypertensive rats subjected to physical training. Exp Physiol. 2008;93(5):589–98.
133. Giani JF, Gironacci MM, Munoz MC, Pena C, Turyn D, Dominici FP. Angiotensin-(1-7) stimulates the phosphorylation of JAK2, IRS-1 and Akt in rat heart in vivo: role of the AT1 and Mas receptors. Am J Physiol Heart Circ Physiol. 2007;293(2):H1154–63.
134. Munoz MC, Burghi V, Miquet JG, Giani JF, Banegas RD, Toblli JE, et al. Downregulation of the ACE2/Ang-(1-7)/Mas axis in transgenic mice overexpressing GH. J Endocrinol. 2014;221(2):215–27.
135. Wang Y, Shoemaker R, Powell D, Su W, Thatcher S, Cassis L. Differential effects of Mas receptor deficiency on cardiac function and blood pressure in obese male and female mice. Am J Physiol Heart Circ Physiol. 2017;312(3):H459–H68.
136. Sampaio WO. Souza dos Santos RA, Faria-Silva R, da Mata Machado LT, Schiffrin EL, Touyz RM. Angiotensin-(1-7) through receptor Mas mediates endothelial nitric oxide synthase activation via Akt-dependent pathways. Hypertension. 2007;49(1):185–92.
137. Keidar S, Kaplan M, Gamliel-Lazarovich A. ACE2 of the heart: From angiotensin I to angiotensin (1-7). Cardiovasc Res. 2007;73(3):463–9.
138. Almeida AP, Frabregas BC, Madureira MM, Santos RJ, Campagnole-Santos MJ, Santos RA. Angiotensin-(1-7) potentiates the coronary vasodilatatory effect of bradykinin in the isolated rat heart. Braz J Med Biol Res. 2000;33(6):709–13.
139. Castro CH, Santos RA, Ferreira AJ, Bader M, Alenina N, Almeida AP. Evidence for a functional interaction of the angiotensin-(1-7) receptor Mas with AT1 and AT2 receptors in the mouse heart. Hypertension. 2005;46(4):937–42.
140. Castro-Chaves P, Pintalhao M, Fontes-Carvalho R, Cerqueira R, Leite-Moreira AF. Acute modulation of myocardial function by angiotensin 1–7. Peptides. 2009;30(9):1714–9.
141. Chang RL, Lin JW, Kuo WW, Hsieh DJ, Yeh YL, Shen CY, et al. Angiotensin-(1-7) attenuated long-term hypoxia-stimulated cardiomyocyte apoptosis by inhibiting HIF-1alpha nuclear translocation via Mas receptor regulation. Growth Factors. 2016;34(1–2):11–8.
142. Costa MA, Lopez Verrilli MA, Gomez KA, Nakagawa P, Pena C, Arranz C, et al. Angiotensin-(1-7) upregulates cardiac nitric oxide synthase in spontaneously hypertensive rats. Am J Physiol Heart Circ Physiol. 2010;299(4):H1205–11.
143. de Almeida PW, de Freitas LR, de Morais Gomes ER, Rocha-Resende C, Roman-Campos D, Gondim AN, et al. Functional cross-talk between aldosterone and angiotensin-(1-7) in ventricular myocytes. Hypertension. 2013;61(2):425–30.
144. De Mello WC. Angiotensin (1-7) re-establishes heart cell communication previously impaired by cell swelling: implications for myocardial ischemia. Exp Cell Res. 2014;323(2):359–65.
145. De Mello WC. Intracellular angiotensin (1-7) increases the inward calcium current in cardiomyocytes. On the role of PKA activation. Mol Cell Biochem. 2015;407(1–2):9–16.
146. De Mello WC, Ferrario CM, Jessup JA. Beneficial versus harmful effects of Angiotensin (1-7) on impulse propagation and cardiac arrhythmias in the failing heart. J Renin Angiotensin Aldosterone Syst. 2007;8(2):74–80.
147. Dias-Peixoto MF, Santos RA, Gomes ER, Alves MN, Almeida PW, Greco L, et al. Molecular mechanisms involved in the angiotensin-(1-7)/Mas signaling pathway in cardiomyocytes. Hypertension. 2008;52(3):542–8.
148. Ferrario CM, Jessup J, Chappell MC, Averill DB, Brosnihan KB, Tallant EA, et al. Effect of angiotensin-converting enzyme inhibition and angiotensin II receptor blockers on cardiac angiotensin-converting enzyme 2. Circulation. 2005;111(20):2605–10.
149. Ferreira AJ, Shenoy V, Qi Y, Fraga-Silva RA, Santos RA, Katovich MJ, et al. Angiotensin-converting enzyme 2 activation protects against hypertension-induced cardiac fibrosis involving extracellular signal-regulated kinases. Exp Physiol. 2011;96(3):287–94.
150. Gallagher PE, Ferrario CM, Tallant EA. Regulation of ACE2 in cardiac myocytes and fibroblasts. Am J Physiol Heart Circ Physiol. 2008;295(6):H2373–9.

151. Gava E, de Castro CH, Ferreira AJ, Colleta H, Melo MB, Alenina N, et al. Angiotensin-(1-7) receptor Mas is an essential modulator of extracellular matrix protein expression in the heart. Regul Pept. 2012;175(1–3):30–42.
152. Giani JF, Gironacci MM, Munoz MC, Turyn D, Dominici FP. Angiotensin-(1-7) has a dual role on growth-promoting signalling pathways in rat heart in vivo by stimulating STAT3 and STAT5a/b phosphorylation and inhibiting angiotensin II-stimulated ERK1/2 and Rho kinase activity. Exp Physiol. 2008;93(5):570–8.
153. Giani JF, Miquet JG, Munoz MC, Burghi V, Toblli JE, Masternak MM, et al. Upregulation of the angiotensin-converting enzyme 2/angiotensin-(1-7)/Mas receptor axis in the heart and the kidney of growth hormone receptor knock-out mice. Growth Horm IGF Res. 2012;22(6):224–33.
154. Giani JF, Munoz MC, Mayer MA, Veiras LC, Arranz C, Taira CA, et al. Angiotensin-(1-7) improves cardiac remodeling and inhibits growth-promoting pathways in the heart of fructose-fed rats. Am J Physiol Heart Circ Physiol. 2010;298(3):H1003–13.
155. Gomes ER, Lara AA, Almeida PW, Guimaraes D, Resende RR, Campagnole-Santos MJ, et al. Angiotensin-(1-7) prevents cardiomyocyte pathological remodeling through a nitric oxide/guanosine 3′,5′-cyclic monophosphate-dependent pathway. Hypertension. 2010;55(1):153–60.
156. Goulter AB, Goddard MJ, Allen JC, Clark KL. ACE2 gene expression is up-regulated in the human failing heart. BMC Med. 2004;2:19.
157. Grobe JL, Mecca AP, Mao H, Katovich MJ. Chronic angiotensin-(1-7) prevents cardiac fibrosis in DOCA-salt model of hypertension. Am J Physiol Heart Circ Physiol. 2006;290(6):H2417–23.
158. Guimaraes GG, Santos SH, Oliveira ML, Pimenta-Velloso EP, Motta DF, Martins AS, et al. Exercise induces renin-angiotensin system unbalance and high collagen expression in the heart of Mas-deficient mice. Peptides. 2012;38(1):54–61.
159. Hao P, Yang J, Liu Y, Zhang M, Zhang K, Gao F, et al. Combination of angiotensin-(1-7) with perindopril is better than single therapy in ameliorating diabetic cardiomyopathy. Sci Rep. 2015;5:8794.
160. He JG, Chen SL, Huang YY, Chen YL, Dong YG, Ma H. The nonpeptide AVE0991 attenuates myocardial hypertrophy as induced by angiotensin II through downregulation of transforming growth factor-beta1/Smad2 expression. Heart Vessels. 2010;25(5):438–43.
161. Ishiyama Y, Gallagher PE, Averill DB, Tallant EA, Brosnihan KB, Ferrario CM. Upregulation of angiotensin-converting enzyme 2 after myocardial infarction by blockade of angiotensin II receptors. Hypertension. 2004;43(5):970–6.
162. Iwata M, Cowling RT, Gurantz D, Moore C, Zhang S, Yuan JX, et al. Angiotensin-(1-7) binds to specific receptors on cardiac fibroblasts to initiate antifibrotic and antitrophic effects. Am J Physiol Heart Circ Physiol. 2005;289(6):H2356–63.
163. Jesus ICG, Scalzo S, Alves F, Marques K, Rocha-Resende C, Bader M, et al. Alamandine acts via MrgD to induce AMPK/NO activation against ANG II hypertrophy in cardiomyocytes. Am J Physiol Cell Physiol. 2018;314(6):C702–C11.
164. Kozlovski VI, Lomnicka M, Fedorowicz A, Chlopicki S. On the mechanism of coronary vasodilation induced by angiotensin-(1-7) in the isolated guinea pig heart. Basic Clin Pharmacol Toxicol. 2007;100(6):361–5.
165. Kumagai H, Khosla M, Ferrario C, Fouad-Tarazi FM. Biological activity of angiotensin-(1-7) heptapeptide in the hamster heart. Hypertension. 1990;15(2 Suppl):I29–33.
166. Lei ZY, Huang YS, Xiao R, Zhang BQ, Zhang Q. Effects of angiotensin (1-7) and enalaprilat on function of isolated rat heart perfused by burn serum. Zhonghua Shao Shang Za Zhi. 2009;25(3):180–3.
167. Li Y, Wu J, He Q, Shou Z, Zhang P, Pen W, et al. Angiotensin (1-7) prevent heart dysfunction and left ventricular remodeling caused by renal dysfunction in 5/6 nephrectomy mice. Hypertens Res. 2009;32(5):369–74.
168. Liang B, Li Y, Han Z, Xue J, Zhang Y, Jia S, et al. ACE2-Ang (1–7) axis is induced in pressure overloaded rat model. Int J Clin Exp Pathol. 2015;8(2):1443–50.

169. Liao XX, Guo RX, Ma H, Wang LC, Chen ZH, Yang CT, et al. Effects of angiotensin-(1-7) on oxidative stress and functional changes of isolated rat hearts induced by ischemia-reperfusion. Nan Fang Yi Ke Da Xue Xue Bao. 2008;28(8):1345–8.

170. Lin CS, Pan CH, Wen CH, Yang TH, Kuan TC. Regulation of angiotensin converting enzyme II by angiotensin peptides in human cardiofibroblasts. Peptides. 2010;31(7):1334–40.

171. Gallagher PE, Chappell MC, Ferrario CM, Tallant EA. Distinct roles for ANG II and ANG-(1-7) in the regulation of angiotensin-converting enzyme 2 in rat astrocytes. Am J Physiol Cell Physiol. 2006;290(2):C420–6.

172. Liu E, Yang S, Xu Z, Li J, Yang W, Li G. Angiotensin-(1-7) prevents atrial fibrosis and atrial fibrillation in long-term atrial tachycardia dogs. Regul Pept. 2010;162(1–3):73–8.

173. Liu E, Xu Z, Li J, Yang S, Yang W, Li G. Enalapril, irbesartan, and angiotensin-(1-7) prevent atrial tachycardia-induced ionic remodeling. Int J Cardiol. 2011;146(3):364–70.

174. Liu HZ, Gao CY, Wang XQ, Fu HX, Yang HH, Wang XP, et al. Angiotensin(1–7) attenuates left ventricular dysfunction and myocardial apoptosis on rat model of adriamycin-induced dilated cardiomyopathy. Zhonghua Xin Xue Guan Bing Za Zhi. 2012;40(3):219–24.

175. Luo D, Zhuang X, Luo C, Long M, Deng C, Liao X, et al. Continuous angiotensin-(1-7) infusion improves myocardial calcium transient and calcium transient alternans in ischemia-induced cardiac dysfunction rats. Biochem Biophys Res Commun. 2015;467(4):645–50.

176. Martins Lima A, Xavier CH, Ferreira AJ, Raizada MK, Wallukat G, Velloso EP, et al. Activation of angiotensin-converting enzyme 2/angiotensin-(1-7)/Mas axis attenuates the cardiac reactivity to acute emotional stress. Am J Physiol Heart Circ Physiol. 2013;305(7):H1057–67.

177. McCollum LT, Gallagher PE, Tallant EA. Angiotensin-(1-7) abrogates mitogen-stimulated proliferation of cardiac fibroblasts. Peptides. 2012;34(2):380–8.

178. Meng W, Zhao W, Zhao T, Liu C, Chen Y, Liu H, et al. Autocrine and paracrine function of Angiotensin 1-7 in tissue repair during hypertension. Am J Hypertens. 2014;27(6):775–82.

179. Niu X, Xue Y, Li X, He Y, Zhao X, Xu M, et al. Effects of angiotensin-(1-7) on the proliferation and collagen synthesis of arginine vasopressin-stimulated rat cardiac fibroblasts: role of mas receptor-calcineurin-NF-kappaB signaling pathway. J Cardiovasc Pharmacol. 2014;64(6):536–42.

180. Pachauri P, Garabadu D, Goyal A, Upadhyay PK. Angiotensin (1-7) facilitates cardioprotection of ischemic preconditioning on ischemia-reperfusion-challenged rat heart. Mol Cell Biochem. 2017;430(1–2):99–113.

181. Pan CH, Wen CH, Lin CS. Interplay of angiotensin II and angiotensin(1–7) in the regulation of matrix metalloproteinases of human cardiocytes. Exp Physiol. 2008;93(5):599–612.

182. Patel VB, Takawale A, Ramprasath T, Das SK, Basu R, Grant MB, et al. Antagonism of angiotensin 1–7 prevents the therapeutic effects of recombinant human ACE2. J Mol Med (Berl). 2015;93(9):1003–13.

183. Peltonen T, Napankangas J, Ohtonen P, Aro J, Peltonen J, Soini Y, et al. (Pro)renin receptors and angiotensin converting enzyme 2/angiotensin-(1-7)/Mas receptor axis in human aortic valve stenosis. Atherosclerosis. 2011;216(1):35–43.

184. Porsti I, Bara AT, Busse R, Hecker M. Release of nitric oxide by angiotensin-(1-7) from porcine coronary endothelium: implications for a novel angiotensin receptor. Br J Pharmacol. 1994;111(3):652–4.

185. Qi Y, Shenoy V, Wong F, Li H, Afzal A, Mocco J, et al. Lentivirus-mediated overexpression of angiotensin-(1-7) attenuated ischaemia-induced cardiac pathophysiology. Exp Physiol. 2011;96(9):863–74.

186. Raffai G, Khang G, Vanhoutte PM. Angiotensin-(1-7) augments endothelium-dependent relaxations of porcine coronary arteries to bradykinin by inhibiting angiotensin-converting enzyme 1. J Cardiovasc Pharmacol. 2014;63(5):453–60.

187. Santos RA, Castro CH, Gava E, Pinheiro SV, Almeida AP, Paula RD, et al. Impairment of in vitro and in vivo heart function in angiotensin-(1-7) receptor MAS knockout mice. Hypertension. 2006;47(5):996–1002.

188. Shah A, Gul R, Yuan K, Gao S, Oh YB, Kim UH, et al. Angiotensin-(1-7) stimulates high atrial pacing-induced ANP secretion via Mas/PI3-kinase/Akt axis and Na+/H+ exchanger. Am J Physiol Heart Circ Physiol. 2010;298(5):H1365–74.
189. Souza AP, Sobrinho DB, Almeida JF, Alves GM, Macedo LM, Porto JE, et al. Angiotensin II type 1 receptor blockade restores angiotensin-(1-7)-induced coronary vasodilation in hypertrophic rat hearts. Clin Sci (Lond). 2013;125(9):449–59.
190. Teixeira LB, Parreiras ESLT, Bruder-Nascimento T, Duarte DA, Simoes SC, Costa RM, et al. Ang-(1-7) is an endogenous beta-arrestin-biased agonist of the AT1 receptor with protective action in cardiac hypertrophy. Sci Rep. 2017;7(1):11903.
191. Trask AJ, Groban L, Westwood BM, Varagic J, Ganten D, Gallagher PE, et al. Inhibition of angiotensin-converting enzyme 2 exacerbates cardiac hypertrophy and fibrosis in Ren-2 hypertensive rats. Am J Hypertens. 2010;23(6):687–93.
192. Velkoska E, Dean RG, Griggs K, Burchill L, Burrell LM. Angiotensin-(1-7) infusion is associated with increased blood pressure and adverse cardiac remodelling in rats with subtotal nephrectomy. Clin Sci (Lond). 2011;120(8):335–45.
193. Wang L, Luo D, Liao X, He J, Liu C, Yang C, et al. Ang-(1-7) offers cytoprotection against ischemia-reperfusion injury by restoring intracellular calcium homeostasis. J Cardiovasc Pharmacol. 2014;63(3):259–64.
194. Wang LP, Fan SJ, Li SM, Wang XJ, Gao JL, Yang XH. Protective role of ACE2-Ang-(1-7)-Mas in myocardial fibrosis by downregulating KCa3.1 channel via ERK1/2 pathway. Pflugers Arch. 2016;468(11–12):2041–51.
195. Wang X, Ye Y, Gong H, Wu J, Yuan J, Wang S, et al. The effects of different angiotensin II type 1 receptor blockers on the regulation of the ACE-AngII-AT1 and ACE2-Ang(1-7)-Mas axes in pressure overload-induced cardiac remodeling in male mice. J Mol Cell Cardiol. 2016;97:180–90.
196. Watts JA, Gellar MA, Stuart L, Obraztsova M, Marchick MR, Kline JA. Effects of angiotensin (1-7) upon right ventricular function in experimental rat pulmonary embolism. Histol Histopathol. 2011;26(10):1287–94.
197. Zhao P, Li F, Gao W, Wang J, Fu L, Chen Y, et al. Angiotensin1-7 protects cardiomyocytes from hypoxia/reoxygenation-induced oxidative stress by preventing ROS-associated mitochondrial dysfunction and activating the Akt signaling pathway. Acta Histochem. 2015;117(8):803–10.
198. Zhao J, Liu E, Li G, Qi L, Li J, Yang W. Effects of the angiotensin-(1-7)/Mas/PI3K/Akt/nitric oxide axis and the possible role of atrial natriuretic peptide in an acute atrial tachycardia canine model. J Renin Angiotensin Aldosterone Syst. 2015;16(4):1069–77.
199. Zhao J, Liu T, Liu E, Li G, Qi L, Li J. The potential role of atrial natriuretic peptide in the effects of Angiotensin-(1-7) in a chronic atrial tachycardia canine model. J Renin Angiotensin Aldosterone Syst. 2016;17(1):1470320315627409.

Blood Vessels

Ang-(1-7) and Vessels

Walyria O. Sampaio and Rhian M. Touyz

Ang-(1-7), Vasodilation and Hemodynamics Effects

Vasodilation is the first described and most well-known vascular action of Ang-(1-7) [1]. These effects are evident in various vascular beds, including peripheral, conduit, coronary, renal, and cerebral arteries. Ang-(1-7)-induced vasorelaxation has been observed in aortic rings of Sprague–Dawley [2] and mRen-2 transgenic rats [3], canine [4] and porcine coronary arteries [5], canine middle cerebral artery [6], piglet pial arterioles [7] and feline systemic vasculature [7], rabbit renal afferent arterioles [8] and mesenteric microvessels of normotensive, and hypertensive rats [9, 10]. Ang-(1-7) also has a synergistic effect on bradykinin-mediated vasodilation, as observed in rat (de coronary vessels [11], rat and human kidney vessels [12, 13], mesenteric arteries of normal and salt-fed rats [14, 15] and pancreatic microcirculation [16]. Ang-(1-7) also potentiates the dose-dependent relaxing effects of ghrelin in pulmonary artery rings of rats [17]. In diabetic rats, chronic administration of Ang-(1-7) contributes to reduce the carotid resistance and to increase blood flow [18] (Fig. 1).

The specific local modulation of vascular tone and blood flow distribution is an essential effect of Ang-(1-7) on hemodynamics control. This specificity is well described by blood flow distribution studies with microspheres [19, 20]. Interestingly, simultaneous actions on blood flow distribution and cardiac output were observed, significantly impacting blood pressure regulation. Ang-(1-7) increases vascular conductance in the mesenteric, cerebral, cutaneous, and renal territories, decreasing

W. O. Sampaio (✉)
University of Itaúna, Biological Sciences Institute, Itaúna, MG, Brazil

R. M. Touyz (✉)
Institute of Cardiovascular and Medical Sciences, BHF Glasgow Cardiovascular Research Centre, University of Glasgow, Glasgow, UK
e-mail: rhian.touyz@glasgow.ac.uk

© Springer Nature Switzerland AG 2019
R. A. S. Santos (ed.), *Angiotensin-(1-7)*,
https://doi.org/10.1007/978-3-030-22696-1_7

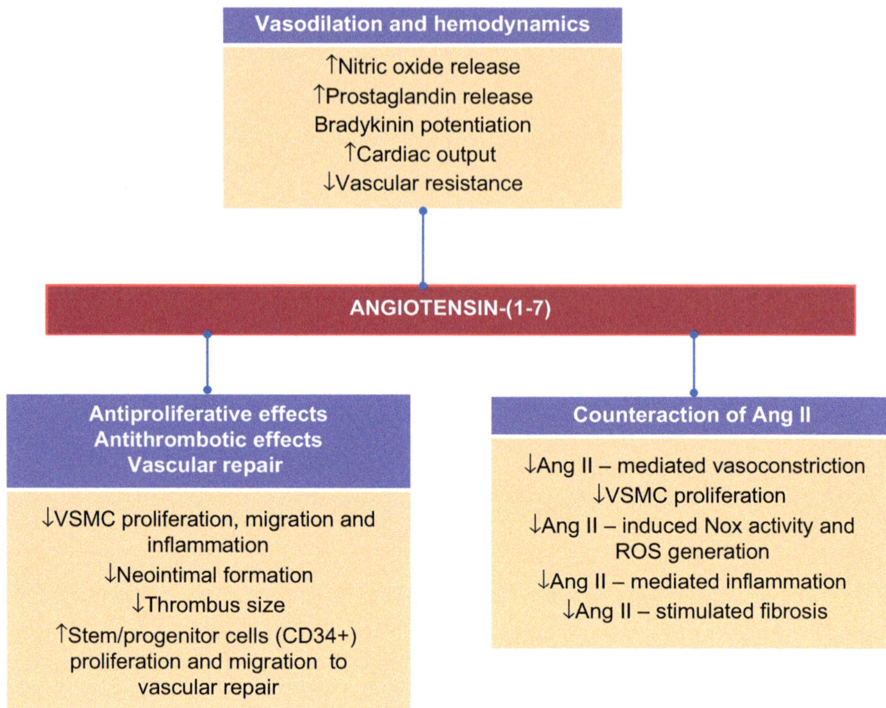

Fig. 1 Diagram demonstrating major vascular effects of Ang-(1-7). In general, Ang-(1-7) opposes actions of Ang II

total peripheral resistance (TPR) by 26%. Additionally, Ang-(1-7) increases cardiac output (CO) by 30% thereby improving hemodynamic status and blood pressure regulation [19]. Likewise, these regional actions are found in transgenic rats with a lifetime increase in circulating Ang- (1-7)-(TGRL-3292) [20]. This model is characterized by an increase in vascular conductance in the kidneys, lungs, adrenals, spleen, brain, testis, and brown fat tissue [20]. Further studies using Mas-deficient mice confirmed the important participation of Ang-(1-7) on flow distribution. As opposed to TGRL-3292 rats, Mas-deficient mice exhibit increased vascular resistance of various territories such as kidney, lung, adrenal gland, mesentery, spleen, and brown adipose tissue, which leads to increased total peripheral resistance. On the other hand, the lack of Mas-receptor decreased cardiac index [21].

Although several studies confirm the vasodilatory action of Ang-(1-7) in animal models and isolated vessels, this effect is not so clear in the human vasculature and further studies are still warranted. Nevertheless, recent studies strongly indicate the participation of Ang-(1-7)/Mas cascade in human hemodynamics, especially in pathological conditions. Patients with acute heart failure have higher serum ACE2 concentrations, lower serum Ang-(1-7) levels, and lower serum ACE activity and plasma aldosterone concentrations than healthy volunteers [22]. Van Twist et al. [23] observed a significant dose-dependent increase in blood flow to the kidney during intrarenal infusion of Ang-(1-7) in hypertensive patients. These authors [23] also demonstrated that the effect of Ang-(1-7) infusion in renal blood flow was reduced in stenotic kidneys. This effect on renal blood flow was weakened in patients on a low-salt diet, probably due to the fact that low-salt diet leads to an increase in circulating angiotensin peptides, including Ang-(1-7) [24]. Interestingly, similar finding was reported in rat renal vessels [25]. Mendonça et al. [26] demonstrated that Ang-(1-7) significantly attenuates Ang II-induced vasoconstriction in human mammary arteries from patients undergoing coronary revascularization probably through a direct effect on vascular smooth muscle cells (VSMCs), since this action was not abolished by A-779, PD123177 or endothelium removal.

Conflicting data in humans regarding the pathophysiological role of Ang-(1-7) may relate to methodological discrepancies or vascular territory selectivity to Ang-(1-7). Initially, it was demonstrated that the infusion of Ang-(1-7) in patients chronically treated with ACE inhibitors (ACEi) had no effect on forearm blood flow, while the infusion of bradykinin caused vasodilation [27]. This finding suggested that Ang-(1-7) had no participation in the hemodynamic actions of ACEi. However, in that study, effects of ACE inhibition-induced increases in circulating levels of Ang-(1-7) were not considered. Thus, during ACE blockade, the use of a Mas receptor antagonist A-779 would have been more appropriate to evaluate the vasodilatory effects mediated through the Ang-(1-7)/Mas axis. Similarly, no effect was found in the forearm blood flow of normotensive patients, in which Ang-(1-7) did not alter vasodilation produced by bradykinin infusion [28]. Sasaki et al. [29] obtained divergent results and demonstrated a dose-dependent vasodilation in forearm circulation of normotensive subjects and patients with essential hypertension. Ueda et al. [30] also reported a dose-dependent potentiation of bradykinin vasodilation by Ang-(1-7) in forearm resistance vessels of normotensive healthy men, confirming the bradykinin-potentiating effect of Ang-(1-7) described in animal models [9, 10, 31]. In the forearm of normotensive patients as well as in mammary arteries in vitro Ang-(1-7) attenuates the vasoconstrictor effect of Ang II, but not noradrenaline [29, 30]. Ang-(1-7) also modulates Ang II in renal vessels in vitro, but does not appear to have a pronounced effect in normal physiological regulation of renal vascular function in vivo [32].

Antiproliferative Effects

In the vasculature, Ang-(1-7) has antiproliferative effects that oppose the mitogenic actions of Ang II. In VSMCs, Ang-(1-7) induces release of prostaglandins (PGI2 and PGE2) and decreases Ang II–stimulated activation of MAP kinases (ERK1/2)2 [33, 34]. Ang-(1-7), through Mas receptor, also attenuates Ang II-induced proliferation, migration, and inflammation of VSMCs. These effects are mediated through processes that inactivate ROS-induced PI3K/Akt and MAPK/ERK signaling [35]. Ang-(1-7)/Mas receptor also inhibits transactivation of epidermal growth factor receptor tyrosine kinases (ErbB2, ErbB3, and ErbB4) mediated by high-glucose, Ang II and norepinephrine. This results in decreased activation of downstream signaling pathways (ROCK, p38MAP kinase, ERK1/2, eNOS, and IkB-α) involved in diabetic vasculopathy [36]. Mas agonist inhibition with AVE0991 has comparable effects in attenuating Ang II-induced VSMCs proliferation as Ang-(1-7). This action was associated with Mas-mediated inhibition of heme oxygenase (HO-1)/p38MAPK signaling pathway [37]. After vascular injury, Ang-(1-7) is also able to reduce neo-intimal formation [38]. Similar effects were observed in a rat stenting model, where Ang-(1-7) treatment produced a significant reduction in neointimal thickness, neo-intimal area, and percentage stenosis [39]. These effects are further evidenced with Mas receptor deletion, which causes marked increase in aortic intima and in intimal thickening, indicating the importance of Ang-(1-7)/Mas vasoprotection in athero-sclerosis [40]. Through this mechanism, a synergistic anti-atherosclerotic action of Ang-(1-7) and losartan was demonstrated in ApoE−/− mice, a model of atheroscle-rosis; Ang-(1-7) and losartan treatment improved endothelial function, attenuated macrophage infiltration, and inhibited VSMCs proliferation and migration [41]. In addition, direct effects of Ang-(1-7) on vessel wall smooth muscle restored the decreased expression of lineage markers, including smooth muscle (SM) α-actin, SM22α, calponin, and smoothelin, in VSMCs and retarded the osteogenic transition of these cells by reducing the expression of bone-associated proteins in rats with vascular calcification [42]. Similar antiproliferative characteristics have also been observed in cardiac fibroblasts [43] and tumor cells [44, 45] suggesting that Ang-(1-7) has anti-fibrotic and antitumor actions. Moreover, Ang-(1-7) has been sug-gested to be anti-angiogenic [46]. Together, these findings suggest that targeting the Ang-(1-7)/Mas pathway may be an effective anti-cancer approach. Clinical studies testing this notion have already been initiated.

Antithrombotic Effects

Activation of the Ang-(1-7)/Mas axis also has significant antithrombotic effects, as evidenced in studies in Mas knockout mice. The deletion of Mas receptor increased venous thrombus size and shortened the bleeding time in these ani-mals [47]. Additionally, an orally active form of Ang-(1-7), where Ang-(1-7) was incorporated in cyclodextrin (Ang-(1-7)-CyD), caused an increase in the plasma concentration of Ang-(1-7) accompanied by antithrombotic actions in spontane-ously hypertensive rats (SHR) [48]. Interestingly, the anticoagulant effect was observed at doses of 10 or 30 µg/kg of Ang-(1-7), while higher doses of Ang-(1-7)-CyD [100 µg/kg of Ang-(1-7)] did not inhibit thrombus formation. Both acute and chronic treatment had antithrombotic actions (60% and 67% thrombus weight inhibition for acute and chronic administration, respectively). This antithrombotic effect of Ang-(1-7)-CyD was not observed in $Mas^{-/-}$ mice [47, 49]. Moreover, the Mas antagonist, A-779, reduced the time to arterial thrombosis and tail bleeding time in Bradykinin B2 receptor-deleted mice (Bdkrb2−/−). NO and prostacyclin release in platelets, stimulated via Ang-(1-7)/Mas, probably mediated the anti-thrombogenic [50].

Ang-(1-7), Progenitor Cells and Vascular Repair

In addition to protecting the vasculature through its anti-Ang II effects, Ang-(1-7) may promote vascular repair by directly influencing reparative stem/progenitor cells, particularly $CD34^+$ cells. These cells, when exposed to ischemic or hypoxic stress, proliferate and migrate to the injured areas and accelerate vascular repair thereby preventing tissue damage [51]. CD34+ cells express ACE2 and Mas receptor and are responsive to Ang-(1-7) [52]. Recent studies in human CD34+ cells demonstrated that Ang-(1-7) or its analogue NorLeu3-Ang-(1-7), through Mas receptor, stimulate migration and proliferation, which are the features of vasoreparative potential [53].

In experimental models of diabetes, activation of the Ang-(1-7)/Mas pathway increased levels of circulating pro-repair bone marrow progenitor cells (Lineage$^-$Sca-1$^+$c-Kit$^+$(LSK) cells), processes that involve Rho kinase [54]. Genetic ablation of MasR prevented ischemia-induced mobilization of LSK cells and impaired blood flow recovery, which was associated with decreased proliferation and migration of LSK cells [54]. Together, these results suggest that Ang-(1-7)/MasR is functionally active in progenitor cells and may have therapeutic potential in vascular disease.

Signaling Underlying Ang-(1-7) Vascular Actions

Most of the Ang-(1-7)/Mas vasoprotective effects occur via nitric oxide (NO) and prostaglandin release, important mediators in vasodilation. Moreover, Ang-(1-7) directly counterregulates the intracellular signaling pathways elicited by Ang II. Figure 2 summarizes the main signaling pathways through which Ang-(1-7) induces its cellular actions.

NO Release

Ang-(1-7) binding to its Mas receptor induces phosphorylation of Akt through a wortmannin-sensitive manner, which in turn, regulates phosphorylation/dephosphorylation of Ser1177/Thr495 eNOS in human endothelial cells [55]. In resting conditions, eNOS is phosphorylated on Thr495 and only weakly phosphorylated on Scr1177. The simultaneous phosphorylation of Ser1177 and Thr495 modulates the active state of eNOS in endothelial cells and increases NO production [56]. Similarly, Ang-(1-7) stimulates NO release via Akt in Mas-transfected Chinese hamster ovary cells [55]. Ang-(1-7) also induces activation of downstream components such as Forkhead box protein O1(FOXO1) transcription factor, an important negative modulator of Akt signaling [57]. These results highlight the complex modulatory effect of Ang-(1-7), which likely coordinates negative feedback loops through finely tuned feedback loops.

Ang-(1-7)-mediated (PI3K)/AKT activation has been shown to counteract the negative effects of Ang II on insulin signaling in endothelial cells and is involved

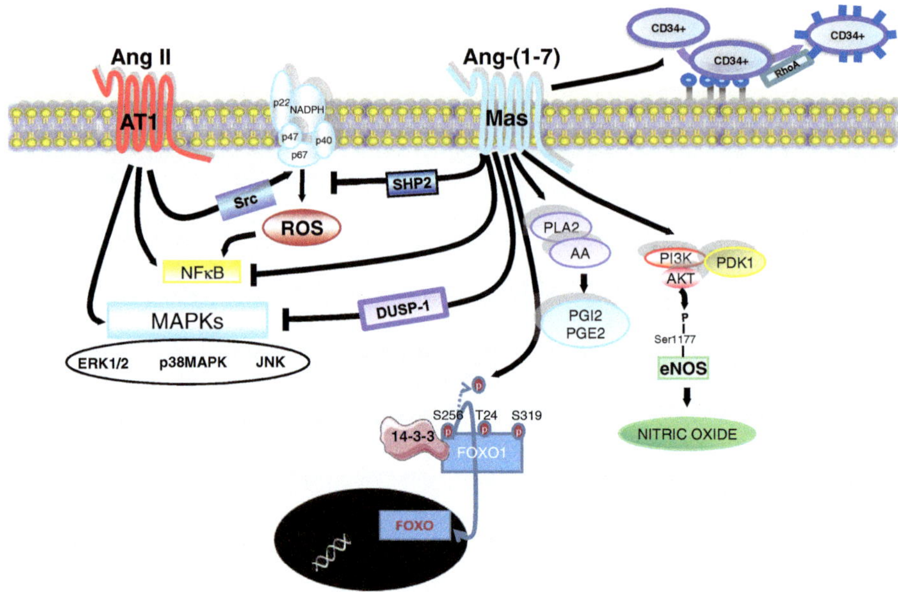

Fig. 2 Signaling pathways stimulated by Ang-(1-7) in the vessels - Ang-(1-7) induces nitric oxide (NO) release, via post-translational regulation of endothelial nitric oxide synthase (eNOS) by Akt, which phosphorylates eNOS Ser 1177. Paradoxically, Ang-(1-7) also activates the transcription factor FOXO1, dephosphorylating FOXO1 Ser256. Furthermore, Ang-(1-7) inhibits several pathways induced by Ang II, including the Src-mediated NAD(P)H oxidase, MAPKs (ERK1/2, p38, JNK), and inflammatory pathways such as NF-κB. The inactivation of NAD(P)H oxidase (Nox) and MAPKs by Ang-(1-7) involves SHP2 and DUSP1 phosphatases, respectively. In addition to these protective intracellular actions, Ang-(1-7) activates stem/progenitor cells CD34+ proliferation and migration, promoting vascular repair

in the survival and proliferation of CD34(+) cells from diabetic individuals [52, 58]. In fructose-fed rats, this pathway seems to mediate the improvement in insulin sensitivity induced by Ang-(1-7) in liver, skeletal muscle, and adipose tissue [59].

Prostaglandin Release

Early studies performed in rabbit isolated vas deferens [60], astrocytes [61], gliomas cells [62], and porcine aortic endothelial cells [63] demonstrated that Ang-(1-7) selectivity increases prostaglandin synthesis. The release of prostaglandins seems to be involved in renal function, including sodium transport [64] and natriuresis [65]. Chronic infusion of Ang-(1-7) in SHR increased urinary excretion of prostaglandin E_2 and 6-keto-prostaglandin $F_{1\alpha}$, and induced diuresis, natriuresis, and a drop in blood pressure [66]. Prostaglandins also participate in antihypertensive effects of

Ang-(1-7) during treatment with angiotensin-converting enzyme (ACE) inhibitors. In rabbit VSMCs, the augmentation of prostacyclin was also found in Ang-(1-7)-stimulated cells. This effect seemed to be mediated via CaM kinase II/ MAP kinase activation, enhancing $cPLA_2$ activity and release of arachidonic acid (AA) for prostacyclin formation [67]. The prostacyclin release also mediates antiproliferative actions of Ang-(1-7) in VSMCs [33].

Cyclooxygenase-related products are also involved in potentiation of bradykinin-induced vasodilation by Ang-(1-7) in coronary [68] and mesentery arteries [69, 70].

Despite effects of Ang-1-7 on cyclooxygenase-2 (Cox-2) expression, the functional effect seems to depend on the inflammatory and oxidative status. In cardiac fibroblast, Ang-(1-7) blocked the ET-1-induced COX2 and prostaglandin synthase, improving the balance between proliferative and anti-proliferative prostaglandins [43] in cerebral arteries. On the other hand, Ang-(1-7) treatment increases the expression of COX2 and decreases expression of matrix metalloproteinase-9 (MPP-9), reducing formation and rupture of intracranial aneurysms in elastase and Ang II-infused mice [71].

Counterregulation of Ang II Oxidative and Inflammatory Signaling

Ang II is a potent vasoconstrictor, pro-oxidative, pro-inflammatory, and pro-proliferative peptide in the vasculature. Ang-(1-7) counterregulates these effects by negatively modulating the intracellular signaling pathways underlying these deleterious Ang II-mediated effects. In human endothelial cells, Ang-(1-7) opposes Ang II-stimulated pro-oxidative and proliferative signaling, and reduces phosphorylation of c-Src and activation of NAD(P)H oxidase. This effect is mediated by phosphorylation of SHP-2, preventing Ang II–induced SHP-2 dephosphorylation and promoting interaction between SHP-2 and c-Src, which leads to c-Src inhibition. The Ang-(1-7) antagonist, A-779, inhibited these actions, demonstrating that these effects are mediated through receptor Mas [72]. Similar results were obtained in human brain microvascular endothelial cells, where Ang-(1-7)/MasR axis reduced Ang II-induced oxidative stress and cell dysfunction, via blockage of Nox2/ROS and activation of PI3K/NO [73]. In human umbilical vein endothelial cells (HUVECs) simulated with Ang II and insulin, Ang-(1-7) restored the insulin-induced Akt/eNOS/NO production as well as inhibited the serine phosphorylation of IRS1 induced by Ang II [58]. The inhibition of Ang II signaling was also demonstrated in cremaster microvessels, where Ang-(1-7) raised DUSP1 to decrease MAP kinase/Smad/CTGF signaling, decreasing Ang II-stimulated fibrosis in these resistance arterioles [74]. The effect of Ang-(1-7) on ERK1/2 inhibition via DUSP-1 activation was also observed in cardiac fibroblasts. (McCollum). Ang-(1-7) also blocked Ang II-induced proliferation, migration, and inflammation of VSMCs by inactivating ROS-mediated PI3K/Akt and MAPK/ERK signaling [75]. On the other hand, a proteomic study comparing Ang II and Ang-(1-7) signaling in rat microvascular endothelial cells indicated that activation of ERK1/2 and p38MAPK is a convergent pathway of Ang II/AT1R and Ang-(1-7) to control angiogenesis. Regarding inflammation control, Ang-(1-7) attenuates Ang II-induced ICAM-1, VCAM-1, and MCP-1 expression by suppressing p38MAPK signaling and translocation of NF-kappaB in HUVEC [75, 76].

Several in vivo studies in various experimental models also showed the importance of the counterrregulatory actions of Ang-(1-7) in vascular function. In Mas-deficient mice from two different genetic backgrounds, C57Bl/6 and FVB/N, endothelial function was impaired [77, 78]. Furthermore, worsening of 2 kidney–1 clip Goldblatt hypertension was observed in Mas knockout mice [79]. Conversely, short-term infusion of Ang-(1-7) improves endothelial function and increases the hypotensive effect of intra-arterial acetylcholine administration in normotensive rats [80]. In diet-induced obese mice, chronic treatment with Ang 1-7 induced a significant improvement in endothelial function and reversed the elevated aortic expression of NAD(P)H oxidase subunits (p22(phox) and p47(phox))and plasma TBARS [81]. Similar results were also observed in diabetic rats, in which carotid blood flow was restored by chronic treatment with Ang-(1-7) probably through *Mas*-mediated antioxidant effects opposing AT1R-activated NAD(P)H oxidase [18].

Conclusion

Ang-(1-7) is a biologically active peptide derived from Ang II, and functions, in large part, in opposition to Ang II. In the vasculature, Ang-(1-7), through Mas receptor, promotes vasodilation and is anti-proliferative, anti-fibrotic, and anti-inflammatory, thereby maintaining vascular integrity and promoting vascular health. These effects are mediated through multiple pathways, including activation of Akt, increased NO generation, decreased ROS generation, and inhibition of Ang II-stimulated signaling pathways. Ang-(1-7) also influences vascular repair by activating CD34+ cells and other progenitor/stem cells involved in tissue repair. While experimental and pre-clinical studies strongly support a vasoprotective role for Ang-(1-7), this has yet to be translated to the clinic [82].

References

1. Santos RA, Brosnihan KB, Jacobsen DW, DiCorleto PE, Ferrario CM. Production of angiotensin-(1-7) by human vascular endothelium. Hypertension. 1992;19:II56–61.
2. le Tran Y, Forster C. Angiotensin-(1-7) and the rat aorta: modulation by the endothelium. J Cardiovasc Pharmacol. 1997;30(5):676–82.
3. Lemos VS, Cortes SF, Silva DM, Campgnole-Santos MJ, Santos RA. Angiotensin-(1-7) is involved in the endothelium-dependent modulation of phenylefrine-induced contraction in the aorta of m-Ren transgenic rats. Br J Pharmacol. 2002;135:1743–8.
4. Brosnihan KB, Li P, Ferrario CM. Angiotensin-(1-7) dilates canine coronary arteries through kinins and nitric oxide. Hypertension. 1996;27(3 Pt2):523–8.
5. Pörsti I, Bara AT, Busse R, Hecker M. Release of nitric oxide by angiotensin-(1-7) from porcine coronary endothelium: implications for a novel angiotensin receptor. Br J Pharmacol. 1994;111:652–4.

6. Feterik K, Smith L, Katusic ZS. Angiotensin-(1-7) causes endothelium-dependent relaxation in canine middle cerebral artery. Brain Res. 2000;873:75–82.
7. Meng W, Busija DW. Comparative effects of angiotensin-(1-7) and angiotensin II on piglet pial arterioles. Stroke. 1993;24:2041–5.
8. Ren Y, Garvin JL, Carretero OA. Vasodilator action of angiotensin-(1-7) on isolated rabbit afferent arterioles. Hypertension. 2002;39:799–802.
9. Oliveira MA, Fortes ZB, Santos RAS, Khosla MC, Carvalho MHC. Synergistic effect of angiotensin-(1-7) on bradykinin arteriolar dilation in vivo. Peptides. 1999;20:1195–201.
10. Fernandes L, Fortes ZB, Nigro D, Tostes RCA, Santos RAS, Carvalho MHC. Potentiation of bradykinin by angiotensin-(1-7) on arterioles of spontaneously hypertensive rats studied in vivo. Hypertension. 2001;37:703–9.
11. de Moraes PL, Kangussu LM, Castro CH, Almeida AP, Santos RAS, Ferreira AJ. Vasodilator effect of angiotensin-(1-7) on vascular coronary bed of rats: role of Mas, ACE and ACE2. Protein Pept Lett. 2017;24(9):869–75.
12. Santos RAS, Passaglio KT, Pesquero JB. Bader M, Simões e Silva AC. Interactions between kinins and angiotensin-(1-7) in kidney and blood vessels. Hypertension. 2001;38:660–4.
13. van Twist DJ, Houben AJ, de Haan MW, Mostard GJ, Kroon AA, de Leeuw PW. Angiotensin-(1-7)-induced renal vasodilation in hypertensive humans is attenuated by low sodium intake and angiotensin II co-infusion. Hypertension. 2013;62:789–93.
14. Grace JA, Klein S, Herath CB, Granzow M, Schierwagen R, Masing N, Walther T, Sauerbruch T, Burrell LM, Angus PW, Trebicka J. Activation of the MAS receptor by angiotensin-(1-7) in the renin–angiotensin system mediates mesenteric vasodilatation in cirrhosis. Gastroenterology. 2013;145(4):874–84.
15. Raffai G, Lombard JH. Angiotensin-(1-7) selectively induces relaxation and modulates endothelium-dependent dilation in mesenteric arteries of salt-fed rats. J Vasc Res. 2016;53(1–2):105–18.
16. Yuan L, Li Y, Li G, Song Y, Gong X. Ang(1-7) treatment attenuates β-cell dysfunction by improving pancreatic microcirculation in a rat model of type 2 diabetes. J Endocrinol Investig. 2013;36(11):931–7.
17. Dincă M, Dumitriu IL, Gurzu MB, Slătineanu SM, Foia L, Vâţă L, Cojocaru E, Petrescu G. Ghrelin and Ang 1-7 have cumulative vasodilatory effects on pulmonary vessels. Rev Med Chir Soc Med Nat Iasi. 2010;114(3):803–7.
18. Pernomian L, Gomes MS, Restini CB, de Oliveira AM. MAS-mediated antioxidant effects restore the functionality of angiotensin converting enzyme 2-angiotensin-(1-7)-MAS axis in diabetic rat carotid. Biomed Res Int. 2014;2014:640329.
19. Sampaio WO, Nascimento AA, Santos RA. Systemic and regional hemodynamic effects of angiotensin-(1-7) in rats. Am J Physiol Heart Circ Physiol. 2003;284(6):H1985–94.
20. Botelho-Santos GA, Sampaio WO, Reudelhuber TL, Bader M, Campagnole-Santos MJ, Santos RAS. Expression of an angiotensin-(1-7)-producing fusion protein in rats induced marked changes in regional vascular resistance. Am J Physiol Heart Circ Physiol. 2007;292(5):H2485–90.
21. Botelho-Santos GA, Bader M, Alenina N, Santos RA. Altered regional blood flow distribution in Mas-deficient mice. Ther Adv Cardiovasc Dis. 2012;6(5):201–11.
22. Hisatake S, Kiuchi S, Kabuki T, Oka T, Dobashi S, Ikeda T. Serum angiotensin-converting enzyme 2 concentration and angiotensin-(1-7) concentration in patients with acute heart failure patients requiring emergency hospitalization. Heart Vessel. 2017;32(3):303–8.
23. Van Twist DJ, Houben AJ, De Haan MW, Mostard GJ, De Leeuw PW, Kroon AA. Angiotensin-(1-7)-induced renal vasodilation is reduced in human kidneys with renal artery stenosis. J Hypertens. 2014;32(12):2428–32; discussion 2432
24. Kocks MJ, Lely AT, Boomsma F, Jong PE, Navis G. Sodium status and angiotensin-converting enzyme inhibition: effects on plasma angiotensin-(1-7) in healthy man. J Hypertens. 2005;23:597–602.
25. van der Wouden EA, Ochodnický P, van Dokkum RP, Roks AJ, Deelman LE, de Zeeuw D, et al. The role of angiotensin(1-7) in renal vasculature of the rat. J Hypertens. 2006;24:1971–8.

26. Mendonça L, Mendes-Ferreira P, Bento-Leite A, Cerqueira R, Amorim MJ, Pinho P, Brás-Silva C, Leite-Moreira AF, Castro-Chaves P. Angiotensin-(1-7) modulates angiotensin II-induced vasoconstriction in human mammary artery. Cardiovasc Drugs Ther. 2014;28(6):513–22.

27. Davie AP, McMurray JJ. Effect of angiotensin-(1-7) and bradykinin in patients with heart failure treated with an ACE inhibitor. Hypertension. 1999;34(3):457–60.

28. Wilsdorf T, Gainer JV, Murphey LJ, Vaughan DE, Brown NJ. Angiotensin-(1-7) does not affect vasodilator or TPA responses to bradykinin in human forearm. Hypertension. 2001;37:1136–40.

29. Sasaki S, Higashi Y, Nakagawa K, Matsuura H, Kajiyama G. OshimaT. Effects of angiotensin-(1-7) on forearm circulation in normotensive subjects and patients with essential hypertension. Hypertension. 2001;38:90–4.

30. Ueda S, Masumori-Maemoto S, Wada A, Ishii M, Brosnihan KB, Umemura S. Angiotensin(1-7) potentiates bradykinin-induced vasodilatation in man. J Hypertens. 2001;19:2001–9.

31. Almeida AP, Frábregas BC, Madureira MM, Santos RJ, Campagnole-Santos MJ, Santos RA. Angiotensin-(1-7) potentiates the coronary vasodilatatory effect of bradykinin in the isolated rat heart. Braz J Med Biol Res. 2000;33(6):709–13.

32. Roks AJ, Nijholt J, van Buiten A, van Gilst WH, de Zeeuw D, Henning RH. Low sodium diet inhibits the local counter-regulator effect of angiotensin-(1-7) on angiotensin II. J Hypertens. 2004;22:2355–61.

33. Tallant EA, Clark MA. Molecular mechanisms of inhibition of vascular growth by angiotensin-(1-7). Hypertension. 2003;42(4):574–9.

34. Freeman EJ, Chisolm GM, Ferrario CM, Tallant EA. Angiotensin-(1-7) inhibits vascular smooth muscle cell growth. Hypertension. 1996;28:104–8.

35. Zhang F, Ren X, Zhao M, Zhou B, Han Y. Angiotensin-(1-7) abrogates angiotensin II-induced proliferation, migration and inflammation in VSMCs through inactivation of ROS-mediated PI3K/Akt and MAPK/ERK signaling pathways. Sci Rep. 2016;6:34621.

36. Akhtar S, Chandrasekhar B, Attur S, Dhaunsi GS, Yousif MH, Benter IF. Transactivation of ErbB family of receptor tyrosine kinases is inhibited by angiotensin-(1-7) via its Mas receptor. PLoS One. 2015;10(11):e0141657.

37. Sheng-Long C, Yan-Xin W, Yi-Yi H, Ming F, Jian-Gui H, Yi-Li C, Wen-Jing X, Hong M. AVE0991, a nonpeptide compound, attenuates angiotensin II-induced vascular smooth muscle cell proliferation via induction of Heme Oxygenase-1 and downregulation of p-38 MAPK phosphorylation. Int J Hypertens. 2012;2012:958298.

38. Tallant EA, Diz DI, Ferrario CM. State-of-the-Art lecture. Antiproliferative actions of angiotensin-(1-7) in vascular smooth muscle. Hypertension. 1999;34(4 Pt 2):950–7.

39. Langeveld B, van Gilst WH, Tio RA, Zijlstra F, Roks AJ. Angiotensin-(1-7) attenuates neointimal formation after stent implantation in the rat. Hypertension. 2005;45(1):138–41.

40. Alsaadon H, Kruzliak P, Smardencas A, Hayes A, Bader M, Angus P, Herath C, Zulli A. Increased aortic intimal proliferation due to MasR deletion in vitro. Int J Exp Pathol. 2015;96(3):183–7.

41. Yang J, Sun Y, Dong M, Yang X, Meng X, Niu R, Guan J, Zhang Y, Zhang C. Comparison of angiotensin-(1-7), losartan and their combination on atherosclerotic plaque formation in apolipoprotein E knockout mice. Atherosclerosis. 2015;240(2):544–9.

42. Sui YB, Chang JR, Chen WJ, Zhao L, Zhang BH, Yu YR, et al. Angiotensin-(1-7) inhibits vascular calcification in rats. Peptides. 2013;42:25–34.

43. McCollum LT, Gallagher PE, Tallant EA. Angiotensin-(1-7) abrogates mitogen-stimulated proliferation of cardiac fibroblasts. Peptides. 2012;34(2):380–8.

44. Gallagher PE, Tallant EA. Inhibition of human lung cancer cell growth by angiotensin-(1-7). Carcinogenesis. 2004;25:2045–52.

45. Ni L, Feng Y, Wan H, Ma Q, Fan L, Qian Y, et al. Angiotensin-(1-7) inhibits the migration and invasion of A549 human lung adenocarcinoma cells through inactivation of the PI3K/Akt and MAPK signaling pathways. Oncol Rep. 2012;27:783–90.

46. Xu J, Fan J, Wu F, Huang Q, Guo M, Lv Z, Han J, Duan L, Hu G, Chen L, Liao T, Ma W, Tao X, Jin Y. The ACE2/Angiotensin-(1-7)/Mas receptor axis: pleiotropic roles in cancer. Front Physiol. 2017;8:276.

47. Fraga-Silva RA, Pinheiro SVB, Gonçalves ACC, Alenina N, Bader M, Santos RA. The antithrombotic effect of angiotensin-(1-7) involves Mas-mediated NO release from platelets. Mol Med. 2008;14(1–2):28–35.

48. Fraga-Silva RA, Costa-Fraga FP, De Sousa FB, Alenina N, Bader M, Sinisterra RD, et al. An orally active formulation of angiotensin-(1-7) produces an antithrombotic effect. Clinics. 2011;66(5):837–41.

49. Fraga-Silva RA, Da Silva DG, Montecucco F, Mach F, Stergiopulos N, Silva RF, et al. The angiotensin-converting enzyme 2/angiotensin-(1-7)/Mas receptor axis: a potential target for treating thrombotic diseases. Thromb Haemost. 2012;108(6):1089–96.

50. Fang C, Stavrou E, Schmaier AA, Grobe N, Morris M, Chen A, et al. Angiotensin- (1-7) and Mas decrease thrombosis in Bdkrb2−/− mice by increasing NO and prostacyclin to reduce platelet spreading and glycoprotein VI activation. Blood. 2013;121(15):3023–32.

51. Mackie AR, Losordo DW. CD34-positive stem cells: in the treatment of heart and vascular disease in human beings. Texas Heart Inst J. 2011;38:474–85.

52. Jarajapu YP, Bhatwadekar AD, Caballero S, Hazra S, Shenoy V, Medina R, Kent D, Stitt AW, Thut C, Finney EM, Raizada MK, Grant MB. Activation of the ACE2/angiotensin-(1-7)/Mas receptor axis enhances the reparative function of dysfunctional diabetic endothelial progenitors. Diabetes. 2013;62(4):1258–69.

53. Singh N, Joshi S, Guo L, Baker MB, Li Y, Castellano RK, Raizada MK, Jarajapu YP. ACE2/Ang-(1-7)/Mas axis stimulates vascular repair-relevant functions of CD34+ cells. Am J Physiol Heart Circ Physiol. 2015;309(10):H1697–707.

54. Vasam G, Joshi S, Thatcher SE, Bartelmez SH, Cassis LA, Jarajapu YP. Reversal of bone marrow mobilopathy and enhanced vascular repair by angiotensin-(1-7) in diabetes. Diabetes. 2017;66(2):505–18.

55. Sampaio WO, Santos RAS, Faria-Silva R, Machado LTM, Schiffrin EL, Touyz RM. Angiotensin-(1-7) through receptor Mas mediates endothelial nitric oxide synthase activation via Akt-dependent pathways. Hypertension. 2007;49:185–92.

56. Fleming I, Fisslthaler B, Dimmeler S, Kemp BE, Busse R. Phosphorylation of Thr(495) regulates Ca(2+)/calmodulin-dependent endothelial nitric oxide synthase activity. Circ Res. 2001;88:E68–75.

57. Verano-Braga T, Schwämmle V, Sylvester M, Passos-Silva DG, Peluso AA, Etelvino GM, et al. Time-resolved quantitative phosphoproteomics: new insights into angiotensin-(1-7) signaling networks in human endothelial cells. J Proteome Res. 2012;11(6):3370–81.

58. Tassone EJ, Sciacqua A, Andreozzi F, Presta I, Perticone M, Carnevale D, Casaburo M, Hribal ML, Sesti G, Perticone F. Angiotensin (1-7) counteracts the negative effect of angiotensin II on insulin signalling in HUVECs. Cardiovasc Res. 2013;99(1):129–36.

59. Muñoz MC, Giani JF, Burghi V, Mayer MA, Carranza A, Taira CA, Dominici FP. The Mas receptor mediates modulation of insulin signaling by angiotensin-(1-7). Regul Pept. 2012;177(1–3):1–11.

60. Trachte GJ, Meixner K, Ferrario CM, Khosla MC. Prostaglandin production in response to angiotensin-(1-7) in rabbit isolated vasa deferentia. Prostaglandins. 1990;39(4):385–94.

61. Jaiswal N, Tallant EA, Diz DI, Khosla MC, Ferrario CM. Subtype 2 angiotensin receptors mediate prostaglandin synthesis in human astrocytes. Hypertension. 1991;17:1115–20.

62. Jaiswal N, Diz DI, Tallant EA, Khosla MC, Ferrerio CM. Characterization of angiotensin receptors mediating prostaglandin synthesis in CG glioma cells. Am J Phys. 1991;260:R1000–6.

63. Jaiswal N, Diz DI, Chappell MC, Khosla MC, Ferrario CM. Stimulation of endothelial cell prostaglandin production by angiotensin peptides. Characterization of receptors. Hypertension. 1992;19(2 Suppl):II49–55.

64. Andreatta-van Leyen S, Romero MF, Khosla MC, Ferrario CM, Douglas JG. Modulation of phospholipase A2 activity and sodium transport by angiotensin-(1-7). Kidney Int. 1993;44:932–6.

65. Hilchey SD, Bell-Quilley CP. Association between the natriuretic action of angiotensin-(1-7) and selective stimulation of renal prostaglandin I2 release. Hypertension. 1995;25(6):1238–44.

66. Benter IF, Ferrario CM, Morris M, Diz DI. Antihypertensive actions of angiotensin-(1-7) in spontaneously hypertensive rats. Am J Phys. 1995;269:H313–9.

67. Muthalif MM, Benter IF, Uddin MR, Harper JL, Malik KU. Signal transduction mechanisms involved in angiotensin-(1-7)-stimulated arachidonic acid release and prostanoid synthesis in rabbit aortic smooth muscle cells. J Pharmacol Exp Ther. 1998;284(1):388–98.
68. Ferreira AJ, Santos RA, Almeida AP. Angiotensin-(1-7) improves the post-ischemic function in isolated perfused rat hearts. Braz J Med Biol Res. 2002;35(9):1083–90.
69. Oliveira MA, Fortes ZB, Santos RA, Kosla MC, De Carvalho MH. Synergistic effect of angiotensin-(1-7) on bradykinin arteriolar dilation in vivo. Peptides. 1999;20(10):1195–201.
70. Oliveira MA, Carvalho MH, Nigro D, Passaglia Rde C, Fortes ZB. Elevated glucose blocks angiotensin-(1-7) and bradykinin interaction: the role of cyclooxygenase products. Peptides. 2003;24(3):449–54.
71. Peña Silva RA, Kung DK, Mitchell IJ, Alenina N, Bader M, Santos RA, Faraci FM, Heistad DD, Hasan DM. Angiotensin 1-7 reduces mortality and rupture of intracranial aneurysms in mice. Hypertension. 2014;64(2):362–8.
72. Sampaio WO, Castro CH, Santos RA, Schiffrin EL, Touyz RM. Angiotensin-(1-7) counterregulates angiotensin II signaling in human endothelial cells. Hypertension. 2007;50:1093–8.
73. Xiao X, Zhang C, Ma X, Miao H, Wang J, Liu L, Chen S, Zeng R, Chen Y, Bihl JC. Angiotensin-(1-7) counteracts angiotensin II-induced dysfunction in cerebral endothelial cells via modulating Nox2/ROS and PI3K/NO pathways. Exp Cell Res. 2015;336(1):58–65.
74. Carver KA, Smith TL, Gallagher PE, Tallant EA. Angiotensin-(1-7) prevents angiotensin II-induced fibrosis in cremaster microvessels. Microcirculation. 2015;22(1):19–27.
75. Zhang F, Ren J, Chan K, Chen H. Angiotensin-(1-7) regulates angiotensin II-induced VCAM-1 expression on vascular endothelial cells. Biochem Biophys Res Commun. 2013;430(2):642–6.
76. Liang B, Wang X, Zhang N, Yang H, Bai R, Liu M, Bian Y, Xiao C, Yang Z. Angiotensin-(1-7) attenuates angiotensin II-induced ICAM-1, VCAM-1, and MCP-1 expression via the MAS receptor through suppression of P38 and NF-κB pathways in HUVECs. Cell Physiol Biochem. 2015;35(6):2472–82.
77. Xu P, Costa-Goncalves AC, Todiras M, Rabelo LA, Sampaio WO, Moura MM, et al. Endothelial dysfunction and elevated blood pressure in MAS gene-deleted mice. Hypertension. 2008;51:574–80.
78. Rabelo LA, Xu P, Todiras M, Sampaio WO, Buttgereit J, Bader M, et al. Ablation of angiotensin- (1-7) receptor Mas in C57Bl/6 mice causes endothelial dysfunction. J Am Soc Hypertens. 2008;2:418–24.
79. Rakušan D, Bürgelová M, Vaněčková I, Vaňourková Z, Husková Z, Skaroupková P, et al. Knockout of angiotensin- (1-7) receptor Mas worsens the course of two-kidney, one-clip Goldblatt hypertension: roles of nitric oxide deficiency and enhanced vascular responsiveness to angiotensin II. Kidney Blood Press Res. 2010;33:476–88.
80. Faria-Silva R, Duarte FV, Santos RA. Short-term angiotensin(1-7) receptor MAS stimulation improves endothelial function in normotensive rats. Hypertension. 2005;46(4):948–52.
81. Beyer AM, Guo DF, Rahmouni K. Prolonged treatment with angiotensin 1-7 improves endothelial function in diet-induced obesity. J Hypertens. 2013;31(4):730–8.
82. Touyz RM, Montezano AC. Angiotensin-(1-7) and vascular function: the clinical context. Hypertension. 2018;71(1):68–9.

Kidney

Ana Cristina Simões e Silva
and Robson Augusto Souza Santos

Introduction

Over the past two decades, considerable advances have been made in our understanding of the renin–angiotensin system (RAS) [1, 2]. Besides the classical RAS axis formed by angiotensin converting enzyme (ACE), Angiotensin (Ang) II, and Ang type 1 receptor (AT_1), the biological relevance of another Ang fragment, the heptapeptide Ang-(1-7), has been widely recognized [1–3]. Ang-(1-7) acts through a specific G-protein-coupled receptor, the Mas receptor [4], and is mainly formed by the action of the ACE homolog enzyme, ACE2, which converts Ang II into Ang-(1-7) [5, 6]. In general, this named alternative RAS axis formed by ACE2, Ang-(1-7), and Mas receptor antagonizes the actions of ACE/Ang II/AT_1 axis in several organs and systems. Therefore, the RAS is generally conceived as a dual function system in which the final effects are the consequence of the balance between both RAS axes.

However, the complexity of this system has increased even more in very recent years by the discovery of novel RAS mediators such as Ang-(1-9) [7, 8] and Alamandine [9, 10], a second receptor for Ang-(1-7) [11] and by the identification of a role for the enzyme neprylisin (NEP) [12]. NEP forms Ang-(1-7) in murine

A. C. Simões e Silva (✉)
Department of Pediatrics, Interdisciplinary Laboratory of Medical Investigation, Faculty of Medicine, UFMG, Belo Horizonte, MG, Brazil

Interdisciplinary Laboratory of Medical Investigation, Faculty of Medicine, Federal University of Minas Gerais (UFMG), Belo Horizonte, MG, Brazil

R. A. S. Santos
National Institute of Science and Technology in Nano Biopharmaceutics –
Department of Physiology and Biophysics, Biological Sciences Institute,
Federal University of Minas Gerais, Belo Horizonte, MG, Brazil

Department of Physiology and Pharmacology, Biological Sciences Institute,
Federal University of Minas Gerais, Belo Horizonte, MG, Brazil

© Springer Nature Switzerland AG 2019

R. A. S. Santos (ed.), *Angiotensin-(1-7)*,
https://doi.org/10.1007/978-3-030-22696-1_8

and human kidney [12]. Ang-(1-9) is formed from Ang I hydrolysis by the action of ACE2 [13]. Ang-(1-9) can also be cleaved by ACE resulting in Ang-(1-7) [13]. The interaction of Ang-(1-9) with the AT_2 receptor reduces blood pressure (BP) and reverses/ameliorates cardiovascular injury in animal models of hypertension [7, 14]. Some authors proposed that the activation of both counter-regulatory RAS axes, ACE2/Ang-(1-7)/Mas and ACE2/Ang-(1-9)/AT_2, can oppose the effects of ACE/Ang II/AT1 axis and prevent or reverse organ damage in experimental models of renal and heart diseases [8]. Alamandine or Ala1-Ang-(1-7) was isolated from human plasma and rat and mouse heart [9, 10]. Alamandine is a product of decarboxylation of the N-terminal Asp residue of Ang II to form Ala1-Ang II or Angiotensin A, which is subsequently converted into Ala1-Ang-(1-7) or Almandine by ACE2 [9]. The only difference in chemical structure of Alamandine and Ang-(1-7) is the substitution of the N-terminal Asp residue by Ala. Alamandine binds to an specific receptor, the Mas-related G-protein-coupled receptor, member D (MrgD), and exerts antihypertensive, antifibrotic, and cardiovascular actions comparable to Ang-(1-7) [9, 10]. In regard to MrgD receptor, Tetzner et al. [11] identified this receptor also as a second receptor for Ang-(1-7). Ang-(1-7) failed to increase cAMP concentration in primary mesangial cells with genetic deficiency of both Mas and MrgD receptors [11]. Knockout mice for MrgD showed an impaired hemodynamic response to Ang-(1-7) administration. Moreover, the Ang type 2 (AT_2) receptor blocker, the compound PD123319, was able to block both Mas and MrgD receptors [11]. The assessment of angiotensin metabolism in kidney homogenates resulted in the identification of NEP as a major source of renal Ang-(1-7) in mice and humans [12]. These findings were supported by matrix-assisted laser desorption ionization imaging technique, showing NEP-mediated Ang-(1-7) formation in whole mice kidney [12]. In addition, pharmacological inhibition of NEP led to a strong decrease in Ang-(1-7) levels in murine kidneys. Further studies are necessary to clarify the precise meaning of these new discoveries in human physiology.

In the present chapter, we focus on recent findings related to the role of the ACE2/Ang-(1-7)/Mas axis in regulating renal function and in the physiopathology of renal diseases in experimental models and in patients.

ACE2/Ang-(1-7)/Mas Receptor Axis in Renal Physiology

Many studies have addressed the complexity of renal actions of Ang-(1-7) [15–34]. In vitro studies reported a diuretic/natriuretic action of Ang-(1-7) [16–23]. This effect was also observed in experimental models, mostly by inhibition of sodium reabsorption at proximal tubule [16, 17, 19, 20]. Ang-(1-7) inhibited Na-K-ATPase activity in renal cortex [19] and in isolated convoluted proximal tubules [20]. In renal tubular epithelial cells, Ang-(1-7) activated phospholipase A2 leading to inhibition of transcellular transport of sodium [16]. In vitro studies also indicated that Ang-(1-7) modulates the stimulatory effect of Ang II on the Na-ATPase activity in proximal tubule through an A779-sensitive receptor [23]. Conversely, several studies showed an antidiuretic/antinatriuretic effect induced by Ang-(1-7), especially in water-loaded animals [24–34]. The administration of Ang-(1-7) and of the oral

Mas receptor agonist, AVE0991, exerted potent antidiuretic effect in water-loaded rats [25, 27–29] and mice [32] via Mas receptor activation [4]. In vitro, Ang-(1-7) increased water transport in inner medullary collecting duct by means of the interaction between Mas and vasopressin type 2 receptors [31]. In line with these findings, the administration of Mas receptor antagonists, A-779 and D-Pro^7, elicited diuretic effects due to elevation of glomerular filtration rate and water excretion [27, 30, 33]. Despite divergent data concerning tubular actions of Ang-(1-7), the studies support its role in the regulation of glomerular filtration, water, and sodium handling. In this regard, Castelo-Branco et al. [35] recently reported that high intratubular concentrations (10^{-6} M) of Ang-(1-7) inhibit the Na^+-H^+ exchanger 3 (NHE3) in the proximal tubules of hypertensive rats, whereas low doses stimulate it. Because this protein is a Na^+-H^+ exchanger that mediates the proximal renal reabsorption of fluid and plays an important role in the maintenance of systemic extracellular volume, blood pressure, and pH, intratubular Ang-(1-7) at high concentrations (10^{-6} M) seems to be able to control, at least in part, hypertension caused by high plasma level of Ang II in hypertensive animals [35]. This effect was mediated by Mas and AT_2 receptors [35]. In another recent study, O'Neil et al. [36] showed that the enhanced urinary excretion of sodium observed in animals on a low-sodium diet during intrarenal Ang-(1-7) infusion is associated with increased intrarenal Ang-(1-7) levels at the beginning of experimental protocol. This association suggests that the natriuretic response to exogenous Ang-(1-7) in the surroundings of a stimulated RAS was increased because the endogenous Ang-(1-7) concentration in the kidney was higher in these conditions [36]. The authors also detected that Ang-(1-7)-induced natriuresis and diuresis are inhibited by both AT_1 and Mas receptor blockade [36]. The precise mechanism by which both receptors mediate natriuretic effect of Ang-(1-7) remains unclear. To sum up, differences between species, local, and systemic concentrations of Ang-(1-7), nephron segment, level of RAS activation, and sodium and water status can be responsible for variable results in regard to renal effects of Ang-(1-7) [37].

Close to important tubular actions, Ang-(1-7) also takes part in renal hemodynamic regulation, mostly by facing the effects of Ang II. When the RAS is excessively stimulated, the increased production of Ang II at renal tissue, acting on AT_1 receptors, may result in both systemic and glomerular capillary hypertension, which in turn contributes hemodynamic injury to the vascular endothelium and glomerulus [38–40]. In contrast, Ren et al. [41] detected that Ang-(1-7) induces dilatation of preconstricted renal afferent arterioles in rabbits and Sampaio et al. [42] observed that the infusion of Ang-(1-7) at low concentrations increases renal blood flow in rats. Ang-(1-7) also blunted the effects of Ang II, including vasoconstriction and stimulation of noradrenaline release in rat-isolated kidney [43]. More recently, Youssif et al. (2017) showed that Ang-(1-7) exhibits potent vasorelaxant action in isolated renal artery and this effect depends on an intact endothelium and on the stimulation of nitric oxide (NO) and guanylate cyclase pathways [44]. The authors also verified that Ang-(1-7)-dependent vasorelaxation was sensitive to antagonists against Mas, AT_1, AT_2, and Bradykinin type 2 (B2) receptors [44].

The mechanisms by which Ang-(1-7) counteracts the renal effects of Ang II are not fully elucidated. Some mechanisms are the competition for the binding of

Ang II to AT_1 receptors, modulation of signaling transduction, and interference with the synthesis of AT_1 receptors [45–50]. In this regard, Kostenis et al. [51] reported that Mas receptor hetero-oligomerizes with AT_1 receptor and inhibits the intracellular Ca^{+2} mobilization effect of Ang II [51]. In addition, prior exposure to Ang-(1-7) caused a mild decrease in the number of AT_1 receptors in the cortical tubulo-interstitial area of the kidney [50]. Another important mechanism that may play a role in physiological effects of Ang-(1-7) is the intra-mitochondrial RAS, which also includes ACE2/Ang-(1-7)/Mas receptor axis [52]. Renal actions of an intramitochondrial ACE2/Ang-(1-7)/Mas receptor axis may encompass the release of NO, activation of anti-apoptotic pathways, and/or the reduction of oxidative stress [52]. Furthermore, the loss of Ang-(1-7) tone within the kidney may accelerate deleterious mitochondrial pathways that increase oxidative stress and enhance apoptosis under pathological conditions [52].

Figure 1 shows the effects of Ang-(1-7) in renal physiology.

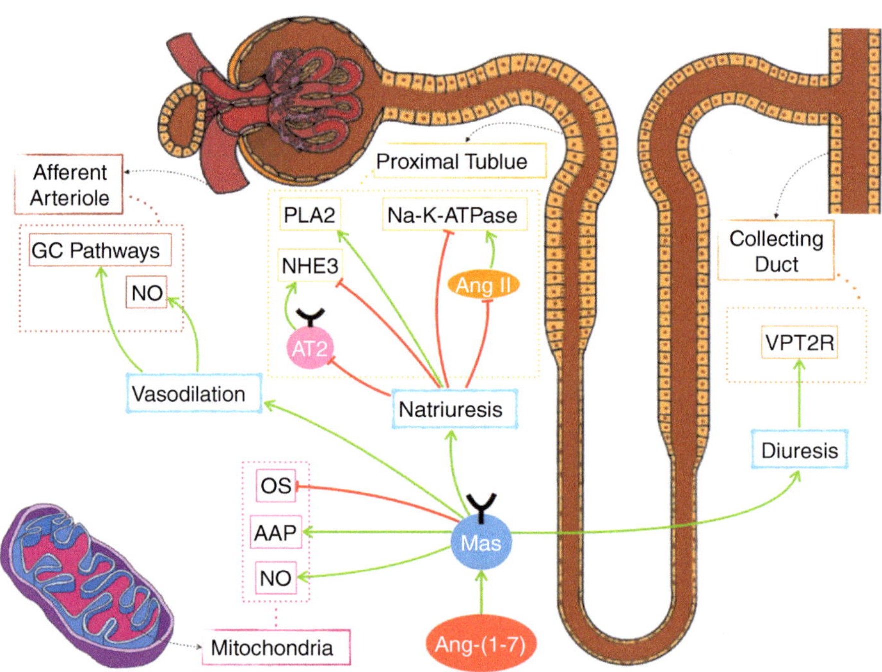

Fig. 1 Effects of Angiotensin-(1-7) [Ang-(1-7)] in renal physiology. (*Ang* angiotensin, *AT1* Ang type 1 receptor, *GC* guanylate cyclase; *NO* nitric oxide, *PLA2* phospholipase A2, *NHE3* sodium hydrogen exchanger, *OS* oxidative stress, *VPT2R* vasopressin receptor type 2)

Physiopathological Role of ACE2/Ang-(1-7)/Mas in Renal Diseases

Experimental Studies

Experimental models of renal diseases showed a protective role for Ang-(1-7) [53–61]. The infusion of Ang-(1-7) reduced glomerulosclerosis by opposing Ang II effects in experimental glomerulonephritis [54]. In adriamycin-induced nephropathy, the oral administration of the Mas agonist, the compound AVE 0991, improved renal function parameters, reduced urinary protein loss, and attenuated histological changes [59]. The administration of AVE 0991 had renoprotective effects in experimental acute renal injury, as seen by improvement of function, decreased tissue injury, prevention of local and remote leukocyte infiltration, and reduced release of the chemokine CXCL1 [56]. By employing an experimental model of chronic intermittent hypoxia, Lu et al. [61] showed that Ang-(1-7) infusion reduces blood pressure and protects the kidneys against tissue injury. These effects seem to be mediated, at least in part, by reducing inflammation, oxidative stress, and fibrosis [61]. The administration of Ang-(1-7) for 12 weeks in 5/6 nephrectomized male C57Bl/6 mice reduced blood pressure, attenuated elevations in plasma urea and creatinine, and preserved cardiac function [62]. On the other hand, despite reducing blood pressure to the same extent as Ang-(1-7), the antihypertensive agent, hydralazine, was not able to improve renal function and to avoid cardiac changes, thereby suggesting that renoprotection obtained with Ang-(1-7) was not mediated merely by the control of hypertension [62]. The infusion with Ang-(1-7) also prevented renal lesion in a model of unilateral ureteral obstruction by suppressing renal apoptosis and fibrosis, possibly through the inhibition of TGF-β1/Smad signaling and recovery of G2/M cell cycle arrest, and the subsequent suppression of AT$_1$ receptor expression [60]. Moreover, exogenous Ang-(1-7) increased ACE2 expression, which could potentially mediate an increase in endogenous Ang-(1-7) in a positive feedback mechanism via Mas receptor.

Ang-(1-7) also produced beneficial effects in experimental diabetes nephropathy by reducing urinary protein excretion without affecting blood pressure in male adult STZ diabetic rats compared with untreated diabetic animals [63]. The same research group treated diabetic spontaneous hypertensive rats with identical dose of Ang-(1-7) and obtained for a second time a significant reduction in urinary protein excretion with no changes in mean arterial pressure [64]. Furthermore, Ang-(1-7) normalized the vascular responses to vasoconstrictors and prevented renal NOX-induced oxidative stress [64]. In an experimental model of type 2 diabetes, the KK-A^y/Ta mouse, the treatment with Ang-(1-7) opposed the Ang II-induced glomerular injury by reducing mesangial expansion, TGF-β and fibronectin mRNA, and NOX activity [65]. In Zucker diabetic fatty rats, Ang-(1-7) diminished triglyceridemia, proteinuria, and systolic blood pressure together with restoration of creatinine clearance [66].

Additionally, Ang-(1-7) reduced renal fibrosis, attenuated renal oxidative stress, and decreased renal immunostaining of inflammatory markers to values similar to those displayed by control animals [66]. Mori et al. [67] reported that Ang-(1-7) treatment exerts renoprotective effects on diabetic nephropathy, associated with reduction of oxidative stress, inflammation, fibrosis, and lipotoxicity. Another mechanism of renoprotection attributed to Ang-(1-7) in diabetic nephropathy was the inhibition of mitochondrial fission in high-glucose-induced podocytes by upregulation of micro-RNA-30a and downregulation of the apoptosis proteins Drp1 and p53 [68]. More recently, Zhao et al. [69] investigated the interactions of ACE2/Ang-(1-7)/Mas receptor axis and nuclear factor erythroid 2-related factor 2 (Nrf2) in renal proximal tubule cells and in the development of systemic hypertension and kidney injury in diabetic Akita mice. Genetic deletion of Nrf2 or pharmacological inhibition of this factor in Akita mice attenuated hypertension, renal injury, tubulointerstitial fibrosis, and urinary albumin/creatinine ratio. These renoprotective effects were associated with increased expression of ACE2 and Mas receptor in renal proximal tubule cells, elevated urinary levels of Ang-(1-7), and downregulated expression of angiotensinogen, ACE, and profibrotic genes in Akita mice [69].

Acquired or genetic ACE2 deficiency exacerbated renal damage and albuminuria in experimental models, possibly facilitating the damaging effects of Ang II [70–75]. Chronic administration of MLN-4760, an ACE2 inhibitor, produced albuminuria and matrix protein deposition in control or diabetic mice [73]. Renal expression of ACE2 was reduced in the renal cortex of mice that underwent subtotal nephrectomy and in a rat model of renal ischemia/reperfusion [75, 76]. In a model of unilateral ureteral obstruction, the deletion of ACE2 gene resulted in a fourfold increase in the ratio of intrarenal Ang II/Ang-(1-7) and these changes were associated with the development of progressive tubulointerstitial fibrosis and inflammation with high levels of TNF-α, IL-1β, and MCP-1 [57]. Enhanced renal fibrosis and inflammation were attributed to marked increase in intrarenal Ang II signaling (AT$_1$/ERK1/2), TGF-β_1/Smad2/3, and NF-κB signaling pathways [57]. Dual RAS blockade normalized ACE2 expression and prevented hypertension, albuminuria, tubulointersticial fibrosis, and tubular apoptosis in Akita angiotensinogen-transgenic mice [77]. Genetic deficiency of ACE2 activity in mice fosters oxidative stress via AT$_1$-dependent effect in the kidney [78]. Accordingly, daily treatment with recombinant ACE2 ameliorated renal fibrosis in apolipoprotein E-deficient mice via modulation of mTOR/ERK signaling and via augmentation of Ang-(1-7)/Ang II ratio [79]. Taken together, these studies suggested that ACE2/Ang-(1-7)/Mas axis modulates oxidative stress, inflammation, apoptosis, and fibrosis at renal tissue.

The interactions between Ang II receptors, AT$_1$ and AT$_2$, and Mas receptor in renal tissue are very complex. AT$_1$ and Mas receptors were codistributed in renal mesangial cells of rats and Ang-(1-7), through the binding to Mas, counteracted the stimulatory effects of Ang II on ERK1/2 and TGF-β_1 pathways mediated by AT$_1$ receptors [58]. Moreover, Ng et al. [80] showed that Mas receptor expression is reduced in the kidneys of rats with chronic kidney disease (CKD), and in cultured human proximal tubular cells, indoxyl sulfate, a uremic toxin, downregulated renal expression of Mas receptor and upregulated TGF-β_1. Ma et al. [68] evaluated the renal expression of various RAS components and examined the renal injury after placing mice with genetic deletion of AT_2 receptor gene on high fat diet for 16 weeks. The authors found that AT_2 knockout mice have decreased cortical ACE2 activity, Mas expression, and Ang-(1-7) levels in renal tissue [68]. These animals also exhibited increased expression of ACE and AT_1 receptor in renal cortex and higher levels of Ang II [68]. These changes in RAS components were accompanied by increased systolic blood pressure, higher indices of kidney injury, mesangial matrix expansion score, and microalbuminuria [68]. More recently, Patel et al. [81] investigated whether renal AT_2 and Mas receptor physically interact and are interdependent to stimulate cell signaling and promote natriuresis in obese rats. The study showed that AT_2 and Mas receptor are colocalized in kidney sections of obese Zucker rats and in human proximal tubule epithelial cells [81]. In addition, both receptors are functionally interdependent in terms of stimulating NO and promoting diuretic/natriuretic response [81].

On the other hand, few reports have suggested that Ang-(1-7) may exacerbate renal injury paradoxically in certain experimental conditions, suggesting that dose or route of administration, state of activation of the local RAS, cell-specific signaling, or non-Mas-mediated pathways may contribute to these deleterious responses [37]. Esteban et al. [82] reported that renal deficiency of Mas diminished renal damage in unilateral ureteral obstruction and in ischemia/reperfusion injury, and that the infusion of Ang-(1-7) to wild-type mice elicited an inflammatory response. Velkoska et al. [83] verified that a 10-day infusion of Ang-(1-7) in rats with subtotal nephrectomy was associated with deleterious effects on blood pressure and heart function. The same research group recently reported that the combined administration of ramipril and Ang-(1-7) prevents the increase blood pressure and cardiac fibrosis produced by the isolate infusion of Ang-(1-7) in rats submitted to subtotal nephrectomy [84]. Cell-specific signaling pathways associated with Ang-(1-7) in the kidney could play a role in the variable response. For instance, Ang-(1-7) displays growth inhibitory properties and antagonizes the effects of Ang II in the proximal tubule [85], whereas, in human mesangial cells, the heptapeptide seems to stimulate cell growth pathways by increasing arachidonic acid release and by MAPK phosphorylation [86]. Results obtained in rat mesangial cells are also divergent. While Liu et al. [87] reported that Ang-(1-7) stimulates ERK1/2 phosphorylation via Mas activation, Oudit et al. [88] showed that Ang-(1-7), acting via Mas receptor, inhibits high glucose-stimulated NOX activation. In addition, in primary cultures of mouse mesangial cells, Moon et al. [65] showed that Ang-(1-7) attenuated Ang II-induced MAPK phosphorylation and expression of TGF-β1, fibronectin, and collagen IV. It must be said, however, that majority of studies suggest an overall renoprotective effect of administering Ang-(1-7) in vivo.

Clinical Studies

Ang-(1-7) can be measured by diverse methods in plasma and urine samples from healthy subjects and patients with several clinical conditions [89–98]. The concentration of Ang-(1-7) may differ in plasma and urine samples of the same subject, since untreated adults with primary hypertension exhibited lower urinary levels of Ang-(1-7) than normotensive controls [90]. Significant differences in circulating levels of Ang II and Ang-(1-7) were detected in pediatric hypertensive patients [92]. Children with renovascular disease had plasma Ang II levels higher than plasma Ang-(1-7), whereas patients with primary hypertension had a selective elevation of plasma Ang-(1-7) [92]. In pediatric patients with chronic kidney disease (CKD), higher levels of Ang-(1-7) and Ang II were also detected in hypertensive patients when compared to normotensives at the same CKD stage [93]. Patients at end-stage renal disease presented an even more pronounced elevation of Ang-(1-7) levels, suggesting a deviation in RAS metabolism toward Ang-(1-7) synthesis [93]. Whether the elevation in plasma Ang-(1-7) provides a counter-regulatory mechanism against Ang II-mediated vasoconstriction or patients with primary hypertension and CKD may have disturbances of Mas receptor signaling remains to be determined. In line with these findings, Rocha et al. [98] recently showed that fetuses with posterior ureteral valves (PUV) had higher urinary concentrations of ACE2 and of Ang-(1-7) than healthy neonates, whereas ACE levels were lower in the urine of PUV fetuses than in the urine of healthy neonates. The increase of components of the protective RAS axis may represent a regulatory response to the intense inflammatory process triggered by PUV [99]. Another possible explanation for the elevation of ACE2 and Ang-(1-7) in the urine of fetuses with PUV could be a dysfunction or reduced expression of the Mas receptor at kidney tissue in PUV fetuses. In this regard, Ng et al. [80] reported that Mas receptor expression is reduced in the kidneys of CKD rats and the administration of the uremic toxin indoxyl sulfate induced downregulation of Mas receptor probably via upregulation of TGF-β1 in proximal tubules.

Mizuiri et al. [100] demonstrated that renal biopsies from patients with IgA nephropathy had significantly reduced glomerular and tubulointerstitial immunostaining for ACE2 compared with healthy controls. On the other hand, glomerular ACE staining was increased. These findings raise the possibility that an upward shift in the intrarenal ACE/ACE2 ratio favoring increased synthesis of Ang II and reduced Ang-(1-7) might lead to progressive nephron loss in this condition [100]. Circulating ACE2 activity was measured in kidney transplant patients and positively correlated with age, serum creatinine, and gama-glutamyl transferase levels [101]. Accordingly, Anguiano et al. [96] showed in adult patients with CKD, without previous history of cardiovascular disease, that plasma ACE2 activity directly correlated with classical cardiovascular risk factors including older age, diabetes, and male gender. The authors hypothesized that circulating ACE2 is altered in CKD patients at risk for cardiovascular event [96]. Angiotensin peptides [Ang I, Ang II, Ang-(1-7), Ang-(1-5), Ang-(2-8), Ang-(3-8)], renin, and aldosterone were measured in 12 hemodialysis patients, who received a kidney transplant and had excellent graft function 6–12 months thereafter [97]. Peptides were simultaneously measured by a mass spectrometry-based method. After kidney transplant, patients increased the positive correlation between renin and Ang II levels. However, plasma Ang-(1-7) was undetectable in hemodialysis and in transplanted patients [97].

It has also been suggested that the beneficial effects of ACE inhibitors and of Ang receptor blockers may be due, at least in part, to an activation of ACE2/Ang-(1-7)/Mas axis, since the chronic treatment with these drugs increases plasma concentrations of Ang-(1-7) [91, 93, 102]. Kocks et al. [102] showed that during ACE inhibition, administration of a low sodium diet did not affect plasma levels of Ang II but induced a significant elevation in Ang-(1-7) concentration. Furthermore, in a murine model of adriamycin-induced nephropathy, renoprotective effects of Losartan were blunted in mice with genetic deletion of Mas receptor, indicating that Mas receptor activation is essential for renal actions of AT_1 receptor antagonism [59]. In this regard, Iwanami et al. [103] showed that the hypotensive and anti-hypertrophic effects of the AT_1 receptor blocker, azilsartan, may also involve activation of the ACE2/Ang-(1-7)/Mas axis. Another relevant aspect is the complex interaction between Mas and AT_1 receptor. Kostenis et al. [51] showed that Mas can hetero-oligomerize with AT_1 and, by so doing, inhibits the actions of Ang II. Thus, Mas may act as a physiological antagonist of AT_1 receptor signaling.

Concluding Remarks

Experimental models of renal diseases suggest that the activation of the ACE2/Ang-(1-7)/Mas axis has a protective role. The few data provided by human studies also indicate a beneficial role for the activation of this alternative RAS axis. In addition, it has been hypothesized that the beneficial effects of ACEi and ARBs might involve, at least in part, the elevation of plasma Ang-(1-7) levels. Further studies are clearly needed to elucidate the mechanisms by which both RAS axes modulate renal function and take part in the physiopathology of renal diseases. Nevertheless, current knowledge supports the possibility that drugs which mimic or enhance the function of the ACE2/Ang-(1-7)/Mas axis may be beneficial for the treatment of renal diseases.

References

1. Simões e Silva AC, Flynn JT. The renin-angiotensin-aldosterone system in 2011: role in hypertension and chronic kidney disease. Pediatr Nephrol. 2012;27:1835–45.
2. Santos RAS, Ferreira AJ, Simões e Silva AC. Recent advances in the angiotensin converting enzyme 2-Angiotensin-(1-7)-Mas axis. Exp Physiol. 2008;93:519–27.
3. Simões e Silva AC, Teixeira MM. ACE inhibition, ACE2 and Angiotensin-(1-7) in kidney and cardiac inflammation and fibrosis. Pharmacological Res. 2016;107:154–62.
4. Santos RA, Simões e Silva AC, Maric C, Silva DM, Machado RP, de Buhr I, et al. Angiotensin-(1-7) is an endogenous ligand for the G protein-coupled receptor Mas. Proc Natl Acad Sci U S A. 2003;100(14):8258–63.
5. Tipinis SR, Hooper NM, Hyde R, Karran E, Christie G, Turner AJ. A human homolog of angiotensin-converting enzyme. Cloning and functional expression as a captopril-insensitive carboxypeptidase. J Biol Chem. 2000;275(43):33238–43.
6. Donoghue M, Hsieh F, Baronas E, Godbout K, Gosselin M, Stagliano N, et al. A novel angiotensin-converting enzyme-related carboxypeptidase (ACE2) converts angiotensin I to angiotensin 1-9. Circ Res. 2000;87(5):E1–9.

7. Flores-Munoz M, Work LM, Douglas K, Denby L, Dominiczak AF, Graham D, Nicklin SA. Angiotensin-(1-9) attenuates cardiac fibrosis in the stroke-prone spontaneously hypertensive rat via the angiotensin type 2 receptor. Hypertension. 2012;59:300–7.

8. Ocaranza MP, Michea L, Chiong M, Lagos CF, Lavandero S, Jalil JE. Recent insights and therapeutic perspectives of angiotensin-(1-9) in the cardiovascular system. Clin Sci (Lond). 2014;127:549–57.

9. Lautner RQ, Villela DC, Fraga-Silva RA, Silva N, Verano-Braga T, Costa-Fraga F, et al. Discovery and characterization of alamandine: a novel component of the renin-angiotensin system. Circ Res. 2013;112:1104–11.

10. Vilela DC, Passos-Silva DG, Santos RA. Alamandine: a new member of the angiotensin family. Curr Opin Nephrol Hypertens. 2014;23:130–4.

11. Tetzner A, Gebolys K, Meinert C, Klein S, Uhlich A, Trebicka J, et al. G-protein–coupled receptor MrgD is a receptor for Angiotensin-(1-7) involving adenylyl cyclase, cAMP, and phosphokinase. Hypertension. 2016;68:185–94.

12. Domeniq O, Manzel A, Grobe N, Königshausen E, Kaltenecker CC, Kovarik JJ, et al. Neprilysin is a mediator of alternative renin-angiotensin-system activation in the murine and human kidney. Sci Rep. 2016;6:33678. https://doi.org/10.1038/srep33678.

13. Rice GI, Thomas DA, Grant PJ, Turner AJ, Hooper NM. Evaluation of angiotensin-converting enzyme (ACE), its homologue ACE2 and neprilysin in angiotensin peptide metabolism. Biochem J. 2004;383:45–51.

14. Flores-Muñoz M, Smith NJ, Haggerty C, Milligan G, Nicklin SA. Angiotensin1-9 antagonises pro-hypertrophic signalling in cardiomyocytes via the angiotensin type 2 receptor. J Physiol. 2011;589:939–51.

15. Ferrario CM, Jessup J, Gallagher PE, Averill DB, Brosnihan KB, Tallant EA, et al. Effects of renin-angiotensin system blockade on renal angiotensin-(1-7) forming enzymes and receptors. Kidney Int. 2005;68:2189–96.

16. Andreatta-van Leyen S, Romero MF, Khosla MC, Douglas JC. Modulation of phospholipase A2 and sodium transport by angiotensin-(1-7). Kidney Int. 1993;44(5):932–6.

17. Handa RK. Angiotensin-(1-7) can interact with the rat proximal tubule AT(4) receptor system. Am J Physiol. 1999;277(1 Pt2):F75–83.

18. Lara LS, Carvalho T, Leao-Ferreira LR, Lopes AG, Caruso-Neves C. Modulation of the Na+K+ATPase activity by angiotensin-(1-7) in MDCK cells. Reg Peptides. 2005;129(1–3):221–6.

19. Lara LS, Cavalcante F, Axelband F, De Souza AM, Lopes AG, Caruso-Neves C. Involvement of the Gi/o/cGMP/PKG pathway in the AT2-mediated inhibition of outer cortex proximal tubule Na+-ATPase by angiotensin-(1-7). Biochem J. 2006;395(1):183–90.

20. Dellipizzi AM, Hilchley SD, Bell-Quilley CP. Natriuretic action of angiotensin-(1-7). Br J Pharmacol. 1994;111(1):1–3.

21. Vallon V, Richter K, Heyne N, Osswald H. Effect of intratubular application of angiotensin-(1-7) on nephron function. Kidney Blood Press Res. 1997;20(4):233–9.

22. Lopez O, Gironacci M, Rodriguez D, Pena C. Effect of angiotensin-(1-7) on ATPase activities in several tissues. Reg Peptides. 1998;77(1–3):135–9.

23. Bürgelová M, Kramer HJ, Teplan V, Velicková G, Vítko S, Heller J, et al. Intrarenal infusion of angiotensin-(1-7) modulates renal function responses to exogenous angiotensin II in the rat. Kidney Blood Press Res. 2002;25(4):202–10.

24. Santos RAS, Campagnole-Santos MJ, Baracho NCV, Fontes MA, Silva LC, Neves LA, et al. Characterization of a new angiotensin antagonist selective for angiotensin-(1-7): evidence that the actions of angiotensin-(1-7) are mediated by specific angiotensin receptors. Brain Res Bull. 1994;35(4):293–8.

25. Santos RAS, Baracho NC. Angiotensin-(1-7) is a potent antidiuretic peptide in rats. Braz J Med Biol Res. 1992;25(6):651–4.

26. Garcia NH, Garvin JL. Angiotensin-(1-7) has a biphasic effect on fluid absorption in the proximal straight tubule. J Am Soc Nephrol. 1994;5(4):1133–8.

27. Santos RAS, Simões e Silva AC, Magaldi AJ, Khosla MC, Cesar KR, Passaglio KT, et al. Evidence for a physiological role of angiotensin-(1-7) in the control of the hydroelectrolyte balance. Hypertension. 1996;27(4):875–84.
28. Simões e Silva AC, Baracho NCV, Passaglio KT, Santos RAS. Renal actions of angiotensin-(1-7). Braz J Med Biol Res. 1997;30(4):503–13.
29. Baracho NC, Simões e Silva AC, Khosla MC, Santos RAS. Effect of selective angiotensin antagonists on the antidiuresis produced by angiotensin-(1-7) in water-loaded rats. Braz J Med Biol Res. 1998;31(9):1221–7.
30. Simões e Silva AC, Bello AP, BaCho NC, Khosla MC, Santos RAS. Diuresis and natriuresis produced by long term administration of a selective angiotensin-(1-7) antagonist in normotensive and hypertensive rats. Reg Peptides. 1998;74(2–3):177–84.
31. Magaldi AJ, Cesar KR, Araujo M, Simões e Silva AC, Santos RA. Angiotensin-(1-7) stimulates water transport in rat inner medullary collecting duct: evidence for novel involvement of vasopressin V2 receptor. Pflugers Archiv. 2003;447(2):223–30.
32. Pinheiro SVB, Simões e Silva AC, Sampaio WO, de Paula RD, Mendes EP, Bontempo ED, et al. The nonpeptide AVE 0991 is an angiotensin-(1-7) receptor Mas agonist in the mouse kidney. Hypertension. 2004;44(4):490–6.
33. Santos RAS, Haibara AS, Campagnole-Santos MJ, Simões e Silva AC, de Paula RD, Pinheiro SV, et al. Characterization of a new selective antagonist for Angiotensin-(1-7), D-Pro7–Angiotensin-(1-7). Hypertension. 2003;41(3 Pt2):737–43.
34. Ferreira AJ, Pinheiro SVB, Castro CH, Silva GA, Simões e Silva AC, Almeida AP, et al. Renal function in transgenic rats expressing an angiotensin-(1-7)-producing fusion protein. Regul Pept. 2007;137(3):128–33.
35. Castelo-Branco RC, Leite-Dellova DCA, Fernandes FB, Malnic G, Mello-Aires M. The effects of angiotensin-(1-7) on the exchanger NHE3 and on [ca^{+2}] i in the proximal tubules of spontaneously hypertensive rats. Am J Physiol Renal Physiol. 2017;313:F450–60.
36. O'Neil J, Healy V, Johns EJ. Intrarenal Mas and AT1 receptors play a role in mediating the excretory actions of renal interstitial angiotensin-(1-7) infusion in anaesthetized rats. Exp Physiol. 2017;102:1700–15.
37. Zimmerman D, Burns KD. Angiotensin-(1-7) in kidney disease: a review of the controversies. Clin Sci (Lond). 2012;123:333–46.
38. Navar LG, Nishiyama A. Why are angiotensin concentrations so high in the kidney? Curr Op Nephrol Hypert. 2004;13(1):107–15.
39. Brewster UC, Perazella MA. The renin-angiotensin-aldosterone system and the kidney: effects on kidney disease. Am J Med. 2004;116(4):263–72.
40. Jaimes EA, Tian RX, Pearse D, Raij L. Up-regulation of glomerular COX-2 by angiotensin II: role of reactive oxygen species. Kidney Int. 2005;68(5):2143–53.
41. Ren Y, Garvin JL, Carretero OA. Vasodilator action of angiotensin-(1-7) on isolated rabbit afferent arterioles. Hypertension. 2002;39(3):799–802.
42. Sampaio WO, Nascimento AA, Santos RAS. Systemic and regional hemodynamic effects of angiotensin-(1-7) in rats. Am J Physiol Heart Circ Physiol. 2003;284(6):H1985–94.
43. Stegbauer J, Oberhauser V, Vonend O, Rump LC. Angiotensin-(1-7) modulates vascular resistance and sympathetic neurotransmission in kidneys of spontaneously hypertensive rats. Cardiovasc Res. 2004;61(2):352–9.
44. Yousif MHM, Benter IF, Diz DI, Chappell MC. Angiotensin-(1-7)-dependent vasorelaxation of the renal artery exhibits unique angiotensin and bradykinin receptor selectivity. Peptides. 2017;90:10–6.
45. Roks AJ, Van Geel PP, Pinto YM, Buikema H, Henning RH, de Zeeuw D, et al. Angiotensin-(1-7) is a modulator of the human renin-angiotensin system. Hypertension. 1999;34(2):296–301.
46. Mahon JM, Carr RD, Nicol AK, Henderson IW. Angiotensin-(1-7) is an antagonist at the type I angiotensin II receptor. J Hypertens. 1994;12:1377–81.

47. Chansel D, Vandermeersch S, Oko A, Curat C, Ardaillou R. Effects of angiotensin IV and angiotensin-(1-7) on basal and angiotensin II-stimulated cytosolic Ca2+ in mesangial cells. Eur J Pharmacol. 2001;414:165–75.

48. Oudot A, Vergely C, Ecarnot-Laubriet A, Rochette L. Pharmacological concentration of angiotensin-(1-7) activates NADPH oxidase after ischemia-reperfusion in rat heart through AT1 receptor stimulation. Regul Pept. 2005;127(1–3):101–10.

49. Zhu Z, Zhong J, Zhu S, Liu D, Van Der Giet M, Tepel M. Angiotensin-(1-7) inhibits angiotensin II-induced signal transduction. J Cardiovasc Pharmacol. 2002;40:693–700.

50. Clark MA, Tallant EA, Tommasi E, Bosch S, Diz DI. Angiotensin-(1-7) reduces renal angiotensin II receptors through a cyclooxygenase-dependent mechanism. J Cardiovasc Pharmacol. 2003;41:276–83.

51. Kostenis E, Milligan G, Christopoulos A, Sanchez-Ferrer CF, Heringer-Walther S, Sexton PM, et al. G-protein-coupled receptor Mas is a physiological antagonist of the angiotensin II type 1 receptor. Circulation. 2005;111(14):1806–13.

52. Wilson BA, Nautiyal M, Gwathmey TM, Rose JC, Chappell MC. Evidence for mitochondrial angiotensin-(1-7) system in the kidney. Am J Physiol Renal Physiol. 2016;310:F637–45.

53. Pinheiro SV, Ferreira AJ, Kitten GT, da Silveira KD, da Silva DA, Santos SH, et al. Genetic deletion of the angiotensin-(1-7) receptor Mas leads to glomerular hyperfiltration and microalbuminuria. Kidney Int. 2009;75(11):1184–93.

54. Zhang J, Noble NA, Border WA, Huang Y. Infusion of angiotensin-(1-7) reduces glomerulosclerosis through counteracting angiotensin II in experimental glomerulonephritis. Am J Physiol Renal Physiol. 2010;298(3):F579–88.

55. Giani JF, Munoz MC, Pons RA, Cao G, Toblli JE, Turyn D, et al. Angiotensin-(1-7) reduces proteinuria and diminishes structural damage in renal tissue of stroke-prone spontaneously hypertensive rats. Am J Physiol Renal Physiol. 2011;300(1):F272–82.

56. Barroso LC, Silveira KD, Lima CX, Borges V, Bader M, Rachid M, et al. Renoprotective effects of AVE0991, a nonpeptide Mas receptor agonist, in experimental acute renal injury. Int J Hypertens. 2012:808726.

57. Liu Z, Huang XR, Chen HY, Penninger JM, Lan HY. Loss of angiotensin-converting enzyme 2 enhances TGF-beta/Smad-mediated renal fibrosis and NF-kappaB-driven renal inflammation in a mouse model of obstructive nephropathy. Lab Investig. 2012;92(5):650–61.

58. Xue H, Zhou L, Yuan P, Wang Z, Ni J, Yao T, et al. Counteraction between angiotensin II and angiotensin-(1-7) via activating angiotensin type I and Mas receptor on rat renal mesangial cells. Regul Pept. 2012;177(1–3):12–20.

59. Silveira KD, Barroso LC, Vieira AT, Cisalpino D, Lima CX, Bader M, et al. Beneficial effects of the activation of the angiotensin-(1-7) Mas receptor in a murine model of adriamycin-induced nephropathy. PLoS One. 2013;8:e66082.

60. Kim CS, Kim IJ, Bae EH, Ma SK, Lee J, Kim SW. Angiotenisn-(1-7) attenuates kidney injury due to obstructive nephropathy in rats. PLoS One. 2015;10:e0142664. https://doi.org/10.1371/journal.pone.0142664.

61. Lu W, Kang J, Hu K, Tang S, Zhou X, Yu S, Xu L. Angiotensin-(1-7) relieved renal injury induced by chronic intermittent hypoxia in rats by reducing inflammation, oxidative stress and fibrosis. Braz J Med Biol Res. 2017;50:e5594. https://doi.org/10.1590/1414-431X20165594.

62. Li Y, Wu J, He Q, Shou Z, Zhang P, Pen W, et al. Angiotensin (1-7) prevent heart dysfunction and left ventricular remodeling caused by renal dysfunction in 5/6 nephrectomy mice. Hypertens Res. 2009;32(5):369–74.

63. Benter IF, Yousif MH, Cojocel C, Al-Maghrebi M, Diz DI. Angiotensin-(1-7) prevents diabetes-induced cardiovascular dysfunction. Am J Physiol Heart Circ Physiol. 2007;292(1):H666–72.

64. Benter IF, Yousif MH, Dhaunsi GS, Kaur J, Chappell MC, Diz DI. Angiotensin-(1-7) prevents activation of NADPH oxidase and renal vascular dysfunction in diabetic hypertensive rats. Am J Nephrol. 2008;28(1):25–33.

65. Moon JY, Tanimoto M, Gohda T, Hagiwara S, Yamazaki T, Ohara I, et al. Attenuating effect of angiotensin-(1-7) on angiotensin II-mediated NAD(P)H oxidase activation in type 2 diabetic nephropathy of KK-A(y)/Ta mice. Am J Physiol Renal Physiol. 2011;300(6):F1271–82.

66. Giani JF, Burghi V, Veiras LC, Tomat A, Munoz MC, Cao G, et al. Angiotensin-(1-7) attenuates diabetic nephropathy in Zucker diabetic fatty rats. Am J Physiol Renal Physiol. 2012;302(12):F1606–15.
67. Mori J, Patel VB, Ramprasath T, Alrob OA, Desaulniers J, Scholey JW, et al. Angiotensin 1-7 mediates renoprotection against diabetic nephropathy by reducing oxidative stress, inflammation and lipotoxicity. Am J Physiol Renal Physiol. 2014;306(8):F812–21.
68. Ma L, Han C, Peng T, Li N, Zhang B, Zhen X, Yang X. Ang-(1e7) inhibited mitochondrial fission in high-glucose-induced podocytes by upregulation of miR 30a and downregulation of Drp1 and p53. J Chin Med Association. 2016;79:597–604.
69. Zhao S, Ghosh A, Lo C-S, Chenier I, Scoley JW, Filep JG, et al. Nrf2 deficiency upregulates intrarenal angiotensin-converting Enzyme-2 and angiotensin 1-7 receptor expression and attenuates hypertension and nephropathy in diabetic mice. Endocrinology. 2018;159:836–52.
70. Oudit GY, Herzenberg AM, Kassiri Z, Wong D, Reich H, Khokha R, et al. Loss of angiotensin-converting enzyme-2 leads to the late development of angiotensin II-dependent glomerulosclerosis. Am J Pathol. 2006;168(6):1808–20.
71. Ye M, Wysocki J, William J, Soler MJ, Cokic I, Batlle D. Glomerular localization and expression of angiotensin-converting enzyme 2 and angiotensin-converting enzyme: implications for albuminuria in diabetes. J Am Soc Nephrol. 2006;17(11):3067–75.
72. Soler MJ, Wysocki J, Ye M, Lloveras J, Kanwar Y, Batlle D. ACE2 inhibition worsens glomerular injury in association with increased ACE expression in streptozotocin-induced diabetic mice. Kidney Int. 2007;72(5):614–23.
73. Wong DW, Oudit GY, Reich H, Kassiri Z, Zhou J, Liu QC, et al. Loss of angiotensin-converting enzyme-2 (Ace2) accelerates diabetic kidney injury. Am J Pathol. 2007;171(2):438–51.
74. Wysocki J, Ye M, Soler MJ, Gurley SB, Xiao HD, Bernstein KE, et al. ACE and ACE2 activity in diabetic mice. Diabetes. 2007;55(7):2132–9.
75. Dilauro M, Zimpelmann J, Robertson SJ, Genest D, Burns KD. Effect of ACE2 and angiotensin-(1-7) in a mouse model of early chronic kidney disease. Am J Physiol Renal Physiol. 2010;298(6):F1523–32.
76. Silveira KD, Pompermayer Bosco KS, Diniz LR, Carmona AK, Cassali GD, Bruna-Romero O, et al. ACE2-angiotensin-(1-7)-Mas axis in renal ischaemia/reperfusion injury in rats. Clin Sci (Lond). 2010;119(9):385–94.
77. Lo CS, Liu F, Shi Y, Maachi H, Chenier I, Godin N, et al. Dual RAS blockade normalizes angiotensin-converting enzyme-2 expression and prevents hypertension and tubular apoptosis in Akita angiotensinogen-transgenic mice. Am J Physiol Renal Physiol. 2012;302(7):F840–52.
78. Wysocki J, Ortiz-Melo DI, Mattocks NK, Xu K, Prescott J, Evora K, et al. ACE2 deficiency increases NADPH-mediated oxidative stress in the kidney. Physiol Rep. 2014;2(3):e00264.
79. Chen L-J, Xu Y-L, Song B, Yu H-M, Oudit GY, Xu R, et al. Angiotensin-converting enzyme 2 ameliorates renal fibrosis by blocking the activation of mTOR/ERK signaling in apolipoprotein E-deficient mice. Peptides. 2016;79:49–57.
80. Ng HY, Yisireyili M, Saito S, Lee CT, Adelibieke Y, Nishijima F, Niwa T. Indoxyl sulfate downregulates expression of Mas receptor via OAT3/AhR/Stat3 pathway in proximal tubular cells. PLoS One. 2014;9(3):e91517.
81. Patel SN, Ali Q, Samuel P, Steckelings UM, Hussain T. Angiotensin II type 2 receptor and receptor Mas are Colocalized and functionally interdependent in obese Zucker rat kidney. Hypertension. 2017;70:831–8.
82. Esteban V, Heringer-Walther S, Sterner-Kock A, de Bruin R, van den Engel S, Wang Y, et al. Angiotensin-(1-7) and the g protein-coupled receptor MAS are key players in renal inflammation. PLoS One. 2009;4(4):e5406.
83. Velkoska E, Dean RG, Griggs K, Burchill L, Burrell LM. Angiotensin-(1-7) infusion is associated with increased blood pressure and adverse cardiac remodelling in rats with subtotal nephrectomy. Clin Sci (Lond). 2011;120(8):335–45.
84. Burrell LM, Gayed D, Griggs K, Patel SK, Velkoska E. Adverse cardiac effects of exogenous angiotensin 1-7 in rats with subtotal nephrectomy are prevented by ACE inhibition. PLoS One. 2017;12:e0171975. https://doi.org/10.1371/journal.pone.0171975.

85. Su Z, Zimpelmann J, Burns KD. Angiotensin-(1-7) inhibits angiotensin II-stimulated phosphorylation of MAP kinases in proximal tubular cells. Kidney Int. 2006;69(12):2212–8.
86. Zimpelmann J, Burns KD. Angiotensin-(1-7) activates growth-stimulatory pathways in human mesangial cells. Am J Physiol Renal Physiol. 2009;296(2):F337–46.
87. Liu GC, Oudit GY, Fang F, Zhou J, Scholey JW. Angiotensin-(1-7)-induced activation of ERK1/2 is cAMP/protein kinase A-dependent in glomerular mesangial cells. Am J Physiol Renal Physiol. 2012;302(6):F784–90.
88. Oudit GY, Liu GC, Zhong J, Basu R, Chow FL, Zhou J, et al. Human recombinant ACE2 reduces the progression of diabetic nephropathy. Diabetes. 2010;59(2):529–38.
89. Luque M, Martin P, Martell N, Fernandez C, Brosnihan KB, Ferrario CM. Effects of captopril related to increased levels of prostacyclin and angiotensin-(1-7) in essential hypertension. J Hypertens. 1996;14(6):799–805.
90. Ferrario CM, Martell N, Yunis C, Flack JM, Chappell MC, Brosnihan KB, et al. Characterization of angiotensin-(1-7) in the urine of normal and essential hypertensive subjects. Am J Hypertens. 1998;11(2):137–46.
91. Azizi M, Menard J. Combined blockade of the renin-angiotensin system with angiotensin-converting enzyme inhibitors and angiotensin II type 1 receptor antagonists. Circulation. 2004;109(21):2492–9.
92. Simões e Silva AC, Diniz JS, Regueira Filho A, Santos RA. The renin angiotensin system in childhood hypertension: selective increase of angiotensin-(1-7) in essential hypertension. J Pediatr. 2004;145(1):93–8.
93. Simões e Silva AC, Diniz JS, Pereira RM, Pinheiro SV, Santos RA. Circulating renin Angiotensin system in childhood chronic renal failure: marked increase of Angiotensin-(1-7) in end-stage renal disease. Pediatr Res. 2006;60(6):734–9.
94. Nogueira AI, Santos RA, Simões e Silva AC, Cabral AC, Vieira RL, Drumond TC, et al. The pregnancy-induced increase of plasma angiotensin-(1-7) is blunted in gestational diabetes. Regul Pept. 2007;141:55–60.
95. Vilas-Boas WW, Ribeiro-Oliveira A Jr, Pereira RM, Ribeiro Rda C, Almeida J, Nadu AP, et al. Relationship between angiotensin-(1-7) and angiotensin II correlates with hemodynamic changes in human liver cirrhosis. World J Gastroenterol. 2009;15(20):2512–9.
96. Anguiano L, Riera M, Pascual J, Valdivielso JM, Barrios C, Betriu A, et al. Circulating angiotensin-converting enzyme 2 activity in patients with chronic kidney disease without previous history of cardiovascular disease. Nephrol Dial Transplant. 2015;30:1176–85.
97. Antlanger M, Domenig O, Kovarik JJ, Kaltenecker CC, Kopecky C, Poglitsch M, Säemann MD. Molecular remodeling of the renin-angiotensin system after kidney transplantation. J Renin-Angiotensin-Aldosterone System. 2017;18:1–9. https://doi.org/10.1177/1470320317705232.
98. Rocha NP, Bastos FM, Vieira EL, Prestes TR, Silveira KD, Teixeira MM, Simões e Silva AC. The protective arm of the renin-angiotensin system may counteract the intense inflammatory process in fetuses with posterior urethral valves. J Pediatr (RJ). 2018; in press.
99. Vieira EL, Rocha NP, Bastos FM, da Silveira KD, Pereira AK, Oliveira EA, et al. Posterior urethral valve in fetuses: evidence for the role of inflammatory molecules. Pediatr Nephrol. 2017;32:1391–400.
100. Mizuiri S, Hemmi H, Arita M, Aoki T, Ohashi Y, Miyagi M, et al. Increased ACE and decreased ACE2 expression in kidneys from patients with IgA nephropathy. Nephron Clin Pract. 2011;117(1):c57–66.
101. Soler MJ, Riera M, Crespo M, Mir M, Márquez E, Pascula MJ, et al. Circulating angiotensin-converting enzyme 2 activity in kidney transplantation: a longitudinal pilot study. Nephron Clin Pract. 2012;121(3–4):144–50.
102. Kocks MJ, Lely AT, Boomsma F, de Jong PE, Navis G. Sodium status and angiotensin-converting enzyme inhibition: effects on plasma angiotensin-(1-7) in healthy man. J Hypertens. 2005;23(3):597–602.
103. Iwanami J, Mogi M, Tsukuda K, Wang XL, Nakaoka H, Ohshima K, et al. Role of angiotensin-converting enzyme 2/angiotensin-(1-7)/Mas axis in the hypotensive effect of azilsartan. Hypertens Res. 2014;37(7):616–20.

Lung

Giselle S. Magalhães, Maria Jose Campagnole-Santos, and Maria da Glória Rodrigues-Machado

Renin-Angiotensin System Components in the Lungs

There is a considerable body of evidence for the existence of local, tissue-based, renin-angiotensin system (RAS) in which angiotensin (Ang) peptides production is independent of circulating precursors [13, 56]. Expression of angiotensinogen, the type 1 (AT1) and type 2 (AT2) Ang II receptors in rat and human lung tissue support local generation of Ang II [56]. Membrane angiotensin-converting enzyme (ACE), primarily responsible for conversion of Ang I to Ang II in the circulation, is abundantly expressed in vascular endothelium of pulmonary circulation.

Ang II can modulate inflammatory response promoting cytokine production, expression of endothelial adhesion molecules, inflammatory cell migration, epithelial cell apoptosis, oxidative stress, lung fibroblast growth and fibrosis [56]. The majority of these actions are mediated through the AT1 receptor involving complex intracellular signaling pathways [38]. AT1 receptor, coupled to Gaq/11 protein, can stimulate multiple signaling pathways including MAPK/ERK, Rho/ROCK kinase,

G. S. Magalhães
Department of Physiology and Biophysics, Federal University of Minas Gerais, Belo Horizonte, MG, Brazil

Post-Graduation Program in Health Sciences, Medical Sciences Faculty of Minas Gerais, Belo Horizonte, MG, Brazil

M. J. Campagnole-Santos
Departamento de Fisiologia e Biofísica, INCT-Nanobiofar, Instituto de Ciências Biológicas, Universidade Federal de Minas Gerais, Belo Horizonte, MG, Brazil
e-mail: mjcs@icb.ufmg.br

M. da Glória Rodrigues-Machado (⊠)
Post-Graduation Program in Health Sciences, Medical Sciences Faculty of Minas Gerais, Belo Horizonte, MG, Brazil
e-mail: maria.machado@cienciasmedicasmg.edu.br

© Springer Nature Switzerland AG 2019
R. A. S. Santos (ed.), *Angiotensin-(1-7)*,
https://doi.org/10.1007/978-3-030-22696-1_9

PLCb/IP3/diacylglycerol, tyrosine kinases, and NF-kB [3]. ACE/Ang II/AT1 receptor axis is involved in many lung diseases.

ACE2, an ACE homologous enzyme, has emerged as a potent negative regulator of the RAS. ACE2 regulates RAS signaling, reducing Ang II/AT1 receptor signaling and activating the counterregulatory angiotensin-(1-7) [Ang-(1-7)]/Mas receptor pathway. ACE2 protein is expressed in the lungs, mainly in the vascular endothelium, Clara cells, type I and type II alveolar epithelial cells [26, 27, 48], as well as in smooth muscle of small and medium vessels in the mouse lung [92]. Mas receptor, a functional receptor for Ang-(1-7) [76], is present in thin areas of the bronchial epithelium and smooth muscle [52]. ACE2/Ang-(1-7)/Mas receptor pathway often serves to counterregulate the pro-inflammatory, pro-proliferative, and pro-fibrotic effects of the ACE/Ang II/AT1 receptor pathway [77].

Ang-(1-7) and Pulmonary Arterial Hypertension

Pulmonary hypertension (PH) is a disorder characterized by an increase in mean pulmonary arterial pressure (PAP) $\geq$25 mmHg at rest as assessed by right heart catheterization (RHC) [32]. The term pulmonary arterial hypertension (PAH) describes a group of PH patients characterized hemodynamically by the presence of pre-capillary PH, defined by a pulmonary artery wedge pressure (PAWP) $\leq$15 mmHg and a pulmonary vascular resistance (PVR) >3 Wood units (WU) in the absence of other causes of pre-capillary PH, such as PH due to lung diseases, chronic thromboembolic pulmonary hypertension (CTEPH), or other rare diseases [32]. A hallmark of PAH is a vascular remodeling process that increases PVR and subsequent right ventricular hypertrophy and premature death [78]. Regardless of the underlying disease, chronic cor pulmonale is associated with progressive clinical deterioration and a poor prognosis in most cases. Incidence and prevalence of PAH is very similar in USA (2.0 and 10.6 cases of PAH per million inhabitants, respectively) and in UK (1.1 and 6.6 cases of PAH per million inhabitants, respectively) [49, 62].

Clinical classification of PH categorizes multiple clinical conditions into five groups, according to their similar clinical presentation, pathological findings, hemodynamic characteristics, and treatment strategy [20].

Diagnosis of PH is based on clinical suspicion established by symptoms, typically induced by exertion (shortness of breath, fatigue, weakness, angina, and syncope). Symptoms at rest occur only in advanced circumstances. Abdominal distension and ankle edema will develop with progressing right ventricle (RV) failure. Diseases that cause or are associated with PH as well as other concurrent diseases can modify the presentation of PH [20].

Advances in basic and clinical research into PAH have led to improved understanding of disease pathogenesis and identification of novel therapeutic targets [42]. The aim of specific therapies for PH is to reduce PVR and thereby improve RV function. Currently, five classes of drugs have been applied for PAH: endothelin receptor antagonists (ERAs), prostanoids, phosphodiesterase type 5 inhibitors, soluble guanylate cyclase stimulators, and selective prostacyclin receptor agonists

[89]. Despite improvement in patient symptoms and well-being with these agents, mortality rates remain high (~65% survival at 5 years). New therapies are needed targeting alternative pathways that can reverse pulmonary vascular remodeling, inhibit disease progression, and improve survival [23]. The RAS is being intensively studied as an alternative therapeutic target [90].

A large number of studies have shown that the RAS is importantly involved in PAH pathophysiology [12, 38, 86]. Lungs of patients with PAH express high levels of ACE in the intra-acinar arteries, suggesting that locally increased production of Ang II, a potent pulmonary vasoconstrictor with mitogenic actions, may contribute to the process of pulmonary vascular remodeling [69]. Ang II is also capable of inducing an inflammatory response in the vascular wall. Ang II, via the type 1 (AT1) receptors, enhances the production of reactive oxygen species (ROS) through stimulation of NAD(P)H oxidase in the vascular wall, leading to endothelial dysfunction and vascular inflammation by stimulating the redox-sensitive transcription factors (NF-kB) and by upregulating adhesion molecules, cytokines, and chemokines [9]. De Man et al. [12] demonstrated increased serum levels of renin, Ang I, and Ang II and correlations with disease progression and mortality in patients with idiopathic PAH. Taken together, these findings indicate an active role for RAS in the pulmonary hypertensive process.

There is a body of evidence suggesting that ACE2, either by itself or through its catalytic product Ang-(1-7), opposes the proliferative, hypertrophic, and fibrotic effects of Ang II in many organs, including the lungs, pointing for a plausible protective role against PAH. Ang II appears to be the main substrate for ACE2, and is effectively hydrolyzed to Ang-(1-7). ACE2 protein is expressed in various human organs and in the lungs, it is expressed mainly on the vascular endothelium, and type I and type II alveolar epithelial cells [26, 27].

Studies demonstrate that serum ACE2 was decreased in patients with PAH due to congenital heart disease, and mean PAP was negatively correlated with serum levels of ACE2 [11]. Similar results were found for Ang-(1-7), suggesting the decrease in Ang-(1-7) shifts the balance of the RAS toward the ACE/Ang II/AT1 receptor axis, resulting in increases in vascular remodeling, fibrosis and PAH in congenital heart disease patients [10]. Consistent with these findings, several ACE2 activators such as diminazene aceturate (DIZE) [84], xanthenone (XNT, e 1-[(2-dimethylamino) ethylamino]-4-(hydroxymethyl)-7-[(4-methylphenyl)   sulfonyloxy]-9H-xanthene-9-one) [31], resorcinolnaphthalein [44, 45], and NCP-2454 [24] have been reported in various preclinical models of PAH.

In a recent trial, Hemnes et al. [30] assessed the mechanism, safety, and efficacy of ACE2 (single IV infusion of GSK2586881) in the treatment of patients with idiopathic and heritable PAH (18 years) with functional class I-III. PAH patients had a significant decrease in ACE2 activity as reflected by the increased Ang II/Ang-(1-7) ratio in PAH patients compared with controls. After treatment, PAH patients had a decrease in Ang II/Ang-(1-7) ratio, suggesting increased activity of ACE2. In addition, levels of superoxide dismutase (SOD2) protein were approximately 25% lower in PAH plasma compared with controls. After treatment, there was significant induction of plasma SOD2 protein levels by 2 weeks suggesting induction in the enzymatic activity by GSK2586881. Compared with control, patients with PAH had increased levels of cytokines (IL-10, IL-1β, TNF-α, IL-13, IL-8, and IL-4). After GSK2586881 administration, there was suppression of IL-10, IL-1β, IL-2, and TNF-α that could be detected as early as 2 hours after drug administration and was associated with sustained anti-inflammatory effects with reduced levels of IL-1β, IL-6, IL-8, and TNF-α at 2 weeks [30]. Taken together, these data showed that treatment with ACE2 reduced the markers of oxidant and inflammatory mediators and improved the balance between ACE/Ang II/AT1 receptor and ACE2/Ang-(1-7)/Mas receptor axis.

Ang-(1-7) promotes the release of prostanoids from endothelial cells (EC) and smooth muscle cells (SMC) and the release of nitric oxide (NO). In addition, Ang-(1-7) inhibits proliferation of vascular SMC and EC *in vitro* and *in vivo* and opposes the mitogenic effects of Ang II [77]. Drugs that inhibit the synthesis of Ang II (ACE inhibitors) or that antagonize AT1 receptors (Ang II receptor blockers – ARBs) have been shown to decrease right ventricular hypertrophy, decrease medial thickening and peripheral muscularization of small pulmonary arteries in hypoxic animals [65]. In addition, ACE2 [17, 94] or Ang-(1-7) itself, by targeted gene transfer, protects the lungs in a model of pulmonary hypertension [82]. The effects of Ang-(1-7) appear to be associated with upregulation of endothelial nitric oxide synthase (eNOS) activation via AKT pathway [7]. Recently, Zhang et al. [96] showed that phosphorylation of ACE2 by AMPK enhanced the stability of ACE2, which increased Ang-(1-7) and nitric oxide synthase (eNOS)-derived NO bioavailability in endothelial cells.

Shenoy et al. [85] developed a plant-based oral delivery of ACE2 or Ang-(1-7) to protect against gastric enzymatic degradation and facilitates long-term storage at room temperature. Further, fusion to a transmucosal carrier helped effective systemic absorption from the intestine on oral delivery. Rats fed with bioencapsulated ACE2 or Ang-(1-7) presented attenuation in the development of monocrotaline-induced PH and improvement of cardiopulmonary pathophysiology. Furthermore, in the reversal protocol, oral ACE2 or Ang-(1-7) treatment significantly arrested disease progression, along with improvement in right heart function, and decrease in pulmonary vessel wall thickness. In addition, a combination therapy with ACE2 and Ang-(1-7) augmented the beneficial effects against monocrotaline-induced lung injury. According to the authors, these results provided proof-of-concept for a novel low-cost oral ACE2 or Ang-(1-7) delivery system using transplastomic technology for pulmonary disease therapeutics.

Microvesicles derived from mesenchymal stem cells (MSCs) improve the outcome of PAH [43]. Recently, Liu et al. [50] investigated whether the effect of MSC-derived microvesicles on PAH induced by monocrotaline was correlated with RAS. Animals treated with microvesicles from MSCs notably attenuated the pulmonary artery pressure, reversed the RV hypertrophy and pulmonary vessel remodeling, the inflammation score and the collagen fiber volume fraction. In addition, ACE2 mRNA in the lung tissues and plasma levels of Ang-(1-7) were both upregulated in animals treated with MSC microvesicles. These protective effects were diminished by the use of A-779, a selective inhibitor of the Mas receptor (Fig. 1).

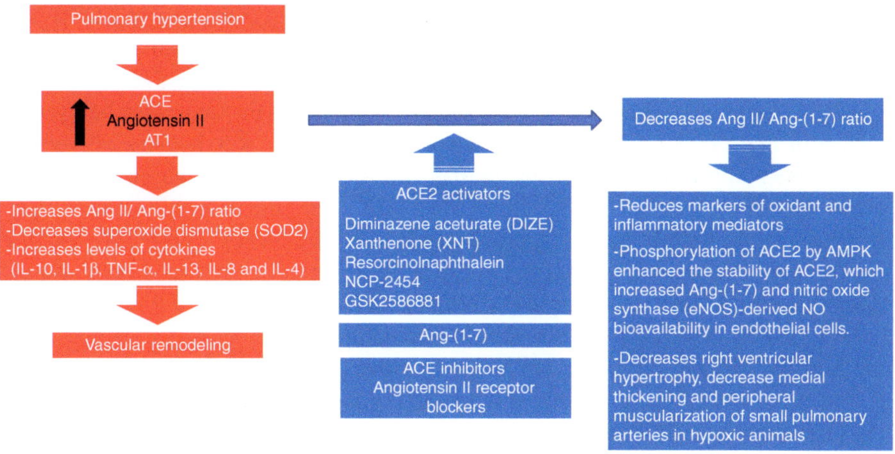

Fig. 1 Effects triggered by treatment with angiotensin-converting enzyme 2 (ACE2), ACE inhibitors and angiotensin II receptor blockers in pulmonary hypertension

Table 1 Origin of the inflammatory insult in ARDS

Pulmonary ARDS	Extrapulmonary ARDS
Pneumonia (bacterial, viral, fungal)	Sepsis syndrome
Aspiration of gastric contents	Non-thoracic trauma
Lung contusion	Transfusion
Inhalation injury	Cardiopulmonary bypass
Near-drowning	Pancreatitis
Fat emboli	Drug overdose
Reperfusion injury	Burn injury
Mechanical ventilation (barotrauma, volutrauma)	

Ang-(1-7) in Acute Respiratory Distress Syndrome

Acute respiratory distress syndrome (ARDS) is a life-threatening form of respiratory failure, that globally accounts for 10% of intensive care unit admissions, representing more than three million patients with ARDS annually [16]. Its first description dates 50 years ago [2]. Since then, ARDS has been redefined several times to ameliorate the accuracy of clinical diagnosis [4, 66, 73]. The last one was the Berlin definition [73] that proposed three categories of ARDS based on the severity of hypoxemia, timing of acute onset, origin of edema, and the chest radiograph or computed tomographic (CT) findings.

ARDS results from a wide spectrum of different risk factors, which can be either local or systemic (Table 1). According to the origin of the inflammatory insult, ARDS can be classified in pulmonary ARDS (ARDSp), as local or direct lung insult and extrapulmonary ARDS (ARDSexp), as systemic or indirect lung injury [21]. There are important clinical differences between ARDSp and ARDSexp in pathology, radiography, respiratory mechanics, response to treatment, and outcomes [21, 80].

ARDS remains a serious clinical problem with the main treatment being supportive in the form of mechanical ventilation. However, if the mechanical ventilation is used improperly, it can exacerbate the tissue damage caused by ARDS, known as ventilator-induced lung injury (VILI). To date, the only intervention demonstrated to improve clinical outcomes in ARDS is the use of a protective ventilatory strategy that uses low tidal volumes (VT) of 6 mL/kg predicted body weight compared with traditionally applied VT of 12 mL/kg [5].

Different animal models of experimental lung injury have been used to investigate mechanisms of lung injury [1]. In 2011, a committee assembled by the American Thoracic Society (ATS) published a workshop report determining the main features that characterize ARDS in animals and then identifying the most relevant measurements to assess these features. Important traits include (1) histological evidence of tissue injury, (2) alteration of the alveolar capillary barrier, (3) the presence of an inflammatory response, and (4) evidence of physiological dysfunction [59].

A body of evidence demonstrates that the RAS is involved in the pathogenesis of ARDS. In addition to its cardiovascular functions, Ang II is involved in inflammatory and fibrogenic processes in the lung [18, 25, 55]. Association between ACE polymorphism and susceptibility, progression, and outcome in ARDS has been demonstrated [36, 57]. Moreover, several studies have shown that inhibition of ARDS by AT1 receptor blockade or inhibition of Ang II formation by ACE has a protective effect on ARDS [34, 72, 81].

Protective effect of losartan has been tested on different models of ARDS. Losartan delayed the onset of ARDS in Wistar rats challenged by *i.t.* instillation of Bordetella bronchiseptica, prevented progressive deterioration of gas exchange and delayed the mortality of infected rats [72]. The signs of inflammation, thickened alveolar septae, and a marked increase in cellularity dominated by polymorphonuclear leukocytes were much less evident in losartan-treated rats. Although this effect was associated with a significant inhibition of lung-neutrophil recruitment, lung bacterial clearance was not impaired but rather, it was significantly improved. Similar results were found with irbesartan. Differently, neither the ACE inhibitor captopril, nor the nonselective peptide inhibitor of Ang II receptors, saralasin, reproduced these effects. The protective effects of losartan on ARDS were attributed, at least in part, to NF-kB and MAPK mechanisms. In a sepsis-induced ARDS using cecal ligation and puncture (CLP), Shen et al. [81] demonstrated that losartan treatment significantly led to inhibition of lung tissue NF-kB activation, attenuated degradation of IkB-alpha, and inhibited phosphorylation of p38MAPK, extracellular signal-regulated kinase 1/2, and c-Jun N-terminal kinase, critical pathways for cytokine release. Similarly, results of Raiden et al. [72] showed that losartan delays the onset of ARDS triggered by a bacterial infection, prevents blood gas deterioration and histopathologic appearance of ARDS, and significantly improved survival after sepsis.

The effects of captopril and losartan have also been tested in fat embolism (FE) and the consequent fat embolism syndrome (FES) that occurs after trauma or surgery and can lead to serious pulmonary injury, including ARDS and death [63]. There was a reduction in pulmonary inflammation, along with a significant decrease in interseptal edema and hemorrhage. Pathologic changes induced by FE in the lumen patency were also diminished with RAS inhibitors. Extending the evidence for the involvement of the RAS in this syndrome, Fletcher et al. [18] demonstrated that aliskiren, a renin inhibitor, protects rat lungs from the histopathological effects of fat embolism.

ACE inhibition or blockade of AT1 receptor favors an increase in Ang-(1-7) levels [77]. In ARDS, an ACE/ACE2 imbalance occurs in favor of increased ACE activity and correlates with lung injury. Previous studies have found that ACE2 mRNA, protein, and enzymatic activity were severely downregulated in human and experimental lung tissue injuries [34, 41]. The decrease in ACE2 expression was importantly involved in severe acute respiratory syndrome (SARS), in which the pathogen, coronavirus (SARS-CoV), triggers severe pneumonia and acute, often lethal, lung failure. Kuba et al. [41] demonstrated that ACE2 is a crucial SARS-CoV receptor in vivo, and both SARS-CoV infections and the Spike protein of the SARS-CoV reduced ACE2 expression, contributing to the severity of lung pathology. In addition, the injection of SARS-CoV Spike into mice worsens acute lung failure *in vivo*. This effect was associated with an increase in Ang II in the lung and it was attenuated by blocking AT1 [41].

In 2005, Imai et al. reported that lack of ACE2 expression (ACE2KO animals) precipitated ARDS, suggesting that ACE2 could present an important role in the prevention of ARDS. ARDS resulted in reduced ACE2 expression and increased Ang II production in ACE2$^{+/+}$ animals as a result of insults. Elastance of the respiratory system, as well as pulmonary edema, was significantly higher in sepsis groups, mainly in ACE2$^{-/-}$ mice. In addition, it was observed thickening of the alveolar wall, edema and pulmonary congestion, infiltration of inflammatory cells and hyaline membrane in sepsis-induced ACE2$^{-/-}$ mice. After 6 hours of observation, all animals in the ACE2$^{+/+}$ group were alive and only 2 of the 10 animals in the ACE$^{-/-}$ group survived. Moreover, intraperitoneal injection of recombinant human ACE2 protein (rhuACE2) in ARDS induced in ACE2$^{-/-}$ mice prevented the increase in elastance of the respiratory system and formation of pulmonary edema. In contrast to ACE2$^{-/-}$ mice, mice with genetic deletion of ACE (ACE$^{-/-}$) are protected against acid aspiration-induced ARDS and inactivation of ACE in ACE2$^{-/-}$ animals attenuates ARDS. Likewise, pharmacological inhibition or genetic deletion of AT1a (AgTr1a$^{-/-}$) receptors significantly attenuated lung function and edema formation. On the other hand, inactivation of AT2 receptors aggravated acute lung injury (ALI) [34].

Recently, bone marrow-derived mesenchymal stem cells (MSCs) overexpressing ACE2 served as a vehicle for gene therapy in lipopolysaccharide (LPS)-induced ARDS mice [28]. MSCs were transduced with ACE2 gene (MSC-ACE2) by a lentiviral vector and then infused into wild-type (WT) and ACE2 knockout

(ACE2−/y) mice following an LPS-induced intratracheal lung injury. MSC-ACE2 improved the lung histopathology, inflammation (decreased the neutrophil counts in the BALF, downregulated the expression of IL-1β and IL-6, and upregulated IL-10 in the lung). Additionally, MSC-ACE2 significantly reduced lung edema, in part by improving lung endothelial permeability, and normalized lung eNOS expression. Increased activity of ACE2 decreased the Ang II and increased the Ang-(1-7) in the lung, thereby inhibiting the detrimental effects of accumulating Ang II.

Protective mechanisms of ACE2 on experimental ARDS are not fully understood. ACE2 regulates RAS signaling, reducing Ang II/AT1 receptor signaling and activating the counterregulatory Ang-(1-7)/Mas receptor pathway. Treatment with lentiviral packaged ACE2 cDNA reduced and ACE2 shRNA increased Ang II/Ang-(1-7) ratio in the bronchoalveolar lavage, LPS-induced lung injury and inflammatory response. These responses were associated with alteration in the phosphorylation of MAPK and were all abolished by A779, a Mas receptor antagonist, suggesting these effects were mediated by Ang-(1-7) [47]. These data indicate that ACE2 protects lung injury via an increase in Ang-(1-7), which in turn stimulates Mas-mediated signaling to inhibit ERK1/2 and NF-κB activation [46, 47]. A recent study indicates that early initiation of therapy after experimental ALI induced by oleic acid and continuous drug delivery are most beneficial for optimal therapeutic efficiency of Ang-(1-7) treatment [88].

The cornerstone of ARDS management remains mechanical ventilation. However, mechanical ventilation with high tidal volumes causes lung hemorrhage and edema and activates inflammatory pathways, process referred as ventilator-induced lung injury (VILI). Jiang et al. [37] demonstrated an increase in lung Ang II levels induced by VILI. Deleterious effects were attenuated by captopril, an ACE inhibitor. These results suggested that local tissue angiotensin mediates these harmful events in VILI. Using the same VILI model of high tidal volumes, Jerng et al. [35] demonstrated that the lung injury score, bronchoalveolar lavage fluid protein concentration, pro-inflammatory cytokines, and NF-kB activities were significantly increased in the high-volume group compared with controls. In addition, the lung Ang II and mRNA levels of angiotensinogen and AT1 and AT2 receptors were also significantly increased in the high-volume group. Pretreatment with captopril or concomitant infusion with losartan or PD123319 in the high-volume group attenuated the lung injury and inflammation. Losartan and a protease-resistant, cyclic form of Ang-(1-7), showed similar lung protective effects, but losartan caused a significant decrease in blood pressure in the LPS-exposed ventilated animals [93].

Intravenous effect of Ang-(1-7) or its non-peptide agonist, AVE0991, was evaluated in ARDS induced by intravenous injection of oleic acid [40]. Ang-(1-7) or AVE0991 infusion 30 minutes after oleic acid administration reversed lung edema, and attenuated increased myeloperoxidase activity, which reflects neutrophil invasion. In addition, administration of Ang-(1-7) or AVE0991 restored arterial pressure and kept throughout experimental protocol (4 h), which falls rapidly by approximately 40% in untreated animals. Ang-(1-7) or its analog AVE0991 also prevented a decrease in pulmonary vascular resistance, characteristic for the acute phase of ARDS. Further, Ang-(1-7) or AVE0991 blocked the increase in TNF-α concentration in bronchoalveolar (BALF). These effects were antagonized by A779 and D-Pro[7]-Ang-(1-7) [40]. Corroborating with the results of Imai et al. [34], treatment with ibesartan, an AT1 blocker, normalized systemic blood arterial pressure, pulmonary arterial resistance, wet-to-dry lung weight ratio, BALF protein concentration, and myeloperoxidase activity in lung tissue. The beneficial effect of ibesartan was prevented by co-treatment with either A779 or d-Pro[7]-Ang-(1-7) on systemic and pulmonary hemodynamics. Thus, the protective effect of recombinant ACE2 or AT1 antagonization in ALI may be related at least in part to increased formation of Ang-(1-7) and stimulation of its specific receptor signaling pathways [40]. In this same study, the effect of Ang-(1-7) was tested in two murine ARDS models, ventilator-induced lung and acid aspiration injury. Ang-(1-7) reversed the effects in both models [40].

Ang-(1-7) has a antiremodeling role in pulmonary fibrosis that occours after ARDS [8]. Recently, Zambelli et al. [95] evaluated the potential for Ang-(1-7) to attenuate ARDS severity and lung fibrosis in a preclinical ARDS model. These authors evaluated if Ang-(1-7) would reduce the severity of early ARDS induced by the combined 'insults' induced by unilateral acid aspiration model followed by high stretch mechanical ventilation. Ang-(1-7) acute infusion showed a significant improvement of arterial oxygenation and inflammatory response (in terms of polymorphonuclear recruitment into alveoli) in acute ARDS. In other protocol, two weeks of Ang-(1-7) infusion increased blood oxygen saturation and the right lung from treated rats showed a significant reduction in collagen deposition. Thus, the inhibitory effect of Ang-(1-7) on inflammatory cells recruitment seen in the acute phase may be related to the reduction of fibrosis in the later phase. The beneficial effects observed by Jiang et al. [37] and Jerng et al. [35] may be related to the formation of Ang-(1-7) from the use of captopril and/ or losartan.

More recently, Khan et al. [39] reported results of a phase II trial examining the safety and efficacy of using GSK2586881, a recombinant human ACE2 (rhACE2), in 18 and 80 years old patients with ARDS, which had been mechanically ventilated for less than 72 h. The use of twice-daily doses of GSK2586881 infusion (0.4 mg/kg) for 3 days resulted in a decrease in plasma Ang II associated with an increase in Ang-(1-7) and Ang-(1-5) that remained elevated for 48 h. There was also a trend to decrease in IL-6. Although no episodes of hypotension were associated with infusion of GSK2586881, no significant improvement in oxygenation was observed in patients (ratio of partial pressure of arterial oxygen to fraction of inspired oxygen-PaO/FIO$_2$, oxygenation index), or Sequential Organ Failure Assessment-SOFA score between treated and placebo groups was observed, which the authors attributed to numerous factors that were not adequately controlled for in this trial [39]. However, this study reinforces the need for further evaluation of the impact of RAS modulation on pulmonary hemodynamics and markers of pulmonary injury (Fig. 2).

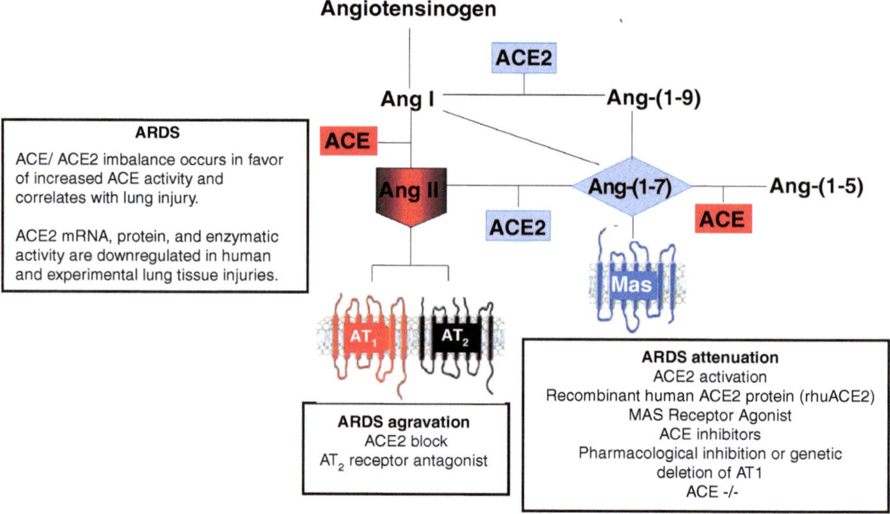

Fig. 2 Alterations in renin-angiotensin system components in acute respiratory distress syndrome (ARDS). (*ACE* angiotensin-converting enzyme, *Ang* angiotensin)

Ang-(1-7) in Asthma

Epidemiological studies show that asthma is currently the most common chronic disease in children, being the major cause of missed days at school and, in adults, loss of working days. In addition, asthma is associated with a significant rate of mortality [74]. The large increase in incidence of asthma is becoming a major global health problem and has encouraged studies aimed at increasing the knowledge of the pathophysiology of asthma, as well as development of new treatments to improve clinical management of the disease, mainly to meet asthma patients who do not respond well to current therapies [51].

Asthma is defined as a reversible airway obstructive disease, caused by airway mucosal edema, inflammation, increased mucus secretion, smooth muscle contraction, and airway hyperreactivity and remodeling [83]. Multiple cells and multiple mediators play a crucial pathophysiological role. The inflammatory response in allergic asthma is characterized by excess production of IgE, mast cell degranulation, and the infiltration of eosinophils and lymphocytes [22, 83]. However, the recruitment and activation of these cells depend on the expression and release of several classes of proteins, such as cytokines, particularly Th2-derived. Inflammatory mediators that increase influx of leukocytes, activity, and survival of eosinophils are positively correlated with asthma severity [14, 19]. Failure to resolve the inflammatory process causes a persistent inflammation with consequent tissue destruction and loss of pulmonary function [14].

There is experimental and clinical evidence indicating that activation of the pulmonary RAS is involved in the pathophysiology of allergic pulmonary disease,

especially through an inappropriate increase in angiotensin II (Ang II) [67, 68].
However, the Ang-(1-7)/Mas receptor axis, recognized as a counterregulatory
peptide system within the RAS, exhibits anti-inflammatory effects and prevents
inappropriate remodeling in different pathophysiological states, such as asthma.
Here, we show the effects of treatment with Ang-(1-7) on the three main changes
observed in chronic asthma: inflammation, pulmonary remodeling, and bronchial
hyperesponsiveness.

Experimental studies try to clear up aspects of the pathophysiology of asthma
mimicking human disease. They classically include two phases: sensitization and
challenge. Sensitization is traditionally performed by intraperitoneal and subcutaneous routes, and the challenges with allergens are performed through aerosol,
intranasal, or intratracheal instillation. Sensibilization increases IgE levels in the
circulation, but does not induce signs of inflammation or pulmonary remodeling.
IgE binds to receptors in eosinophils, mast cells, and basophils. When the challenge
occurs with the same allergen, the allergen provokes an antigenic-antibody reaction
that induces the degranulation of these cells. Degranulation releases inflammatory
mediators that initiate and propagate the process. Ovalbumin (OVA) is a widely
used allergen, because promote to an intense allergic lung inflammation. In addition, the most common species studied in the last two decades is mice, particularly
BALB/c [15, 52].

In an experimental model of acute asthma (BALB/c mice), Ang-(1-7) treatment resulted in inhibition of the OVA-induced increase in total cell counts, eosinophils, lymphocytes, and neutrophils. Ang-(1-7) also significantly reduced the
OVA-induced perivascular and peribronchial inflammation (Fig. 3). Moreover,
Ang-(1-7) attenuated OVA-induced increase in the phosphorylation of IκB-α and
ERK 1/2, suggesting that Ang-(1-7) could mediate an anti-inflammatory pathway
in allergic asthma [15]. In chronic allergic lung inflammation that administration
of Ang-(1-7) or a synthetic analog, AVE 0991 (Mas receptor agonist), decreased
inflammatory cell infiltrate in the peribronchial, perivascular, and alveolar regions
of the lung [52, 75]. Furthermore, Ang-(1-7) treatment decreased chemokines
(CCL2 and CCL5), cytokines (IL-4, IL-5 and GM-CSF), IgE, and two signaling
pathways associated with asthma, the ERK1/2, and possibly the JNK pathways.
Altogether, these results suggest that Ang-(1-7) treatment decreases chemokines
and cytokines essential for the initiation and maintenance of the inflammatory
process, as well as those important for the migration of eosinophils to the site of
injury and reduction of their apoptosis. These effects were associated with the
inhibition of ERK1/2 pathway [52].

It has been demonstrated that genetic Mas deficiency increased chronic allergic pulmonary inflammation. FVB/N mice with genetic deletion of the Mas
receptor subjected to a model of chronic allergic lung inflammation presented a
significant increase in the number of eosinophils in BALF and inflammatory cell
infiltrate in the lung [53]. Furthermore, there was an increase in ERK1/2 phosphorylation and proinflammatory cytokine (IL-13) and chemokines (CCL2/
MCP-1 and CCL5/RANTES) in the lungs of mice asthmatic with genetic deletion of the Mas receptor [53]. Thus, Mas receptor-induced effects are important

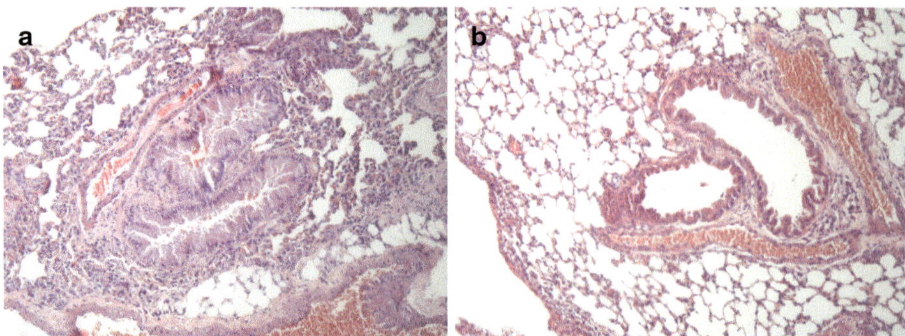

Fig. 3 Representative histological images of lung sections stained with H&E from OVA-sensitized and challenged mice and treated with Ang-(1-7). The OVA produced a pronounced increase in the density of inflammatory cell infiltrate around the airways and blood vessels and alveolar parenchyma (**a**). Treatment with Ang-(1-7) attenuated the inflammatory infiltrate in the peribronchial, perivascular and alveolar regions of the lung (**b**). In addition, OVA mice exhibited significantly greater thickening and inflammation of the alveolar wall and bronchial wall thickness. Ang-(1-7) presented reduced inflammation in the interalveolar space with normal appearance of the alveolar lumen [52]

counterbalancing mechanisms of the RAS for attenuating the inflammatory process in asthma. Moreover, impairment of the Ang-(1-7)/Mas receptor pathway may lead to the deterioration of the pathophysiology of asthma.

Defective apoptosis of eosinophils, the main leukocyte in the pathogenesis of asthma, and delay in its removal lead to lung damage and loss of pulmonary function due to failure in the resolution of inflammation [14, 19]. Recently, we demonstrated a novel action of Ang-(1-7), resolution of allergic lung inflammation [54]. Balb/c mice were sensitized and challenged with OVA and treated with Ang-(1-7) at the peak of the inflammatory process. Treatment with Ang-(1-7) reduced the accumulation of eosinophils in the lung by inducing apoptosis. In addition, Ang-(1-7) treatment reduced the phosphorylation of intracellular signaling pathway, associated with cytokine production and leukocyte survival, the NF-κB. Increase in apoptosis of leukocytes and their clearance by macrophages are essential events to promote resolution of inflammation [70]. Ang-(1-7) treatment increased the clearance of the apoptotic cells by macrophages [54]. This result added important criteria to establish Ang-(1-7) as an endogenous pro-resolutive mediator.

Unregulated or prolonged inflammatory responses in the lungs can lead to tissue damage, pulmonary remodeling, and consequently compromised lung function [33]. There is evidence that lung inflammation and remodelling in both asthmatic patients and in experimental models of asthma are not restricted to the airway and extend into the parenchyma and pulmonary vessels [33]. In addition to leukocytes migrating to the lung, structural cells, airway epithelium and smooth muscle cells secreting a variety of inflammatory mediators and extracellular matrix proteins, can participate in immunomodulation and airway remodelling in asthma [91]. In a model of chronically OVA-sensitized and challenged mice, there was an increase in the deposition of collagen fibres in the airway wall, an increase in the expression of collagen I and III in the lung, along with thickening of the alveolar wall and smooth muscle of the arterioles. In addition, the OVA-mice showed right ventricular hypertrophy, probably due to a functional and structural adaptation in response to chronic pulmonary artery pressure overload [52]

Lung sections from mice that were challenged intranasally with OVA (four consecutive days, with 20 μg OVA) showed severe perivascular and peribronchial fibrosis and marked goblet cell hyper/metaplasia suggesting airway remodeling. In contrast, lung sections from OVA-challenged mice treated with Ang-(1-7) decreased in the perivascular and peribronchial fibrosis and goblet cell hyper/metaplasia [15].

In other studies, mice were sensitized and challenged with OVA three times per week (for four weeks). OVA mice exhibited significantly greater thickening and inflammation of the alveolar wall. The epithelial thickness and collagen deposition in airways and lung parenchyma were increased. In addition, OVA induced an increase in the mRNA expression of collagen I and collagen III. However, OVA-sensitized and challenged animals treated with Ang-(1-7) or AVE0991 presented reduced inflammation in the interalveolar space with normal appearance of alveolar lumen and reduce epithelial thickness [52, 75]. Furthermore, OVA-sensitized and challenged mice treated with Ang-(1-7) presented a marked reduction in collagen deposition in airway walls, lung parenchyma, and mRNA expression of collagen I and III (Fig. 4. [52]).

The model of chronic asthma in mice with lack of the Mas receptor induces an intense degree of lung inflammation and remodeling in a mice strain (FVB/N) less sensitive to an experimental model of asthma. Indeed, FVB/N-WT (wild-type) mice presented an attenuated response to OVA challenge compared with the response observed in Balb/C mice subjected to the same protocol. However, deletion of Mas receptor induces worsening of the development of chronic allergic lung inflammation in mice. These data show that impairment of the Ang-(1-7)/Mas receptor pathway may lead to the deterioration of the pathophysiology of asthma [53].

A recent study, showed that treatment with Ang-(1-7) at the peak of the inflammatory process induced resolution of eosinophilic inflammation in an experimental model of asthma. Balb/c mice were sensitized and challenged with ovalbumin and treated with Ang-(1-7), 24 h after the last OVA challenge. The inclusion of Ang-(1-7) into an oligosaccharide HPβCD cavity protects the peptide during its passage through the gastrointestinal tract. Resolution of inflammation is an active process that allows cessation of inflammation and re-establishment of tissue homeostasis. Therefore, oral treatment with Ang-(1-7) promoted prevention of excessive trafficking of eosinophil to the lung, shutdown intracellular signaling molecules associated with cytokine production and eosinophil survival, apoptosis of recruited eosinophil, and promotion of clearance of apoptotic leukocytes, i.e., efferocytosis. These effects induced the return of pulmonary homeostasis through a decrease in extracellular matrix accumulation and a great reduction in collagen I and III genes expression in the lung [54].

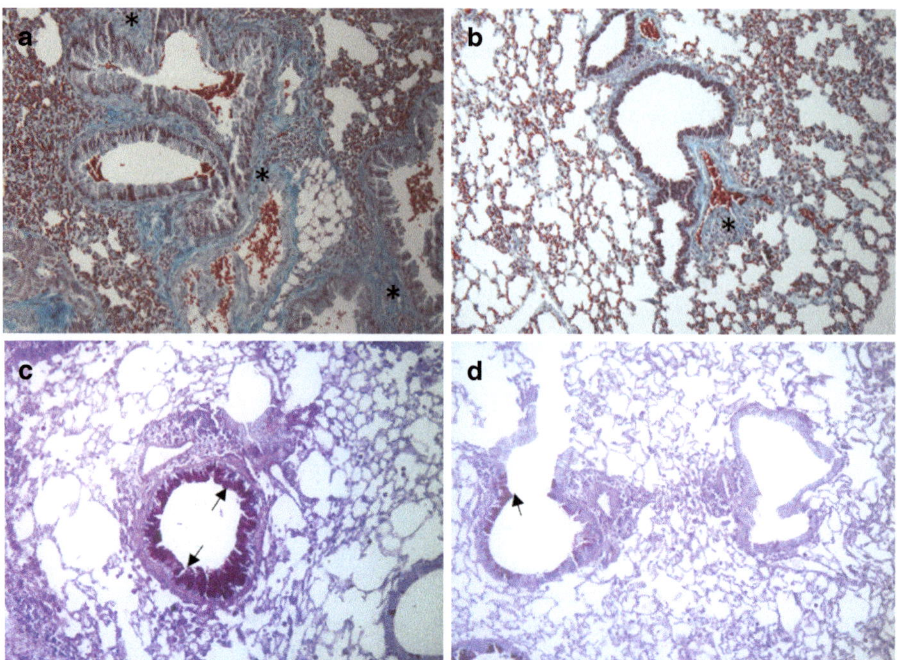

Fig. 4 Representative histological images of lung sections stained with Gomori's trichrome from OVA-sensitized and challenged mice (**a**) and treated with Ang-(1-7) (**b**). OVA-challenged mice presented marked peribronchial and perivascular fibrosis (**a**-asterisks), which was prevented by Ang-(1-7) treatment (**b**). (**c** and **d**) Representative histological images of lung sections stained with periodic acid schiff (PAS) from OVA-sensitized and challenged mice and treated with Ang-(1-7). The OVA-challenged mice presented increased mucus deposition in airways (**c**-arrows). In addition, the treatment with Ang-(1-7) decreased mucus deposition in airways in mice with allergic pulmonary inflammation (**d**) [52]

These data will accelerate the research efforts for the development of new Ang-(1-7)-based pharmacological strategies to control, prevent, and treat chronic inflammation-related diseases, such as asthma. Thus, the observation that Ang-(1-7) is effective through oral route can provide clinical benefits for treatment of allergic asthma, as it can be better tolerated than nebulization or than standard drugs, and it can act sistemically reducing overall inflammation and optimizing health of patients

Ang-(1-7) in Pulmonary Fibrosis

Pulmonary fibrosis (PF) is a fatal lung disease of unknown cause. The disease is characterized by progressive scarring of the lung tissue accompanied by fibroblast proliferation, the sudden onset of lung parenchyma, with thickening of the alveolar septa, hyperplasia of type II pneumocytes (PII), and myofibroblasts, causing narrowing of airways, all leading to a loss of lung function and decreased quality of life

[79]. The estimated prevalence of PF is around 30 cases per 100,000 people, reaching more than 100 individuals per 100,000 people aged 75 years or more [71]. Treatment for PF with anti-inflammatory, immunosuppressive, and antifibrotic agents has not shown promising results to abate the progression of the disease or to improve the quality of life [71]. Therefore, it becomes essential to better understand the disease pathophysiology and to identify novel therapeutic targets/agents for the treatment of PF.

Bleomycin (BLM), used and described method to cause PF in rodents, is a chemotherapeutic used in the treatment of several neoplasias. Challenged with BLM in intratracheal administration causes some lung lesions such as parenchyma inflammation, lesion of the alveolar epithelial cells with reactive hyperplasia, activation and fibroblast to myofibroblast differentiation and pulmonary fibrosis [64]. In addition, the presence of PH secondary to fibrotic lung diseases, called cor pulmonale, indicates poor prognosis with a compromised cardiac function.

Studies demonstrate that Ang II/AT1 receptor is required for the pathogenesis of experimental lung fibrosis. Ang II has a number of profibrotic effects on lung parenchymal, such as induction of growth factors for mesenchymal cells, extracellular matrix deposition, production of cytokines, and increased motility of lung fibroblasts [55, 58]. Recent evidence shows that the counterregulatory molecule Ang-(1-7), the product of the ACE2 acts as an antifibrotic pulmonary survival factor [87].

Shenoy et al. [82] showed that endotracheal instillation of bleomycin evoked a severe fibrotic response, characterized by the accumulation of interstitial lung collagen. In addition, increased lung mRNA levels of an important cytokine that plays a key role in fibrogenesis, the transforming growth factor-β (TGF-β), were also observed. Collagen deposition and TGF-β were significantly decreased by overexpression of ACE2 or Ang-(1-7). Furthermore, this study did detect pulmonary hypertension (PH) and right ventricular hypertrophy (RVH) after bleomycin administration. However, treatment with Ang-(1-7) prevented the development of both PH and RVH. The treatment with Ang-(1-7) or overexpression of ACE2 presented similar beneficial effects, possibly mediated via generation of Ang-(1-7). It is conceivable that the protective effects of ACE2 and Ang-(1-7) on the heart may be secondary to the reduction in the lung fibrosis.

Meng et al. [60] investigated whether the upregulation of the ACE2/Ang-(1-7)/Mas axis protects against BLM-induced pulmonary fibrosis by inhibiting the mitogen-activated protein kinase (MAPK)/NF-κB pathway. In this experimental protocol, male Wistar rats were submitted the PF by BLM and/or AngII. The results showed that Ang-(1-7) regulates the balance of the RAS from the ACE/AngII/AT1R axis toward the ACE2/Ang-(1-7)/Mas axis. The BLM-treated animals presented characteristic histological changes in lung tissue, including areas of inflammatory infiltration, thickening of the alveolar walls, increased interstitial collagen deposition, and a fibroblastic appearance. Chronic infusion with Ang-(1-7) resulted in a protective effect against lung fibrosis. Furthermore, treatment with Ang-(1-7) and lenti-ACE2 protect against BLM- or AngII-induced inflammation and extracellular matrix (ECM) accumulation by inhibiting the MAPK/NF-κB and NF-κB signaling pathways. These results suggest that treatment with Ang-(1-7) decreased activation of MAPKs pathways (ERK1/2, p38, JNK) and NF-κB, which are crucial for lung fibrogenesis [60].

Study in vitro shows that human fetal lung-1 cells were pretreated with compounds that block the activities of AT1 receptor, Mas (A-779), and MAPKs before exposure to Ang II or Ang-(1-7). The human fetal lung-1 cells were infected with lentivirus-mediated ACE2 before exposure to Ang II. Ang-(1-7) and lentivirus-mediated ACE2 inhibited the Ang II-induced MAPK/NF-κB pathway, thereby attenuating inflammation and α-collagen I production, which could be reversed by A-779, Mas receptor antagonist. Ang-(1-7) inhibited Ang II-induced lung fibroblast apoptotic resistance via inhibition of the MAPK/NF-κB pathway and activation of the mitochondrial apoptotic pathway [60].

It is well known that in addition to MAPK and NF-κB activation, the reactive oxygen species (ROS) generated by NADPH oxidase-4 (NOX4) initiates lung fibrosis. ROS generation plays a relevant role in lung fibrosis, and recent studies suggest that NADPH oxidases (NOXs) are key sources of ROS in the fibrotic lung [6]. The NOX4 in mediating fibroblast functions during the lung fibrosis process has been stressed. In addition, mice with genetic deletion of NOX4 are protected against BLM-induced pulmonary fibrosis [29]. Meng et al. [61] showed that NOX4-dependent ROS caused by the activation of the ACE/Ang II/AT1 receptor axis contributes to the development of AngII- or BLM-induced lung fibrosis by fibroblast migration and α-collagen I synthesis. Ang(1-7) and lentiACE2 treatment protect against BLM-induced pulmonary fibrosis by shifting the balance of the RAS toward the ACE2/Ang(1-7)/Mas axis and by inhibiting the generation of ROS. In addition, Ang-(1-7) and llentiACE2 protected against BLM- or Ang II-induced lung fibroblast migration and ECM accumulation by inhibiting the NOX4-derived. These results suggest that the ACE2/Ang(1-7)/Mas axis could be a novel pharmacological antioxidant target for lung fibrosis induced by Ang II-mediated ROS [61].

References

1. Aeffner F, Bolon B, Davis IC. Mouse models of acute respiratory distress syndrome: a review of analytical approaches, pathologic features, and common measurements. Toxicol Pathol. 2015;43(8):1074–92.
2. Ashbaugh DG, Bigelow DB, Petty TL, Levine BE. Acute respiratory distress in adults. Lancet. 1967;2:319–23.
3. Balakumar P, Jagadeesh G. A century old renin-angiotensin system still grows with endless possibilities: AT1 receptor signaling cascades in cardiovascular physiopathology. Cell Signal. 2014;26:2147–60.
4. Bernard GR, Artigas A, Brigham KL, Carlet J, Falke K, Hudson L, Lamy M, Legall JR, Morris A, Spragg R. The American-European consensus conference on ARDS. Definitions, mechanisms, relevant outcomes, and clinical trial coordination. Am J Respir Crit Care Med. 1994;149:818–24.
5. Brower RG, Matthay MA, Morris A, Schoenfeld D, Thompson BT, Wheeler A. The acute respiratory distress syndrome network: ventilation with lower tidal volumes as compared with traditional tidal volumes for acute lung injury and the acute respiratory distress syndrome. N Engl J Med. 2000;342:1301–8.

6. Bocchino M, Agnese S, Fagone E, Svegliati S, Grieco D, Vancheri C, Gabrielli A, Sanduzzi A, Avvedimento EV. Reactive oxygen species are required for maintenance and differentiation of primary lung fibroblasts in idiopathic pulmonary fibrosis. PLoS One. 2010;5:e14003.

7. Chen L, Xiao J, Li Y, Ma H. Ang-(1-7) might prevent the development of monocrotaline induced pulmonary arterial hypertension in rats. Eur Rev Med Pharmacol Sci. 2011;15(1):1–7.

8. Chen Q, Yang Y, Huang Y, Pan C, Liu L, Qiu H. Angiotensin-(1-7) attenuates lung fibrosis by way of Mas receptor in acute lung injury. J Surg Res. 2013;185:740–7.

9. Cheng ZJ, Vapaatalo H, Mervaala E. Angiotensin II and vascular inflammation. Med Sci Monit. 2005;11(6):RA194–205.

10. Dai HL, Dai H, Gong Y, Xiao Z, Guang X, Yin X. Decreased levels of serum angiotensin-(1-7) in patients with pulmonary arterial hypertension due to congenital heart disease. Int J Cardiol. 2014;176:1399–401.

11. Dai HL, Guo Y, Guang XF, Xiao ZC, Zhang M, Yin XL. The changes of serum angiotensin-converting enzyme 2 in patients with pulmonary arterial hypertension due to congenital heart disease. Cardiology. 2013;124:208–12.

12. de Man FS, Tu L, Handoko ML, Rain S, Ruiter G, François C, Schalij I, Dorfmüller P, Simonneau G, Fadel E, Perros F, Boonstra A, Postmus PE, van der Velden J, Vonk-Noordegraaf A, Humbert M, Eddahibi S, Guignabert C. Dysregulated renin-angiotensin-aldosterone system contributes to pulmonary arterial hypertension. Am J Respir Crit Care Med. 2012;186(8):780–9.

13. Duncan JC. Clinical relevance of local renin angiotensin systems. Front Endocrinol (Lausanne). 2014;5:113.

14. Duncan CJA, Lawrie A, Blaylock MG, Douglas JG, Walsh GM. Reduced eosinophil apoptosis in induced sputum correlates with asthma severity. Eur Respir J. 2003;22:484–90.

15. El-Hasmin, et al. Angiotensin-(1-7) inhibits allergic inflammation, via MAS1 receptor, through supression of ERK1/2 and NF-Kb-dependent pathways. Br J Pharmacol. 2012;166(6):1964–76.

16. Fan E, Brodie D, Slutsky AS. Acute respiratory distress syndrome: advances in diagnosis and treatment. JAMA. 2018;319(7):698–710.

17. Ferreira AJ, Shenoy V, Yamazato Y, Sriramula S, Francis J, Yuan L, Castellano RK, Ostrov DA, Oh SP, Katovich MJ, Raizada MK. Evidence for angiotensin-converting enzyme 2 as a therapeutic target for the prevention of pulmonary hypertension. Am J Respir Crit Care Med. 2009;179(11):1048–54.

18. Fletcher AN, Molteni A, Ponnapureddy R, Patel C, Pluym M, Poisner AM. The renin inhibitor aliskiren protects rat lungs from the histopathologic effects of fat embolism. J Trauma Acute Care Surg. 2017;82(2):338–44.

19. Fulkerson PC, Rothember ME. Targeting eosinophyls in allergy, inflammation and beyond. Nat Rev Drug Discov. 2013;12:117–29.

20. Galiè N, Humbert M, Vachiery JL, Gibbs S, Lang I, Torbicki A, Simonneau G, Peacock A, Vonk Noordegraaf A, Beghetti M, Ghofrani A, Gomez Sanchez MA, Hansmann G, Klepetko W, Lancellotti P, Matucci M, McDonagh T, Pierard LA, Trindade PT, Zompatori M, Hoeper M, ESC Scientific Document Group. 2015 ESC/ERS guidelines for the diagnosis and treatment of pulmonary hypertension: the joint task force for the diagnosis and treatment of pulmonary hypertension of the European Society of Cardiology (ESC) and the European Respiratory Society (ERS): endorsed by: Association for European Paediatric and Congenital Cardiology (AEPC), International Society for Heart and Lung Transplantation (ISHLT). Eur Heart J. 2016;37(1):67–119.

21. Gattinoni L, Pelosi P, Suter PM, Pedoto A, Vercesi P, Lissoni A. Acute respiratory distress syndrome caused by pulmonary and extrapulmonary disease. Different syndromes? Am J Respir Crit Care Med. 1998;158(1):3–11.

22. Georas SN, Guo J, De Fanis U, Casolaro V. T–helper cell type-2 regulation in allergic disease. Eur Respir J. 2005;26:1119–37.

23. Ghataorhe P, Rhodes CJ, Harbaum L, Attard M, Wharton J, Wilkins MR. Pulmonary arterial hypertension progress in understanding the disease and prioritizing strategies for drug development. J Intern Med. 2017;282(2):129–41.

24. Haga S, Tsuchiya H, Hirai T, Hamano T, Mimori A, Ishizaka Y. A novel ACE2 activator reduces monocrotaline-induced pulmonary hypertension by suppressing the JAK/STAT and TGF-β cascades with restored caveolin-1 expression. Exp Lung Res. 2015;41(1):21–31.
25. Hagiwara S, Iwasaka H, Matumoto S, Hidaka S, Noguchi T. Effects of an angiotensin-converting enzyme inhibitor on the inflammatory response in in vivo and in vitro models. Crit Care Med. 2009;37:626–33.
26. Hamming I, Cooper ME, Haagmans BL, Hooper NM, Korstanje R, Osterhaus AD, Timens W, Turner AJ, Navis G, van Goor H. The emerging role of ACE2 in physiology and disease. J Pathol. 2007;212(1):1–11.
27. Hamming I, Timens W, Bulthuis ML, Lely AT, Navis G, van Goor H. Tissue distribution of ACE2 protein, the functional receptor for SARS coronavirus. A first step in understanding SARS pathogenesis. J Pathol. 2004;203(2):631–7.
28. He H, Liu L, Chen Q, Liu A, Cai S, Yang Y, Lu X, Qiu H. Mesenchymal stem cells overexpressing angiotensin-converting enzyme 2 rescue lipopolysaccharide-induced lung injury. Cell Transplant. 2015;24(9):1699–715.
29. Hecker L, Vittal R, Jones T, Jagirdar R, Luckhardt TR, Horowitz JC, Pennathur S, Martinez FJ, Thannickal VJ. NADPH oxidase-4 mediates myofibroblast activation and fibrogenic responses to lung injury. Nat Med. 2009;15(9):1077–81.
30. Hemnes AR, Rathinasabapathy A, Austin EA, Brittain EL, Carrier EJ, Chen X, Fessel JP, Fike CD, Fong P, Fortune N, Gerszten RE, Johnson JA, Kaplowitz M, Newman JH, Piana R, Pugh ME, Rice TW, Robbins IM, Wheeler L, Yu C, Loyd JE, West J. A potential therapeutic role for angiotensin-converting enzyme 2 in human pulmonary arterial hypertension. Eur Respir J. 2018;51(6).
31. Hernández Prada JA, Ferreira AJ, Katovich MJ, Shenoy V, Qi Y, Santos RA, Castellano RK, Lampkins AJ, Gubala V, Ostrov DA, Raizada MK. Structure-based identification of small-molecule angiotensin-converting enzyme 2 activators as novel antihypertensive agents. Hypertension. 2008;51(5):1312–7.
32. Hoeper MM, Bogaard HJ, Condliffe R, Frantz R, Khanna D, Kurzyna M, Langleben D, Manes A, Satoh T, Torres F, Wilkins MR, Badesch DB. Definitions and diagnosis of pulmonary hypertension. J Am Coll Cardiol. 2013;62(25 Suppl):D42–50.
33. Holgate ST. Asthma: a simple concept but in reality a complex disease. Eur J Clin Investig. 2011;41:1339–52.
34. Imai Y, Kuba K, Rao S, Huan Y, Guo F, Guan B, Yang P, Sarao R, Wada T, Leong-Poi H, Crackower MA, Fukamizu A, Hui CC, Hein L, Uhlig S, Slutsky AS, Jiang C, Penninger JM. Angiotensin-converting enzyme 2 protects from severe acute lung failure. Nature. 2005;436(7047):112–6.
35. Jerng JS, Hsu YC, Wu HD, Pan HZ, Wang HC, Shun CT, Yu CJ, Yang PC. Role of the renin-angiotensin system in ventilator-induced lung injury: an in vivo study in a rat model. Thorax. 2007;62(6):527–35.
36. Jerng JS, Yu CJ, Wang HC, Chen KY, Cheng SL, Yang PC. Polymorphism of the angiotensin converting enzyme gene affects the outcome of acute respiratory distress syndrome. Crit Care Med. 2006;34:1001–6.
37. Jiang JS, Wang LF, Chou HC, Chen CM. Angiotensin-converting enzyme inhibitor captopril attenuates ventilator-induced lung injury in rats. J Appl Physiol. 2007;102(6):2098–103.
38. Kaparianos A, Argyropoulou E. Local renin-angiotensin II systems, angiotensin-converting enzyme and its homologue ACE2: their potential role in the pathogenesis of chronic obstructive pulmonary diseases, pulmonary hypertension and acute respiratory distress syndrome. Curr Med Chem. 2011;18(23):3506–15.
39. Khan A, Benthin C, Zeno B, Albertson TE, Boyd J, Christie JD, Hall R, Poirier G, Ronco JJ, Tidswell M, Hardes K, Powley WM, Wright TJ, Siederer SK, Fairman DA, Lipson DA, Bayliffe AI, Lazaar AL. A pilot clinical trial of recombinant human angiotensin-converting enzyme 2 in acute respiratory distress syndrome. Crit Care. 2017;21(1):234.
40. Klein N, Gembardt F, Supé S, Kaestle SM, Nickles H, Erfinanda L, Lei X, Yin J, Wang L, Mertens M, Szaszi K, Walther T, Kuebler WM. Angiotensin-(1-7) protects from experimental acute lung injury. Crit Care Med. 2013;41(11):e334–43.

41. Kuba K, Imai Y, Rao S, Gao H, Guo F, Guan B, Huan Y, Yang P, Zhang Y, Deng W, Bao L, Zhang B, Liu G, Wang Z, Chappell M, Liu Y, Zheng D, Leibbrandt A, Wada T, Slutsky AS, Liu D, Qin C, Jiang C, Penninger JM. A crucial role of angiotensin converting enzyme 2 (ACE2) in SARS coronavirus-induced lung injury. Nat Med. 2005;11:875–9.
42. Latus H, Delhaas T, Schranz D, Apitz C. Treatment of pulmonary arterial hypertension in children. Nat Rev Cardiol. 2015;12(4):244–54.
43. Lee C, Mitsialis SA, Aslam M, Vitali SH, Vergadi E, Konstantinou G, Sdrimas K, Fernandez-Gonzalez A, Kourembanas S. Exosomes mediate the cytoprotective action of mesenchymal stromal cells on hypoxia-induced pulmonary hypertension. Circulation. 2012;126(22):2601–011.
44. Li G, Liu Y, Zhu Y, Liu A, Xu Y, Li X, Li Z, Su J, Sun L. ACE2 activation confers endothelial protection and attenuates neointimal lesions in prevention of severe pulmonary arterial hypertension in rats. Lung. 2013;191(4):327–36.
45. Li G, Xu YL, Ling F, Liu AJ, Wang D, Wang Q, Liu YL. Angiotensin-converting enzyme 2 activation protects against pulmonary arterial hypertension through improving early endothelial function and mediating cytokines levels. Chin Med J. 2012;125(8):1381–8.
46. Li Y, Cao Y, Zeng Z, Liang M, Xue Y, Xi C, Zhou M, Jiang W. Angiotensin-converting enzyme 2/ angiotensin-(1-7)/ Mas axis prevents lipopolysaccharide-induced apoptosis of pulmonary microvascular endothelial cells by inhibiting JNK/NF-κB pathways. Sci Rep. 2015;5:8209.
47. Li Y, Zeng Z, Cao Y, Liu Y, Ping F, Liang M, Xue Y, Xi C, Zhou M, Jiang W. Angiotensin-converting enzyme 2 prevents lipopolysaccharide-induced rat acute lung injury via suppressing the ERK1/2 and NF-κB signaling pathways. Sci Rep. 2016;6:27911.
48. Li W, Moore MJ, Vasilieva N, Sui J, Wong SK, Berne MA, Somasundaran M, Sullivan JL, Luzuriaga K, Greenough TC, Choe H, Farzan M. Angiotensin-converting enzyme 2 is a functional receptor for the SARS coronavirus. Nature. 2003;426:450–4.
49. Ling Y, Johnson MK, Kiely DG, et al. Changing demographics, epidemiology, and survival of incident pulmonary arterial hypertension: results from the pulmonary hypertension registry of the United Kingdom and Ireland. Am J Respir Crit Care Med. 2012;186:790–6.
50. Liu Z, Liu J, Xiao M, Wang J, Yao F, Zeng W, Yu L, Guan Y, Wei W, Peng Z, Zhu K, Wang J, Yang Z, Zhong J, Chen J. Mesenchymal stem cell-derived microvesicles alleviate pulmonary arterial hypertension by regulating renin-angiotensin system. J Am Soc Hypertens. 2018;12(6):470–8.
51. Loftus PA, Wise SK. Epidemiology and economic burden of asthma. Int Forum Allergy Rhinol. 2015;5(Suppl 1):S7–10.
52. Magalhães GS, Rodrigues-Machado MG, Motta-Santos D, Silva AR, Caliari MV, Prata LO, Abreu SC, Rocco PR, Barcelos LS, Santos RA, Campagnole-Santos MJ. Angiotensin-(1-7) attenuates airway remodelling and hyperresponsiveness in a model of chronic allergic lung inflammation. Br J Pharmacol. 2015;172(9):2330–23342.
53. Magalhaes GS, Rodrigues-Machado MG, Motta-Santos D, Alenina N, Bader M, Santos RA, Campagnole-Santos MJ. Chronic allergic pulmonary inflammation is aggravated in angiotensin-(1-7) Mas receptor knockout mice. Am J Physiol Lung Cell Mol Physiol. 2016;311(6):L1141–8.
54. Magalhaes GS, Barroso LC, Reis AC, Rodrigues-Machado MG, Gregório JF, Motta-Santos D, Oliveira AC, Perez DA, Barcelos LS, Teixeira MM, Santos RAS, Pinho V, Campagnole-Santos MJ. Angiotensin-(1-7) promotes resolution of eosinophilic inflammation in an experimental model of asthma. Front Immunol. 2018;9:58.
55. Marshall RP, Gohlke P, Chambers RC, Howell DC, Bottoms SE, Unger T, McAnulty RJ, Laurent GJ. Angiotensin II and the fibroproliferative response to acute lung injury. Am J Physiol Lung Cell Mol Physiol. 2004;286:L156–64.
56. Marshall RP. The pulmonary renin-angiotensin system. Curr Pharm Des. 2003;9(9):715–22.
57. Marshall RP, Webb S, Bellingan GJ, Montgomery HE, Chaudhari B, RJ MA, Humphries SE, Hill MR, Laurent GJ. Angiotensin converting enzyme insertion/deletion polymorphism is associated with susceptibility and outcome in acute respiratory distress syndrome. Am J Respir Crit Care Med. 2002;166(5):646–50.

58. Marshall RP, McAnulty RJ, Laurent GJ. Angiotensin II is mitogenic for human lung fibroblasts via activation of the type 1 receptor. Am J Respir Crit Care Med. 2000;161:1999–2004.
59. Matute-Bello G, Downey G, Moore BB, Groshong SD, Matthay MA, Slutsky AS, Kuebler WM. An official American thoracic society workshop report: features and measurements of experimental acute lung injury in animals. Am J Respir Cell Mol Biol. 2011;44:725–38.
60. Meng Y, Yu C-H, Li W, Li T, Luo W, Huang S, Wu P-S, Cai S-X, Li X. Angiotensin-Converting Enzyme 2/Angiotensin-(1-7)/Mas Axis Protects against Lung Fibrosis by Inhibiting the MAPK/NF-κB Pathway. Am J Respir Cell Mol Biol. 2014;50(4):723–36.
61. Meng YTL, Zhou G-s, Chen Y, Yu C-H, Pang M-X, Li W, Li Y, Zhang W-Y, Li X. The Angiotensin-Converting Enzyme 2/Angiotensin (1-7)/Mas Axis Protects Against Lung Fibroblast Migration and Lung Fibrosis by Inhibiting the NOX4-Derived ROS-Mediated RhoA/Rho Kinase Pathway. Antioxid Redox Signal. 2015;20(22(3)):241–58.
62. McGoon MD, Benza RL, Escribano-Subias P, et al. Pulmonary arterial hypertension: epidemiology and registries. J Am Coll Cardiol. 2013;62:D51–9.
63. Mclff TE, Poisner AM, Herndon B, Lankachandra K, Molteni A, Adler F. Mitigating effects of captopril and losartan on lung histopathology in a rat model of fat embolism. J Trauma. 2011;70(5):1186–91.
64. Moeller A, Ask K, Warburton D, Gauldie J, Kolb M. The bleomycin animal model: a useful tool to investigate treatment options for idiopathic pulmonary fibrosis? Int J Biochem Cell Biol. 2008;40:362–82.
65. Morrell NW, Morris KG, Stenmark KR. Role of angiotensin-converting enzyme and angiotensin II in development of hypoxic pulmonary hypertension. Am J Phys. 1995;269(4 Pt 2):H1186–94.
66. Murray JF, Matthay MA, Luce JM, Flick MR. An expanded definition of the adult respiratory distress syndrome. Am Rev Respir Dis. 1988;138:720–3.
67. Myou S, Fujimura M, Kamio Y, Ishiura Y, Kurashima K, Tachibana H, et al. Effect of losartan, a type 1 angiotensin II receptor antagonist, on bronchial hyperresponsiveness to methacholine in patients with bronchial asthma. Am J Respir Crit Care Med. 2000;162:40–4.
68. Myou S, Fujimura M, Kamio Y, Kita T, Watanabe K, Ishiura Y, et al. Effect of candesartan, a type 1 angiotensin II receptor antagonist, on bronchial hyper-responsiveness to methacholine in patients with bronchial asthma. Br J Clin Pharmacol. 2002;54:622–6.
69. Orte C, Polak JM, Haworth SG, Yacoub MH, Morrell NW. Expression of pulmonary vascular angiotensin-converting enzyme in primary and secondary plexiform pulmonary hypertension. J Pathol. 2000;192(3):379–84.
70. Perez DA, Vago JP, Athayde RM, Reis AC, Teixeira MM, Sousa LP, et al. Switching off key signaling survival molecules to switch on the resolution of inflammation. Mediat Inflamm. 2014;2014:829851.
71. Prata LO, et al. ACE2 activator associated with physical exercise potentiates the reduction of pulmonary fibrosis. Exp Biol med. 2017;242(1):8–21.
72. Raiden S, Nahmod K, Nahmod V, Semeniuk G, Pereira Y, Alvarez C, Giordano M, Geffner JR. Nonpeptide antagonists of AT1 receptor for angiotensin II delay the onset of acute respiratory distress syndrome. J Pharmacol Exp Ther. 2002;303:45–51.
73. Ranieri VM, Rubenfeld GD, Thompson BT, Ferguson ND, Caldwell E, Fan E, Camporota L, Slutsky AS. Acute respiratory distress syndrome: the Berlin definition. JAMA. 2012;307:2526–33.
74. Rincon M, Irvin CG. Role of Il-6 in asthma and other inflammatory pulmonary diseases. Int J Biol Sci. 2012;8:1281–90.
75. Rodrigues-Machado MG, et al. AVE 0991, a non-pepitide mimic of angiotensin-(1-) effects, atenuattes pulmonary remodeling in a model of chronic asthma. Br J Pharmacol. 2013;170(4):835–46.
76. Santos RA, Simoes e Silva AC, Maric C, Silva DM, Machado RP, de Buhr I, Heringer-Walther S, Pinheiro SV, Lopes MT, Bader M, Mendes EP, Lemos VS, Campagnole-Santos

MJ, Schultheiss HP, Speth R, Walther T. Angiotensin-(1-7) is an endogenous ligand for the G protein-coupled receptor Mas. Proc Natl Acad Sci U S A. 2003;100(14):8258–63.

77. Santos RAS, Sampaio WO, Alzamora AC, Motta-Santos D, Alenina N, Bader M, Campagnole-Santos MJ. The ACE2/Angiotensin-(1-7)/MAS axis of renin-angiotensin system: focus on angiotensin-(1-7). Physiol Rev. 2018;98(1):505–3.

78. Schannwell CM, Steiner S, Strauer BE. Diagnostics in pulmonary hypertension. J Physiol Pharmacol. 2007;58 Suppl 5(Pt 2):591–602.

79. Selman M, King TE, Pardo A, American Thoracic Society; European Respiratory Society; America College of Chest Physicians. Idiopathic pulmonary fibrosis: prevailing and evolving hypothesis about its pathogenesis and implications for therapy. Ann Intem Med. 2001;134(2):136–51.

80. Shaver CM, Bastarache JA. Clinical and biological heterogeneity in acute respiratory distress syndrome: direct versus indirect lung injury. Clin Chest Med. 2014;35(4):639–53.

81. Shen L, Mo H, Cai L, Kong T, Zheng W, Ye J, Qi J, Xiao Z. Losartan prevents sepsis-induced acute lung injury and decreases activation of nuclear factor kappa B and mitogen-activated protein kinases. Shock. 2009;31(5):500–6.

82. Shenoy V, Ferreira AJ, Qi Y, Fraga-Silva RA, Díez-Freire C, Dooies A, Jun JY, Sriramula S, Mariappan N, Pourang D, Venugopal CS, Francis J, Reudelhuber T, Santos RA, Patel JM, Raizada MK, Katovich MJ. The angiotensin-converting enzyme 2/angiogenesis-(1-7)/Mas axis confers cardiopulmonary protection against lung fibrosis and pulmonary hypertension. Am J Respir Crit Care Med. 2010;182(8):1065–72.

83. Shelhamer JH, Levine SJ, Wu T, Jacoby DB, Kaliner MA, Rennard SL. Airway inflammation. Ann Intern Med. 1995;123:288–304.

84. Shenoy V, Gjymishka A, Jarajapu YP, Qi Y, Afzal A, Rigatto K, Ferreira AJ, Fraga-Silva RA, Kearns P, Douglas JY, Agarwal D, Mubarak KK, Bradford C, Kennedy WR, Jun JY, Rathinasabapathy A, Bruce E, Gupta D, Cardounel AJ, Mocco J, Patel JM, Francis J, Grant MB, Katovich MJ, Raizada MK. Diminazene attenuates pulmonary hypertension and improves angiogenic progenitor cell functions in experimental models. Am J Respir Crit Care Med. 2013;187(6):648–57.

85. Shenoy V, Kwon KC, Rathinasabapathy A, Lin S, Jin G, Song C, Shil P, Nair A, Qi Y, Li Q, Francis J, Katovich MJ, Daniell H, Raizada MK. Oral delivery of Angiotensin-converting enzyme 2 and Angiotensin-(1-7) bioencapsulated in plant cells attenuates pulmonary hypertension. Hypertension. 2014;64(6):1248–59. Erratum in: Hypertension. 2015;65(3):e8.

86. Shenoy V, Qi Y, Katovich MJ, Raizada MK. ACE2 a promising therapeutic target for pulmonary hypertension. Curr Opin Pharmacol. 2011;11(2):150–5.

87. Simoes e Silva AC, Silveira KD, Ferreira AJ, Teixeira MM. ACE2, angiotensin-(1-7) and Mas receptor axis in inflammation and fibrosis. Br J Pharmacol. 2013;169:477–92.

88. Supé S, Kohse F, Gembardt F, Kuebler WM, Walther T, et al. Therapeutic time window for angiotensin-(1-7) in acute lung injury. Br J Pharmacol. 2016;173(10):1618–28.

89. Taichman DB, Ornelas J, Chung L, Klinger JR, Lewis S, Mandel J, Palevsky HI, Rich S, Sood N, Rosenzweig EB, Trow TK, Yung R, Elliott CG, Badesch DB. Pharmacologic therapy for pulmonary arterial hypertension in adults: CHEST guideline and expert panel report. Chest. 2014;146(2):449–75.

90. Tan WSD, Liao W, Zhou S, Mei D, Wong WF. Targeting the renin-angiotensin system as novel therapeutic strategy for pulmonary diseases. Curr Opin Pharmacol. 2018;40:9–17.

91. Van Wetering S, Zuyderduyn S, Ninaber DK, van Sterkenburg MA, Rabe KF, Hiemstra PS. Epithelial differentiation is a determinant in the production of eotaxin-2 and -3 by bronchial epithelial cells in response to IL-4 and IL-13. Mol Immunol. 2007;44:803–11.

92. Wiener RS, Cao YX, Hinds A, Ramirez MI, Williams MC. Angiotensin converting enzyme 2 is primarily epithelial and is developmentally regulated in the mouse lung. J Cell Biochem. 2007;101(5):1278–91.

93. Wösten-van Asperen RM, Lutter R, Specht PA, Moll GN, van Woensel JB, van der Loos CM, van Goor H, Kamilic J, Florquin S, Bos AP. Acute respiratory distress syndrome leads to

reduced ratio of ACE/ACE2 activities and is prevented by angiotensin-(1-7) or an angiotensin II receptor antagonist. J Pathol. 2011;225:618–27.

94. Yamazato Y, Ferreira AJ, Hong KH, Sriramula S, Francis J, Yamazato M, Yuan L, Bradford CN, Shenoy V, Oh SP, Katovich MJ, Raizada MK. Prevention of pulmonary hypertension by angiotensin-converting enzyme 2 gene transfer. Hypertension. 2009;54(2):365–71.

95. Zambelli V, Bellani G, Borsa R, Pozzi F, Grassi A, Scanziani M, Castiglioni V, Masson S, Decio A, Laffey JG, Latini R, Pesenti A. Angiotensin-(1-7) improves oxygenation, while reducing cellular infiltrate and fibrosis in experimental acute respiratory distress syndrome. Inten Care Med Exp. 2015;3(1):44.

96. Zhang J, Dong J, Martin M, He M, Gongol B, Marin TL, Chen L, Shi X, Yin Y, Shang F, Wu Y, Huang HY, Zhang J, Zhang Y, Kang J, Moya EA, Huang HD, Powell FL, Chen Z, Thistlethwaite PA, Yuan ZY, Shyy JY. AMPK phosphorylation of ACE2 in endothelium mitigates pulmonary hypertension. Am J Respir Crit Care Med. 2018;198(4):509–20.

Angiotensin-(1-7): Role in the Endocrine System

Sérgio Henrique Sousa Santos

Abbreviations

ACE	Angiotensin-converting enzyme
AGT	Angiotensinogen
AKT	Protein kinase B
Ang II	Angiotensin II
Ang-(1-5)	Angiotensin-(1-5)
Ang-(1-7)	Angiotensin-(1-7)
ARB	Type 1 angiotensin receptor blockers
AS160	TBC1 domain family member 4
AT_1R	Type 1 AngII receptor
AT_2R	Type 2 AngII receptor
Bcl-2	B-cell lymphoma 2
eCG	Gonadotropic hormone
ERK	Extracellular signal–regulated kinases
GLUT4	Glucose transporter 4
GSK3B	Glycogen synthase kinase 3 beta
IR	Insulin receptor
IRS	Insulin receptor substrate
JNK	JUN N-Terminal Kinase
LH	Luteinizing hormone
MAPK	Mitogen activated protein kinase
MasR	Mas receptor

S. H. S. Santos (✉)
Institute of Agricultural Sciences, Food Engineering College, Universidade Federal de Minas Gerais (UFMG), Montes Claros, MG, Brazil

Laboratory of Health Science, Postgraduate Program in Health Sciences, Universidade Estadual de Montes Claros (UNIMONTES), Montes Claros, MG, Brazil
e-mail: sergiosousas@ufmg.br

mTOR   Mammalian target of rapamycin
NF-KB   Nuclear factor kappa B
NOS   Nitric oxide synthase
p38   P38 mitogen-activated protein kinase
PKC   Protein kinase C
RAS   Renin-angiotensin system
ROS   Reactive oxygen species

Introduction

The renin-angiotensin system (RAS) is one of the most important endocrine systems in maintaining body homeostasis by integrating the systemic effects and also locally adjusting organs functions with important role in modulating energy balance [1–3]. Body catabolism and anabolism are regulated by many endocrine components present in different tissues and also in the circulation. In the last decade, the renin-angiotensin system gained an important visibility with the characterization of Angiotensin-(1-7) [Ang-(1-7)] metabolic and endocrine effects [3].

The RAS starts with angiotensinogen (AGT) expression in different tissues and cell types. Nevertheless, the hepatic cells are regarded as the primary circulating source of AGT in normal homeostasis [4]. Renin enzyme coverts AGT into Ang I, which is rapidly hydrolyzed by angiotensin-converting enzyme (ACE) to the octapeptide, Ang II. Ang II is the oldest and most studied RAS peptide, with potent vasoconstrictive and proliferative effects and broadly elevated in different metabolic diseases such as diabetes, obesity, metabolic syndrome, and liver steatosis [3]. Several other angiotensin peptides are formed from AGT; however, the main one and better described is the Ang-(1-7), that largely opposes Ang II actions thought the Mas receptor [5]. The Ang-(1-7) peptide is also mainly produced as a product from the Ang II degradation by ACE-homologue enzyme (ACE2) [4].

The Ang-(1-7) has been well established as a key hormonal peptide in the last two decades with numerous research findings of new biological effects and therapeutic strategies as described [6]. Currently, it is also well established that ACE2/Ang-(1-7)/ Mas axis is implicated in counteracting different deleterious effects produced by high-level activation of the ACE/Ang II/AT1 axis [7]. Ang-(1-7) and Ang II have been implicated in regulating several hormones and metabolic tissues. Considering the pancreas and energy metabolic endogenous components, the Ang-(1-7) demonstrated to be crucial for regulating insulin resistance, glucagon effects, and energy storage.

Preceding results showed that elevated levels of Ang II by acting through type 1 (AT1) receptors, produce dyslipidemia, glucose intolerance, endothelial dysfunction, atherosclerosis, and several other metabolic disturbances which lead to cardiovascular damage and metabolic disorders [8]. Indeed, Ang-(1-7) broadly opposes these metabolic alterations by activating, mainly, the G-coupled Mas receptor [6]. Besides, Ang-(1-7) interacts with and regulates several local and circulating endocrine effects such as gonadal hormones and adipokines secretion. In recent years, an important cross-modulation of sirtuins (enzymes that regulates metabolism) [9] was demonstrated with Ang-(1-7), improving together the type 2 diabetes. Recently, Muñoz et al. and Santos et al. showed that Mas receptor is an essential component of the insulin receptor-signaling pathway [10, 11]. Altogether, these data indicate that improving Ang-(1-7) signaling and reducing Ang II effects appears to be an important way to treat metabolic abnormalities and open new perspectives for understanding RAS metabolic and hormonal role.

Ang-(1-7) and Pancreatic Hormones

The ACE2/Ang-(1-7)/Mas axis improves pancreatic hormones activities, especially in disease conditions, such as type 2 diabetes. There are several pathways associated with the Ang-(1-7) action mechanisms on regulating insulin and glucagon. A possible mechanism to oppose Ang II effects includes the Ang-(1-7)s ability to increase the phosphorylation of insulin-associated proteins (e.g., Akt, AS160, and GSK-3B). In addition, Ang-(1-7) seems to reduce IRS-1 serine phosphorylation and mammalian target of rapamycin (mTOR)/ JUN N-Terminal Kinase (JNK) activation, while activating ERK1/2 and protein kinase C (PKC), thus improving insulin responsiveness [12]. Ang-(1-7) is also shown to act neurologically in energy metabolism improving insulin sensitivity. It was reported that Ang-(1-7) may potentiate insulin's anorectic actions, regulating energy balance and thus interfering in the pathophysiology of metabolic diseases [13].

Additionally, Ang-(1-7) is shown to cause arterial dilation and microvascular recruitment, which increase muscle insulin delivery and glucose disposal, thus improving whole-body insulin resistance. Insulin delivery to peripheral tissues constitutes an important part of this hormone's actions [14]. Moreover, a study performed using ACE2 knock out (KO) animals, which presents decreased Ang-(1-7) production, evidenced reduced first-phase insulin secretion in response to glucose infusion [15] while ACE2 overexpression was shown to improve first-phase insulin secretion [16]. Additionally, the ACE2/Ang-(1-7)/Mas axis seems to exert a potential beneficial effect in the pancreas via β-cell function improvement [16, 17] while Ang-(1-7) increases insulin metabolism in the pancreas by reducing β-cell oxidative stress and apoptosis [18–20] (possibly modifying the B-cell lymphoma 2 (Bcl-2) family and augmenting pancreatic microcirculation). These results point to the Ang-(1-7) effects ameliorating insulin secretion in addition to previous reported effects on insulin sensitivity.

The Ang-(1-7) effects on glucagon activity are also reported, but remain scarce and controversial in the literature. It was shown that Ang-(1-7) main receptor (Mas receptor) ablation was correlated with increased α-cells/glucagon levels and decreased β-cells, associated with increased fasting glucose levels [21]. The ACE2/Ang-(1-7)/Mas axis is also reported to regulate pancreatic development in mouse with decreased β-cells and increased α-cells, along with observations of impaired insulin secretion and reduced glucose tolerance after prenatal treatment with Mas receptor antagonist [22]. It is discussed that Ang-(1-7) may modulate

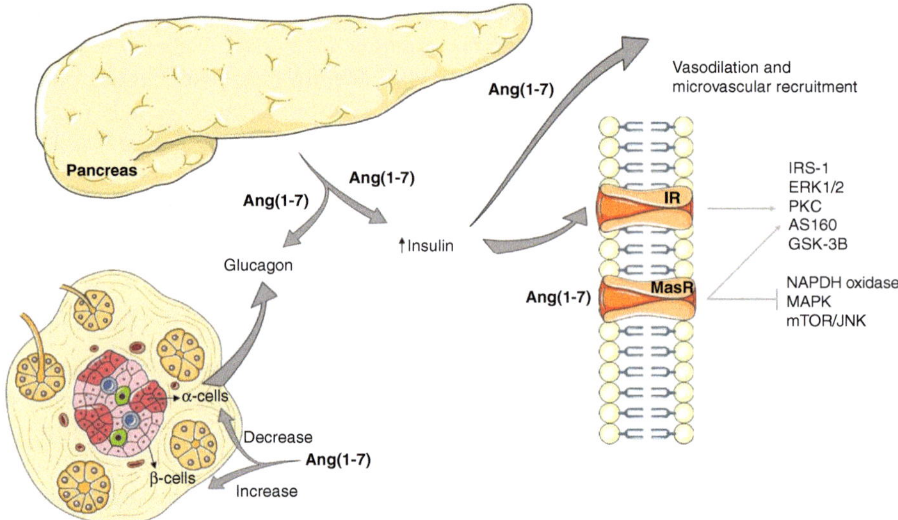

Fig. 1 Ang-(1-7) increases insulin secretion. 2 – Ang-(1-7) induces activation of insulin-associated proteins (IRS-1, ERK1/2, PKC, AS160 and GSK-3B). 3 – Ang-(1-7) decreases NAPDH oxidase, MAPK, and mTOR/JNK activity. 4 – Ang-(1-7) increases vasodilation, which leads to improved insulin delivery and glucose disposal. 5 – Ang-(1-7) decreases α-cells and increases β-cells percentage

endocrine cell differentiation via reactive oxygen species (ROS) [23], possibly via mitogen-activated protein kinase (MAPK) signaling pathway and reduced inducible nitric oxide synthase (iNOS) expression [20]. These findings are important to delineate potential therapeutic alternatives to restore pancreatic islets function and thus recover glucose metabolism [22]. However, the literature also reports that Ang-(1-7) is not able to influence glucagon release by the α-cells nor systemic glucagon levels [24, 25], thus requiring further studies to elucidate possible interactions (Fig. 1).

Angiotensin-(1-7) and Liver

The liver is the main source of circulating angiotensinogen (AGT) and several other important hormones in healthy organisms [2, 3]. Currently, it is well described that imbalance in RAS components produces increased AGT secretion, which is directly associated with augmented Ang II local and systemic levels and several diseases such as insulin resistance, liver steatosis (fatty liver disease), hypertension, and obesity. As previously described, increased Ang II production leads to profibrotic, pro-inflammatory and prooxidant effects modifying several steps of cellular signaling, specially acting through AT1 receptor, interfering in liver diseases such as steatosis, hepatitis and cirrhosis [26, 27]. On the other hand, ACE2/Ang-(1-7)/Mas axis also plays a pivotal protective role in liver disorders [28].

The Ang-(1-7) produces several beneficial effects in liver conditions by reducing inflammation and liver fibrosis [29–31] and regulating liver metabolic function [25]. Using animal models of obesity, Ang-(1-7) oral administration prevented hepatic inflammation by inhibiting resistin/toll-like receptor 4 (TLR4)/MAPK/NF-κB signaling pathway [32]. Bilman et al, showed that an increased circulating Ang-(1-7) is associated with decreased gluconeogenesis in rats [25]. The beneficial effects of augmented ACE2/Ang-(1-7) axis activity is corroborated by data indicating that relative ACE2 deficiency leads to Ang II impaired degradation and accumulation contributing to augmented fibrosis and inflammatory process [33, 34].

Liver steatosis, characterized by excessive lipid accumulation in hepatocytes, is one of the most common hepatic diseases currently, and is regularly associated with obesity, type 2 diabetes mellitus/insulin resistance and dyslipidemia [28, 35, 36]. Rats overexpressing Ang-(1-7) for a lifetime presented decreased concentration of liver triacylglycerol which may result from increased activity of cytosolic lipases and decreased fatty acid uptake in adipose tissue [37]. Supporting this data, ACE2 deletion in mice [reduced Ang-(1-7)] aggravates liver steatosis, which is correlated with increased lipogenic genes expression and decreased fatty acid oxidation-related genes expression in the liver. On the contrary, ACE2 overexpression improved fatty liver disease in db/db mice by enhancing Akt signaling pathway [38]. The Ang-(1-7) pharmacological potential was improved by the production of a new oral formulation characterized by a protected Ang-(1-7) molecule included in acyclic-oligosaccharides (cyclodextrin), which permits an efficient oral treatment, which already showed ability to improve liver steatosis by reducing fat accumulation, lobular inflammation, and fibrosis in rats [39].

In conclusion, the increased ACE2/Ang-(1-7)/Mas activation is able to improve hepatic disorders and reduce Ang II/AT1 deleterious effects. Thus, new approaches aiming to increase Ang-(1-7)/Mas effects could be a novel tool for the treatment and prevention of liver diseases (Fig. 2).

Ang-(1-7) and Adipose Tissue

The adipose tissue is an important endocrine and paracrine organ. The adipocytes exert a significant role in the synthesis and secretion of proinflammatory factors such as cytokines, chemotactic agents, chemoattractants, acute phase proteins, eicosanoids, prostaglandins, hormones, and anti-inflammatory effectors, all called adipokines [40]. Adiponectin and leptin are adipokines and hormones with local and systemic biological effects modulating insulin sensitivity and metabolic syndrome development [41]. The adiponectin is mainly involved in the glycemic control, and high levels of this hormone are associated with reduced insulin resistance, type 2 diabetes, and obesity [42–44]. In addition, leptin showed to be important on energy homeostasis and regulating satiety [45, 46].

As well as adiponectin and leptin, the renin-angiotensin system was also described as an important regulator of the metabolic homeostasis. Some authors suggest a possible interaction among the ACE2/Ang-(1-7)/Mas axis, adiponectin

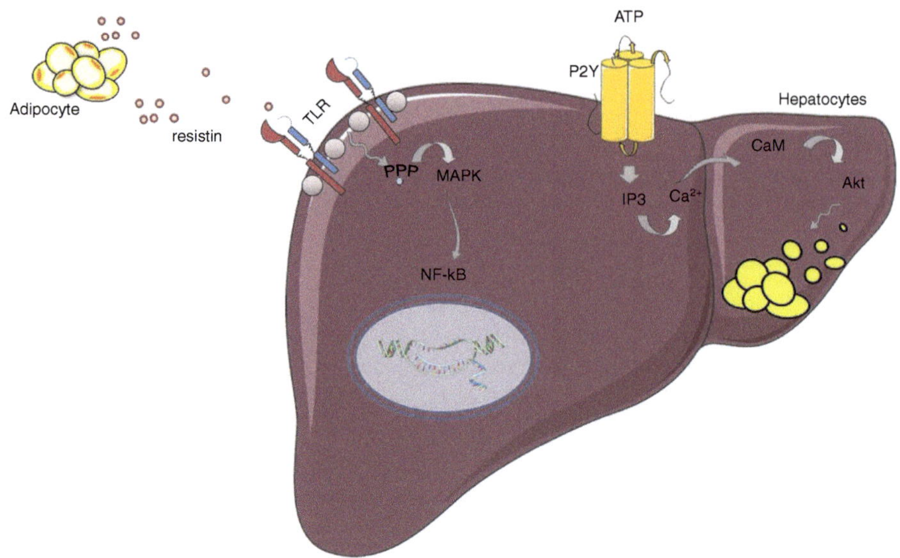

Fig. 2 Ang-(1-7) beneficial effects in hepatic conditions. 1 – Ang-(1-7) reduces hepatic inflammation and fibrosis via resistin/TLR4/MAPK/NF-κB signaling pathway. 2 – Increased circulating Ang- (1-7) is associated with decreased gluconeogenesis. 3 – ACE2 overexpression improves Fatty liver disease via AKT signaling pathway

and leptin [47–50]. Mas receptor knockout mice (Mas−/−) present decreased metabolic efficiency, insulin resistance, dyslipidemia, hyperleptinemia and reduced adiponectin and Glucose transporter type 4 (GLUT4) expression in the adipose tissue [48]. Similarly, it was demonstrated that transgenic rats with increased Ang-(1-7) levels presented increased glucose tolerance, improved insulin sensitivity, and increased adiponectin levels [49].

Recently, the Ang-(1-7) effects on oxidative stress were reported. The reactive oxygen species activity via nicotinamide adenine dinucleotide phosphate (NAPDH) oxidase attenuation was reduced in db/db mice (Leptin receptor deficient mice) with nephropathy after Ang-(1-7) administration [51]. The Ang-(1-7) protective effect against oxidative stress was associated with increased adiponectin levels [47].

Subsequent studies suggested that Ang-(1-7) reduces cardiac dysfunction associated with obesity mainly via attenuation of adiponectin expression and epicardial fat inflammation [52]. ACE2 knockout obese mice presented increased epicardial adipose tissue inflammation, worsened myocardial insulin resistance and consequently impairing cardiac metabolism via increased lipotoxicity and oxidative stress. Ang-(1-7) administration attenuated the deleterious alterations observed in ACE2 knockout mice, especially via maintaining cardiac function [53]. The Ang-(1-7) anti-inflammatory effects were associated with increased myocardial adiponectin levels [52].

The interaction between RAS and leptin levels was described in Mas-KO mice [48] and mice fed a high-sucrose diet [50]. Both models presented increased fat mass, insulin, leptin plasma levels and glucose intolerance. Corroborating these data, transgenic mice with increased Ang-(1-7) levels presented reduced leptin serum levels [54]. The Ang-(1-7)/Mas axis was also recently described as an important modulator of the renal function, exerting renoprotective effects, by decreasing leptin levels synthesized by the perirenal adipose tissue [55].

Ang-(1-7) treatment improved cardiac hypertrophy, myocardial fibrosis, and reduced lipotoxicity by decreasing triglycerides accumulation in the cardiac muscle in db/db mice, which could contribute to improve diastolic dysfunction [56, 57]. These alterations were associated with inhibition of pathological signaling pathway, as mediated by PKC. Erk1/2 phosphorylation was reduced, due to the decreased leptin capacity to stimulate mitogen kinase protein phosphorylation [58]. Ang-(1-7) treatment prevented Erk1/2 loss of phosphorylation, which is cardioprotective [59]. In diabetic cardiomyopathy, leptin was also associated with heart hypertrophy by increasing cardiomyocytes glucose concentration [60]. More recently, it was demonstrated that Ang-(1-7) may reverse this condition via reduced ROS, leptin, p38, and Erk1/2 expression [61]. These findings suggest an important interaction among RAS, adiponectin and leptin and the use of drugs aiming to modulate these molecules, which might compose an efficient alternative to be used on treating metabolic disorders associated with increased body adiposity (Fig. 3).

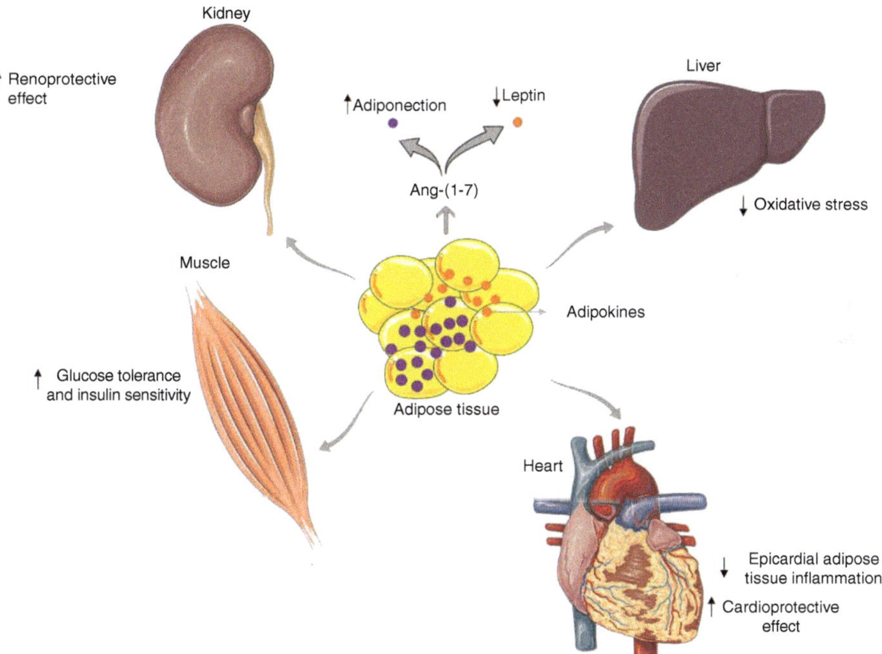

Fig. 3 Ang-(1-7) increases adiponectin and reduces leptin. 1 – Ang-(1-7) has renoprotective effects via attenuation of adipokines produced by the perirenal adipose tissue. 2 – Ang-(1-7) induces glucose and insulin sensitivity improvement. 3 – Ang-(1-7) has cardioprotective effects and reduces inflammation mediated by the epicardial adipose tissue

Ang-(1-7) and Gonadal Hormones

The literature chronologically reports the main RAS components in the female and male reproductive organs in several mammals' species, including rodents and humans [62–66]. The angiotensin-converting enzyme (ACE) has been identified in rat ovaries, in the germinal epithelium adjacent to the corpus luteum, granulosa cells, blood vessels, and stroma [66]. The male reproductive system and testicles also express the RAS classical components that modulates steroidogenesis in Leydig cells and have effects in epididymal and sperm cells function [67]. Some other studies demonstrated that the Mas receptor is present in mice, rats, and humans testicles [68–70] evidencing that Ang-(1-7) is important on female and male reproductive system, as well as on the gonadal hormones effects.

Ang-(1-7) Modulating the Gonadal Effects: Ovary

In the ovary, several studies have reported the involvement of Ang-(1-7) peptide on functions such as folliculogenesis, follicular atresia, ovulation and corpus luteum formation, thus influencing the reproduction biotechnologies efficiency [71–74].

High levels of Ang-(1-7) were found in the rats ovarian (phases: proestrous and estrous) as well as in impubertal rats treated with equine chorionic gonadotropin (eCG), especially in theca cells [74]. This finding indicates the involvement of this peptide in pre- and post-ovulatory events [72]. Reis et al. (2009) reported the Ang-(1-7) presence in rabbit ovaries via immunohistochemistry techniques and found immunoreactivity of this peptide in interstitial cells and oocytes [75–77]. They also observed immunoreactivity for Ang-(1-7) in theca and granulosa cells of preovulatory follicles in pretreated rabbits with equine chorionic gonadotropin (eCG) and in luteal bodies of covered rabbits.

Ang-(1-7) was also detected in women ovarian follicles – primordial, primary, secondary, and antral – stroma, and corpus luteum [78]. Evidence regarding a specific Ang-(1-7) receptor was raised especially after the synthesis of specific antagonist for this peptide [A-779 or D-Ala7-Ang-(1-7)] [79]. In another study, the specific Mas receptor antagonist, A-779, inhibited the germinative vesicle induced by Ang-(1-7) and reduced the oocyte maturation stimulated by the luteinizing hormone (LH) [77]. The Mas receptor presence was also evidenced in a study that added Ang-(1-7) in ovary perfusion medium of rats in vitro, stimulating the estradiol production [72]. Evidence of an Ang-(1-7) specific receptor in rats was also reported [80].

ACE participates in Ang-(1-7) metabolism, as its inhibition might reduce Ang-(1-7) degradation and reduce its conversion into Ang-(1-5) while increasing its bioavailability [81]. Furthermore, ACE inhibitors may also increase Ang I availability, providing higher Ang-(1-7) production via the prolyl-endopeptidase pathway. The cCG treatment in impubertal rats increased the Mas receptor and ACE2 expression, promoting increased Ang-(1-7) immunoreactivity in theca and interstitial cells. These findings support the hypothesis that the ovarian ACE2/Ang-(1-7)/Mas axis is expressed in rats ovarian and is regulated by gonadotropic hormones [77, 82]. The Mas receptor is also involved in vessel relaxation dependent on nitric oxide, induced by the estrogen hormone that induces nitric oxide (NO) production and vasodilation via mechanisms that require Mas receptor activation [83, 84].

Generally, ovarian steroids are not necessary to induce Ang-(1-7) and Mas receptor endometrial expression in rats, as they remain highly expressed in ovariectomized animals. However, estrogen and progestin may modulate the endometrial distribution pattern of this peptide, especially in the glandular compartment [85].

Current evidence also suggests that, in part, the female protective phenotype against hypertension might be due to the ACE2 activity incident within cardiovascular regulatory regions of the brain, potentially mediated by estrogen. Increasing evidence suggests the importance of a central renin-angiotensin pathway, although its localization and mechanisms involved in its expression and regulation still need to be further clarified [86].

Ang (1-7) Modulating the Gonadal Effects: Testicles

The male reproductive structures also express all the RAS components and the gonadotropins regulate the components activity. Of great interest is the expression of the angiotensin-converting enzyme (ACE) in testicles germ cells.

Regarding Ang-(1-7) and MasR, the mRNA expression and distribution for the Mas proto-oncogene were studied by hybridization with marked cDNA probes in mice brain, and its presence was observed in several brain areas [87]. The Mas expression was also described in rats testicles [68], and its expression in mice started only after 18 days of age and increased until the sixth month [69]. Santos et al. (2003), showed for the first time the Ang-(1-7) functional receptor, Mas receptor, localized in mice kidney [88]. In mice and rats testicles, the MasR was evidenced several weeks after birth, coinciding with the puberty period. In mice aging 3 months, the MasR is expressed and localized in Leydig and Sertoli cells, despite highest amounts were observed only in Leydig cells [69, 70]; although Mas Knockout mice did not present fertility alterations [89], the Mas receptor presence in the puberty phase in the testicles, especially in testosterone modulation, suggests that Ang-(1-7) might participate in the testosterone production and thus in the reproductive system.

The human testicle also presents Ang-(1-7), which is more abundant in the Leydig cytoplasm. In the seminiferous tubules, the Ang-(1-7) expression is lower when compared to the interstitial compartment and is predominant in external layers, particularly in Sertoli cells cytoplasm and primary sperm [70].

In order to verify the Ang-(1-7) effects on testicular steroidogenesis, a perfusion system model of isolated organ or incubation system with the peptide was used. In the perfusion model with rats testicles, increased testosterone was observed in the perfusate [90], while in the human testicles incubation, Ang-(1-7) decreased testosterone production [91]. However, even with the in vitro results already reported, this

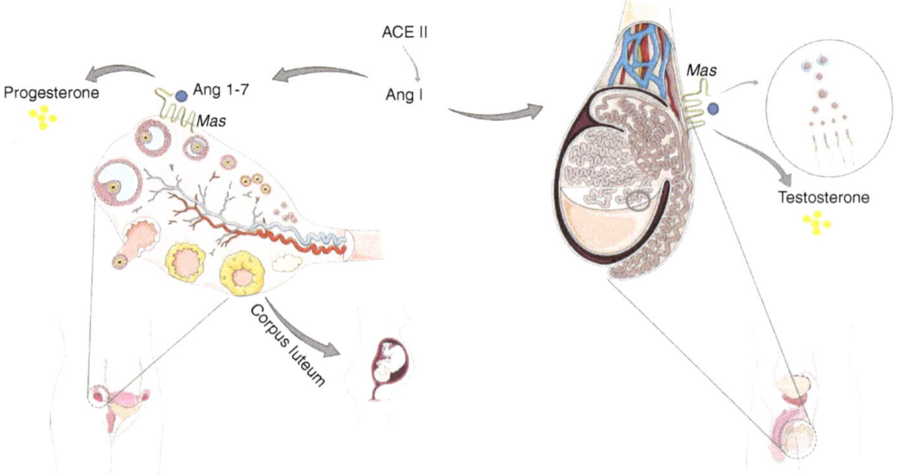

Fig. 4 RAS components and Ang-(1-7) action via Mas Receptor in the reproductive systems. 1 – Ovary: Ang-(1-7) effects on folliculogenesis, ovulation, formation of the corpus luteum responsible for the secretion of progesterone that maintains the ovarian cycle and pregnancy balance. 2 – In the testicles: greater expression of MasR in the Leydig cells responsible for the testosterone secretion that has action in the reproductive tissues (testicles and prostate) and secondary male characteristics

peptide systemic effect on steroidogenesis was not yet verified. In order to verify the metabolism modulation in these effects, Ang-(1-7) was included in polysaccharide cyclodextrin, protecting the peptide against digestive enzymes, allowing its oral administration. Thus, it was possible to verify the Ang-(1-7) effects in reproductive organs, from plasma samples [62].

In this sense, a consistent amount of evidence produced in the last few years suggests that the Ang-(1-7)/Mas pathway is active and demonstrates an important role in female and male reproductive systems (Fig. 4).

Remarks and Perspectives

Recent studies suggest that ACE/Ang-(1-7)/Mas axis modulation may improve metabolic diseases by modulating different signaling pathways involved in the synthesis and secretion of several hormones produced in different sites such as: pancreas, liver, adipose tissue and reproductive system.

Ang-(1-7) showed to be effective on improving type 2 diabetes and obesity in murine models. The Ang-(1-7) or its synthetic analog (AVE-0991) prevented abnormal vascular alterations induced by hyperglycemia in the mesenteric bed, isolated carotid and renal arteries from diabetic rats. Furthermore, treatment with AVE-0991 was capable of restoring cardiac function in diabetogenic conditions, via arterial pressure and contractility parameters normalization [92].

In the liver, Ang-(1-7) showed an essential role improving and treating hepatic diseases by modulating inflammation and fibrosis [29–31] and regulating hepatic metabolic function [92]. The resistin/TLR4/MAPK/NF-κB axis is an important Ang-(1-7) target [32]. Furthermore, the relative ACE2 deficiency leads to Ang II impaired degradation and consequently accumulation of this peptide, contributing to increased fibrosis and inflammation [33, 34].

In the metabolic homeostasis control, adiponectin, leptin, and Ang-(1-7) exert important influence [49]. Ang-(1-7) produced renoprotective effects in experimental models of diabetic nephropathy associated with reduced inflammation, fibrosis, oxidative stress, and lipotoxicity [66–69]. In diabetic cardiomyopathy, leptin was associated with cardiac hypertrophy via increased glucose concentrations in cardiomyocytes [60]. More recently, it was demonstrated that Ang-(1-7) may improve cardiovascular diseases by reducing ROS, leptin, p38 and Erk1/2 expression [61].

In the ovary, several studies reported the Ang-(1-7) involvement in functions such as folliculogenesis, follicular atresia, ovulation and corpus luteum formation, influencing the reproductive biotechnologies efficiency [71–74]. The treatment with eCG in impubertal rats increased Mas receptor and ACE2 expression and Ang-(1-7) imunoreactivity in theca and interstitial cells. These findings support the hypothesis that the ovarian ACE2/Ang-(1-7)/Mas axis is regulated by gonadotrophic hormones [77, 82]. In the testicles, the Mas receptor presence in the pubertal phase suggests that Ang-(1-7) may participate in the testosterone production and thus in the reproductive system [70].

Altogether, these findings are important to delineate potential therapeutic alternatives to restore the metabolic organs and improve metabolism [22], especially considering the pandemic healthcare problem of obesity and diabetes (associated with pancreas disorders) [24, 25]. The increased ACE2/Ang-(1-7)/Mas axis activation is also capable of improving hepatic disorders and reducing the AngII/AT1 deleterious effects. In the adipose tissue, adiponectin and leptin are described as important metabolic regulators [47–50], along with Ang-(1-7) that interferes in obesity-associated signaling pathways. Yet, the Ang-(1-7)/MasR signaling is active and demonstrates an important role in female and male reproductive systems. Thus, new approaches aiming to increase Ang-(1-7) levels might be new important tools for treating and preventing endocrine diseases.

References

1. Patel S, Rauf A, Khan H, Abu-Izneid T. Renin-angiotensin-aldosterone (RAAS): the ubiquitous system for homeostasis and pathologies. Biomed Pharmacother. 2017;94:317–25. PubMed PMID: 28772209.
2. Favre GA, Esnault VL, Van Obberghen E. Modulation of glucose metabolism by the renin-angiotensin-aldosterone system. Am J Physiol Endocrinol Metab. 2015;308(6):E435–49. PubMed PMID: 25564475.
3. Santos SH, Andrade JM. Angiotensin 1-7: a peptide for preventing and treating metabolic syndrome. Peptides. 2014;59:34–41. PubMed PMID: 25017239.

4. Santos SH, Simoes e Silva AC. The therapeutic role of Renin-Angiotensin System blockers in obesity- related renal disorders. Curr Clin Pharmacol. 2014;9(1):2–9. PubMed PMID: 23270435.

5. Santos RAS, Sampaio WO, Alzamora AC, Motta-Santos D, Alenina N, Bader M, et al. The ACE2/Angiotensin-(1-7)/MAS axis of the renin-angiotensin system: focus on angiotensin-(1-7). Physiol Rev. 2018;98(1):505–53. PubMed PMID: 29351514.

6. Rein J, Bader M. Renin-angiotensin system in diabetes. Protein Pept Lett. 2017;24(9):833–40. PubMed PMID: 28758590.

7. Santos SHS. Editorial: renin-angiotensin system: role in chronic diseases. Protein Pept Lett. 2017;24(9):782–3. PubMed PMID: 29210628.

8. Igarashi M, Hirata A, Nozaki H, Kadomoto-Antsuki Y, Tominaga M. Role of angiotensin II type-1 and type-2 receptors on vascular smooth muscle cell growth and glucose metabolism in diabetic rats. Diabetes Res Clin Pract. 2007;75(3):267–77. PubMed PMID: 16934905.

9. Oliveira Andrade JM, Paraiso AF, Garcia ZM, Ferreira AV, Sinisterra RD, Sousa FB, et al. Cross talk between angiotensin-(1-7)/Mas axis and sirtuins in adipose tissue and metabolism of high-fat feed mice. Peptides. 2014;55:158–65. PubMed PMID: 24642355.

10. Munoz MC, Giani JF, Burghi V, Mayer MA, Carranza A, Taira CA, et al. The Mas receptor mediates modulation of insulin signaling by angiotensin-(1-7). Regul Pept. 2012;177(1–3):1–11. PubMed PMID: 22561450.

11. Nunes-Souza V, Alenina N, Qadri F, Penninger JM, Santos RA, Bader M, et al. CD36/sirtuin 1 axis impairment contributes to hepatic steatosis in ACE2-deficient mice. Oxid Med Cell Longev. 2016;2016:6487509. PubMed PMID: 28101297. Pubmed Central PMCID: 5215286.

12. Dominici FP, Burghi V, Munoz MC, Giani JF. Modulation of the action of insulin by angiotensin-(1-7). Clin Sci. 2014;126(9):613–30. PubMed PMID: 24450744.

13. Bruce EB, de Kloet AD. The intricacies of the renin-angiotensin-system in metabolic regulation. Physiol Behav. 2017;178:157–65. PubMed PMID: 27887998. Pubmed Central PMCID: 5600901.

14. Fu Z, Zhao L, Aylor KW, Carey RM, Barrett EJ, Liu Z. Angiotensin-(1-7) recruits muscle microvasculature and enhances insulin's metabolic action via mas receptor. Hypertension. 2014;63(6):1219–27. PubMed PMID: 24711523. Pubmed Central PMCID: 4030711.

15. Niu MJ, Yang JK, Lin SS, Ji XJ, Guo LM. Loss of angiotensin-converting enzyme 2 leads to impaired glucose homeostasis in mice. Endocrine. 2008;34(1–3):56–61. PubMed PMID: 18956256.

16. Bindom SM, Hans CP, Xia H, Boulares AH, Lazartigues E. Angiotensin I-converting enzyme type 2 (ACE2) gene therapy improves glycemic control in diabetic mice. Diabetes. 2010;59(10):2540–8. PubMed PMID: 20660625. Pubmed Central PMCID: 3279528.

17. Shoemaker R, Yiannikouris F, Thatcher S, Cassis L. ACE2 deficiency reduces beta-cell mass and impairs beta-cell proliferation in obese C57BL/6 mice. Am J Physiol Endocrinol Metab. 2015;309(7):E621–31. PubMed PMID: 26389599. Pubmed Central PMCID: 4593774.

18. Chu KY, Leung PS. Angiotensin II Type 1 receptor antagonism mediates uncoupling protein 2-driven oxidative stress and ameliorates pancreatic islet beta-cell function in young Type 2 diabetic mice. Antioxid Redox Signal. 2007;9(7):869–78. PubMed PMID: 17508912.

19. Yuan L, Li Y, Li G, Song Y, Gong X. Ang(1-7) treatment attenuates beta-cell dysfunction by improving pancreatic microcirculation in a rat model of Type 2 diabetes. J Endocrinol Invest. 2013;36(11):931–7. PubMed PMID: 23640708.

20. He J, Yang Z, Yang H, Wang L, Wu H, Fan Y, et al. Regulation of insulin sensitivity, insulin production, and pancreatic beta cell survival by angiotensin-(1-7) in a rat model of streptozotocin-induced diabetes mellitus. Peptides. 2015;64:49–54. PubMed PMID: 25576844.

21. Felix Braga J, Ravizzoni Dartora D, Alenina N, Bader M, Santos RA. Glucagon-producing cells are increased in Mas-deficient mice. Endocr Connect. 2017;6(1):27–32. PubMed PMID: 27998954. Pubmed Central PMCID: 5302165.

22. Wang L, Liang J, Leung PS. The ACE2/Ang-(1-7)/Mas axis regulates the development of pancreatic endocrine cells in mouse embryos. PLoS One. 2015;10(6):e0128216. PubMed PMID: 26029927. Pubmed Central PMCID: 4452480.

23. Zhang F, Liu C, Wang L, Cao X, Wang YY, Yang JK. Antioxidant effect of angiotensin (17) in the protection of pancreatic beta cell function. Mol Med Rep. 2016;14(3):1963–9. PubMed PMID: 27430410. Pubmed Central PMCID: 4991744.
24. Mildner M, Muller-Fielitz H, Stolting I, Johren O, Steckelings M, Raasch W. Glucagon increase after chronic AT1 blockade is more likely related to an indirect leptin-dependent than to a pancreatic alpha-cell-dependent mechanism. Naunyn Schmiedebergs Arch Pharmacol. 2017;390(5):505–18. PubMed PMID: 28144709.
25. Bilman V, Mares-Guia L, Nadu AP, Bader M, Campagnole-Santos MJ, Santos RA, et al. Decreased hepatic gluconeogenesis in transgenic rats with increased circulating angiotensin-(1-7). Peptides. 2012;37(2):247–51. PubMed PMID: 22902596.
26. Klein S, Herath CB, Schierwagen R, Grace J, Haltenhof T, Uschner FE, et al. Hemodynamic effects of the non-peptidic angiotensin-(1-7) agonist AVE0991 in liver cirrhosis. PLoS One. 2015;10(9):e0138732. PubMed PMID: 26406236. Pubmed Central PMCID: 4583473.
27. Liu Y, Li B, Wang X, Li G, Shang R, Yang J, et al. Angiotensin-(1-7) suppresses hepatocellular carcinoma growth and angiogenesis via complex interactions of angiotensin II type 1 receptor, angiotensin II type 2 receptor and Mas receptor. Mol Med. 2015;21:626–36. PubMed PMID: 26225830. Pubmed Central PMCID: 4656199.
28. Moreira de Macedo S, Guimaraes TA, Feltenberger JD, Sousa Santos SH. The role of renin-angiotensin system modulation on treatment and prevention of liver diseases. Peptides. 2014;62:189–96. PubMed PMID: 25453980.
29. Cai SM, Yang RQ, Li Y, Ning ZW, Zhang LL, Zhou GS, et al. Angiotensin-(1-7) improves liver fibrosis by regulating the NLRP3 inflammasome via redox balance,modulation. Antioxid Redox Signal. 2016;24(14):795–812. PubMed PMID: 26728324.
30. Simoes ESAC, Miranda AS, Rocha NP, Teixeira AL. Renin angiotensin system in liver diseases: friend or foe? World J Gastroenterol. 2017;23(19):3396–406. PubMed PMID: 28596676. Pubmed Central PMCID: 5442076.
31. Mak KY, Chin R, Cunningham SC, Habib MR, Torresi J, Sharland AF, et al. ACE2 therapy using Adeno-associated viral vector inhibits liver fibrosis in, mice. Mol Ther. 2015;23(9):1434–43. PubMed PMID: 25997428. Pubmed Central PMCID: 4817885.
32. Santos SH, Andrade JM, Fernandes LR, Sinisterra RD, Sousa FB, Feltenberger JD, et al. Oral angiotensin-(1-7) prevented obesity and hepatic inflammation by inhibition of resistin/TLR4/MAPK/NF-kappaB in rats fed with high-fat diet. Peptides. 2013;46:47–52. PubMed PMID: 23714175.
33. Thomas MC, Pickering RJ, Tsorotes D, Koitka A, Sheehy K, Bernardi S, et al. Genetic Ace2 deficiency accentuates vascular inflammation and atherosclerosis in the ApoE knockout mouse. Circ Res. 2010;107(7):888–97. PubMed PMID: 20671240.
34. Tikellis C, Bialkowski K, Pete J, Sheehy K, Su Q, Johnston C, et al. ACE2 deficiency modifies renoprotection afforded by ACE inhibition in experimental diabetes. Diabetes. 2008;57(4):1018–25. PubMed PMID: 18235039.
35. Cortez-Pinto H, de Moura MC, Day CP. Non-alcoholic steatohepatitis: from cell biology to clinical practice. J Hepatol. 2006;44(1):197–208. PubMed PMID: 16274837.
36. Dowman JK, Armstrong MJ, Tomlinson JW, Newsome PN. Current therapeutic strategies in non-alcoholic fatty liver disease. Diabetes Obes Metab. 2011;13(8):692–702. PubMed PMID: 21449949.
37. Moreira CCL, Lourenco FC, Mario EG, Santos RAS, Botion LM, Chaves VE. Long-term effects of angiotensin-(1-7) on lipid metabolism in the adipose tissue and liver. Peptides. 2017;92:16–22. PubMed PMID: 28438644.
38. Cao X, Yang F, Shi T, Yuan M, Xin Z, Xie R, et al. Angiotensin-converting enzyme 2/angiotensin-(1-7)/Mas axis activates Akt signaling to ameliorate hepatic steatosis. Sci Rep. 2016;6:21592. PubMed PMID: 26883384. Pubmed Central PMCID: 4756304.
39. Feltenberger JD, Andrade JM, Paraiso A, Barros LO, Filho AB, Sinisterra RD, et al. Oral formulation of angiotensin-(1-7) improves lipid metabolism and prevents high-fat diet-induced hepatic steatosis and inflammation in mice. Hypertension. 2013;62(2):324–30. PubMed PMID: 23753417.

40. Bays H, Rodbard HW, Schorr AB, Gonzalez-Campoy JM. Adiposopathy: treating pathogenic adipose tissue to reduce cardiovascular disease risk. Curr Treat Options Cardiovasc Med. 2007;9(4):259–71. PubMed PMID: 17761111.
41. Gil-Campos M, Canete RR, Gil A. Adiponectin, the missing link in insulin resistance and obesity. Clin Nutr. 2004;23(5):963–74. PubMed PMID: 15380884.
42. Fasshauer M, Paschke R. Regulation of adipocytokines and insulin resistance. Diabetologia. 2003;46(12):1594–603. PubMed PMID: 14605806.
43. Lihn AS, Pedersen SB, Richelsen B. Adiponectin: action, regulation and association to insulin sensitivity. Obes Rev. 2005;6(1):13–21. PubMed PMID: 15655035.
44. Stefan N, Stumvoll M. Adiponectin--its role in metabolism and beyond. Hormone Metab Res. 2002;34(9):469–74. PubMed PMID: 12384822.
45. Kershaw EE, Flier JS. Adipose tissue as an endocrine organ. J Clin Endocrinol Metab. 2004;89(6):2548–56. PubMed PMID: 15181022.
46. Villanueva EC, Myers MG Jr. Leptin receptor signaling and the regulation of mammalian physiology. Int J Obes (Lond). 2008;32(Suppl 7):S8–12. PubMed PMID: 19136996. Pubmed Central PMCID: 2648306.
47. Liu C, Lv XH, Li HX, Cao X, Zhang F, Wang L, et al. Angiotensin-(1-7) suppresses oxidative stress and improves glucose uptake via Mas receptor in adipocytes. Acta Diabetol. 2012;49(4):291–9. PubMed PMID: 22042130.
48. Santos SH, Fernandes LR, Mario EG, Ferreira AV, Porto LC, Alvarez-Leite JI, et al. Mas deficiency in FVB/N mice produces marked changes in lipid and glycemic metabolism. Diabetes. 2008;57(2):340–7. PubMed PMID: 18025412.
49. Santos SH, Braga JF, Mario EG, Porto LC, Rodrigues-Machado Mda G, Murari A, et al. Improved lipid and glucose metabolism in transgenic rats with increased circulating angiotensin-(1-7). Arterioscler Thromb Vasc Biol. 2010;30(5):953–61. PubMed PMID: 20203301.
50. Coelho MS, Lopes KL, Freitas Rde A, de Oliveira-Sales EB, Bergasmaschi CT, Campos RR, et al. High sucrose intake in rats is associated with increased ACE2 and angiotensin-(1-7) levels in the adipose tissue. Regul Pept. 2010;162(1–3):61–7. PubMed PMID: 20346375.
51. Mori J, Patel VB, Ramprasath T, Alrob OA, DesAulniers J, Scholey JW, et al. Angiotensin 1-7 mediates renoprotection against diabetic nephropathy by reducing oxidative stress, inflammation, and lipotoxicity. Am J Physiol Renal Physiol. 2014;306(8):F812–21. PubMed PMID: 24553436.
52. Patel VB, Basu R, Oudit GY. ACE2/Ang 1-7 axis: a critical regulator of epicardial adipose tissue inflammation and cardiac dysfunction in obesity. Adipocyte. 2016;5(3):306–11. PubMed PMID: 27617176. Pubmed Central PMCID: 5014009.
53. Loot AE, Roks AJ, Henning RH, Tio RA, Suurmeijer AJ, Boomsma F, et al. Angiotensin-(1-7) attenuates the development of heart failure after myocardial infarction in rats. Circulation. 2002;105(13):1548–50. PubMed PMID: 11927520.
54. Schuchard J, Winkler M, Stolting I, Schuster F, Vogt FM, Barkhausen J, et al. Lack of weight gain after angiotensin AT1 receptor blockade in diet-induced obesity is partly mediated by an angiotensin-(1-7)/Mas-dependent pathway. Br J Pharmacol. 2015;172(15):3764–78. PubMed PMID: 25906670. Pubmed Central PMCID: 4523334.
55. Li H, Li M, Liu P, Wang Y, Zhang H, Li H, et al. Telmisartan ameliorates nephropathy in metabolic syndrome by reducing leptin release from perirenal adipose tissue. Hypertension. 2016;68(2):478–90. PubMed PMID: 27296996.
56. Flagg TP, Cazorla O, Remedi MS, Haim TE, Tones MA, Bahinski A, et al. Ca2+-independent alterations in diastolic sarcomere length and relaxation kinetics in a mouse model of lipotoxic diabetic cardiomyopathy. Circ Res. 2009;104(1):95–103. PubMed PMID: 19023131.
57. van der Meer RW, Hammer S, Smit JW, Frolich M, Bax JJ, Diamant M, et al. Short-term caloric restriction induces accumulation of myocardial triglycerides and decreases left ventricular diastolic function in healthy subjects. Diabetes. 2007;56(12):2849–53. PubMed PMID: 17717279.
58. Zeidan A, Javadov S, Chakrabarti S, Karmazyn M. Leptin-induced cardiomyocyte hypertrophy involves selective caveolae and RhoA/ROCK-dependent p38 MAPK translocation to nuclei. Cardiovasc Res. 2008;77(1):64–72. PubMed PMID: 18006472.

59. Mori J, Patel VB, Abo Alrob O, Basu R, Altamimi T, Desaulniers J, et al. Angiotensin 1-7 ameliorates diabetic cardiomyopathy and diastolic dysfunction in db/db mice by reducing lipotoxicity and inflammation. Circ Heart Fail. 2014;7(2):327–39. PubMed PMID: 24389129.

60. Majumdar P, Chen S, George B, Sen S, Karmazyn M, Chakrabarti S. Leptin and endothelin-1 mediated increased extracellular matrix protein production and cardiomyocyte hypertrophy in diabetic heart disease. Diabetes Metab Res Rev. 2009;25(5):452–63. PubMed PMID: 19391127.

61. Lei Y, Xu Q, Zeng B, Zhang W, Zhen Y, Zhai Y, et al. Angiotensin-(1-7) protects cardiomyocytes against high glucose-induced injuries through inhibiting reactive oxygen species-activated leptin-p38 mitogen-activated protein kinase/extracellular signal-regulated protein kinase 1/2 pathways, but not the leptin-c-Jun N-terminal kinase pathway in vitro. J Diabet Investig. 2017;8(4):434–45. PubMed PMID: 27896943. Pubmed Central PMCID: 5497033.

62. Thomas Unger U, Steckelings M, Santos RASD. The protective arm of the renin angiotensin system (RAS): Elsevier; 2015. https://doi.org/10.1016/C2013-0-23135-4.

63. Raposo-Costa AP, Reis AM. O sistema renina-angiotensina em ovário. Arq Bras Endocrinol Metabol. 2000;44:306–13.

64. Paulson RJ, Do YS, Hsueh WA, Eggena P, Lobo RA. Ovarian renin production in vitro and in vivo: characterization and clinical correlation. Fertil Steril. 1989;51(4):634–8. PubMed PMID: 2647527. Epub 1989/04/01. eng.

65. Itskovitz J, Bruneval P, Soubrier F, Thaler I, Corvol P, Sealey JE. Localization of renin gene expression to monkey ovarian theca cells by in situ hybridization. J Clin Endocrinol Metab. 1992;75(5):1374–80. PubMed PMID: 1430100. Epub 1992/11/01. eng.

66. Husain A, Bumpus FM, De Silva P, Speth RC. Localization of angiotensin II receptors in ovarian follicles and the identification of angiotensin II in rat ovaries. Proc Natl Acad Sci U S A. 1987;84(8):2489–93. PubMed PMID: 3470807. Pubmed Central PMCID: PMC304677. Epub 1987/04/01. eng.

67. Leung PS, Sernia C. The renin-angiotensin system and male reproduction: new functions for old hormones. J Mol Endocrinol. 2003;30(3):263–70. PubMed PMID: 12790798. Epub 2003/06/07. eng.

68. Metzger R, Bader M, Ludwig T, Berberich C, Bunnemann B, Ganten D. Expression of the mouse and rat mas proto-oncogene in the brain and peripheral tissues. FEBS Lett. 1995;357(1):27–32. PubMed PMID: 8001672. Epub 1995/01/02. eng.

69. Alenina N, Baranova T, Smirnow E, Bader M, Lippoldt A, Patkin E, et al. Cell type-specific expression of the Mas proto-oncogene in testis. J Histochem Cytochem. 2002;50(5):691–6. PubMed PMID: 11967280. Epub 2002/04/23. eng.

70. Reis AB, Araujo FC, Pereira VM, Dos Reis AM, Santos RA, Reis FM. Angiotensin (1-7) and its receptor Mas are expressed in the human testis: implications for male infertility. J Mol Histol. 2010;41(1):75–80. PubMed PMID: 20361351. Epub 2010/04/03. eng.

71. Yoshimura Y, Karube M, Aoki H, Oda T, Koyama N, Nagai A, et al. Angiotensin II induces ovulation and oocyte maturation in rabbit ovaries via the AT2 receptor subtype. Endocrinology. 1996;137(4):1204–11. PubMed PMID: 8625890. Epub 1996/04/01. eng.

72. Costa AP, Fagundes-Moura CR, Pereira VM, Silva LF, Vieira MA, Santos RA, et al. Angiotensin-(1-7): a novel peptide in the ovary. Endocrinology. 2003;144(5):1942–8. PubMed PMID: 12697701. Epub 2003/04/17. eng.

73. Li YH, Jiao LH, Liu RH, Chen XL, Wang H, Wang WH. Localization of angiotensin II in pig ovary and its effects on oocyte maturation in vitro. Theriogenology. 2004;61(2-3):447–59. PubMed PMID: 14662143. Epub 2003/12/10. eng.

74. Speth RC, Daubert DL, Grove KL. Angiotensin II: a reproductive hormone too? Regul Pept. 1999;79(1):25–40. PubMed PMID: 9930580. Epub 1999/02/04. eng.

75. dos Reis AM, Viana GEN, Pereira VM, Santos RAS. Angiotensin-(1-7) in the rabbit ovary: a novel local regulator of ovulation. Biol Reprod. 2009;81(Suppl_1):566.

76. Pereira AM, de Souza Junior A, Machado FB, Goncalves GK, Feitosa LC, Reis AM, et al. The effect of angiotensin-converting enzyme inhibition throughout a superovulation protocol in ewes. Res Vet Sci. 2015;103:205–10. PubMed PMID: 26679819. Epub 2015/12/19. eng.

77. Honorato-Sampaio K, Pereira VM, Santos RA, Reis AM. Evidence that angiotensin-(1-7) is an intermediate of gonadotrophin-induced oocyte maturation in the rat preovulatory follicle. Exp Physiol. 2012;97(5):642–50. PubMed PMID: 22247282. Epub 2012/01/17. eng.

78. Reis FM, Bouissou DR, Pereira VM, Camargos AF, dos Reis AM, Santos RA. Angiotensin-(1-7), its receptor Mas, and the angiotensin-converting enzyme type 2 are expressed in the human ovary. Fertil Steril. 2011;95(1):176–81. PubMed PMID: 20674894. Epub 2010/08/03. eng.

79. Santos RA, Campagnole-Santos MJ, Baracho NC, Fontes MA, Silva LC, Neves LA, et al. Characterization of a new angiotensin antagonist selective for angiotensin-(1-7): evidence that the actions of angiotensin-(1-7) are mediated by specific angiotensin receptors. Brain Res Bull. 1994;35(4):293–8. PubMed PMID: 7850477. Epub 1994/01/01. eng.

80. Silva DM, Vianna HR, Cortes SF, Campagnole-Santos MJ, Santos RA, Lemos VS. Evidence for a new angiotensin-(1-7) receptor subtype in the aorta of Sprague-Dawley rats. Peptides. 2007;28(3):702–7. PubMed PMID: 17129638. Epub 2006/11/30. eng.

81. Chappell MC, Pirro NT, Sykes A, Ferrario CM. Metabolism of angiotensin-(1-7) by angiotensin-converting enzyme. Hypertension. 1998;31(1 Pt 2):362–7. PubMed PMID: 9453329. Epub 1998/02/07. eng.

82. Pereira VM, Reis FM, Santos RA, Cassali GD, Santos SH, Honorato-Sampaio K, et al. Gonadotropin stimulation increases the expression of angiotensin-(1--7) and MAS receptor in the rat ovary. Reprod Sci (Thousand Oaks, Calif). 2009;16(12):1165–74. PubMed PMID: 19703990. Epub 2009/08/26. eng.

83. Sobrino A, Vallejo S, Novella S, Lazaro-Franco M, Mompeon A, Bueno-Beti C, et al. Mas receptor is involved in the estrogen-receptor induced nitric oxide-dependent vasorelaxation. Biochem Pharmacol. 2017;129:67–72. PubMed PMID: 28131844. Epub 2017/01/31. eng.

84. Mompeon A, Lazaro-Franco M, Bueno-Beti C, Perez-Cremades D, Vidal-Gomez X, Monsalve E, et al. Estradiol, acting through ERalpha, induces endothelial non-classic renin-angiotensin system increasing angiotensin 1-7 production. Mol Cell Endocrinol. 2016;422:1–8. PubMed PMID: 26562171. Epub 2015/11/13. eng.

85. Vaz-Silva J, Tavares RL, Ferreira MC, Honorato-Sampaio K, Cavallo IK, Santos RA, et al. Tissue specific localization of angiotensin-(1-7) and its receptor Mas in the uterus of ovariectomized rats. J Mol Histol. 2012;43(5):597–602. PubMed PMID: 22684246. Epub 2012/06/12. eng.

86. de Morais SDB, Shanks J, Zucker IH. Integrative physiological aspects of brain RAS in hypertension. Curr Hypertens Rep. 2018;20(2):10. PubMed PMID: 29480460. Epub 2018/02/27. eng.

87. Bunnemann B, Fuxe K, Metzger R, Mullins J, Jackson TR, Hanley MR, et al. Autoradiographic localization of mas proto-oncogene mRNA in adult rat brain using in situ hybridization. Neurosci Lett. 1990;114(2):147–53. PubMed PMID: 2203997. Epub 1990/07/03. eng.

88. Santos RA, Simoes e Silva AC, Maric C, Silva DM, Machado RP, de Buhr I, et al. Angiotensin-(1-7) is an endogenous ligand for the G protein-coupled receptor Mas. Proc Natl Acad Sci U S A. 2003;100(14):8258–63. PubMed PMID: 12829792. Pubmed Central PMCID: PMC166216. Epub 2003/06/28. eng.

89. Walther T, Balschun D, Voigt JP, Fink H, Zuschratter W, Birchmeier C, et al. Sustained long term potentiation and anxiety in mice lacking the Mas protooncogene. J Biol Chem. 1998;273(19):11867–73. PubMed PMID: 9565612. Epub 1998/06/13. eng.

90. Viana GE, Pereira VM, Honorato-Sampaio K, Oliveira CA, Santos RA, Reis AM. Angiotensin-(1-7) induces ovulation and steroidogenesis in perfused rabbit ovaries. Exp Physiol. 2011;96(9):957–65. PubMed PMID: 21666031. Epub 2011/06/15. eng.

91. Reis AB. Expressão da angiotensina 1-7 no testículo humano. Belo Horizonte: Universidade Federal de Minas Gerais; 2006.

92. Benter IF, Yousif MH, Cojocel C, Al-Maghrebi M, Diz DI. Angiotensin-(1-7) prevents diabetes-induced cardiovascular dysfunction. Am J Physiol Heart Circ Physiol. 2007;292(1):H666–72. PubMed PMID: 17213482.

Skeletal Muscle System

María José Acuña, Enrique Brandan,
and Daisy Motta-Santos

Abbreviations

ACE	Angiotensin converting enzyme
ACE2	Angiotensin converting enzyme 2
Ang-(1-7)	Angiotensin-(1-7)
Ang-II	Angiotensin II
AT1	Angiotensin receptor 1
AT2	Angiotensin receptor 2
CTGF/CCN-2	Connective tissue growth factor
DMD	Duchenne muscular dystrophy
ECM	Extracellular matrix

M. J. Acuña (✉)
Centro de Envejecimiento y Regeneración, CARE Chile UC y Departamento de Biología
Celular y Molecular, Facultad de Ciencias Biológicas, Pontificia Universidad Católica de
Chile, Santiago, Chile

Centro de Biología y Química Aplicada (CIBQA), Universidad Bernardo O Higgins,
Santiago, Chile
e-mail: mjose.acuna@ubo.cl

E. Brandan (✉)
Centro de Envejecimiento y Regeneración, CARE Chile UC y Departamento de Biología
Celular y Molecular, Facultad de Ciencias Biológicas, Pontificia Universidad Católica de
Chile, Santiago, Chile
e-mail: ebrandan@bio.puc.cl

D. Motta-Santos (✉)
National Institute of Science and Technology in Nanobiopharmaceutics, UFMG,
Belo Horizonte, MG, Brazil

Sports Department, School of Physical Education, Physiotherapy, and Occupational Therapy
(EEFFTO), UFMG, Belo Horizonte, MG, Brazil
e-mail: daisy@ufmg.br

© Springer Nature Switzerland AG 2019
R. A. S. Santos (ed.), *Angiotensin-(1-7)*,
https://doi.org/10.1007/978-3-030-22696-1_11

MasR	Mas receptor
NMJ	Neuromuscular junctions
RAS	Renin-angiotensin system
SM	Skeletal muscle
TGF-β	Transforming growth factor type-β

Skeletal Muscle Overview

Skeletal muscle (SM) is the most abundant tissue spread through the entire body, plays critical role in keeping the postural control, has a vital function for making movements, both voluntary and nonvoluntary; and is also involved in the balance of energy metabolism.

SM is composed of muscle fibers that are long multinucleated cells surrounded by a specialized form of connective tissue the endomysium, and the fibers are organized as bundles of fibers called fascicles that are surrounded by the perimysium, and groups of fascicles are forming the muscles that are surrounded by the epimysium. Within the muscle fibers, capillaries that are oxygenating and nurturing the tissue are found, and each fiber is innervated by a motor neuron at the neuromuscular junction (NMJ), where the fibers are connected with the peripheral nervous system for the control of the muscle contraction. The quality of the skeletal muscle fibers is strongly influenced by the nerve. Another component of SM are the satellite cells that lie under the fibers basal lamina, these cells are stem cells committed to the muscle lineage. Other cell types present in SM are the fibroadipogenic progenitors (FAPs) and some monocytes residents in the muscle [1–5]. These cell types play important roles in the regeneration of SM. Figure 1 shows a schematic representation of SM architecture.

RAS Components in Skeletal Muscle (SM)

There is increasing evidence that the renin-angiotensin system (RAS) plays a direct role in different tissues and the SM does not seem to be an exception. Several works have shown that components of both arms of the RAS are present in this tissue. Among the components of the classic RAS axis, it has been described that there is angiotensin- converting enzyme (ACE) expression and activity [6–8] and immunohistochemical analyses show that (ACE) is present in the endothelium and in the neuromuscular junction [8, 9] and its levels are increased in SM from Duchenne muscular dystrophy (DMD) patients [9], a myopathy that occurs by the lack of the dystrophin protein and is characterized by muscle weakness and fibrosis [10].

ACE exerts a key role in SM metabolism; in humans, there is a polymorphism on ACE gene, ACE I and D (insertion or deletion of a sequence in intron 16 of the ACE gene), and the ACE I polymorphism is characterized by reduction of ACE levels

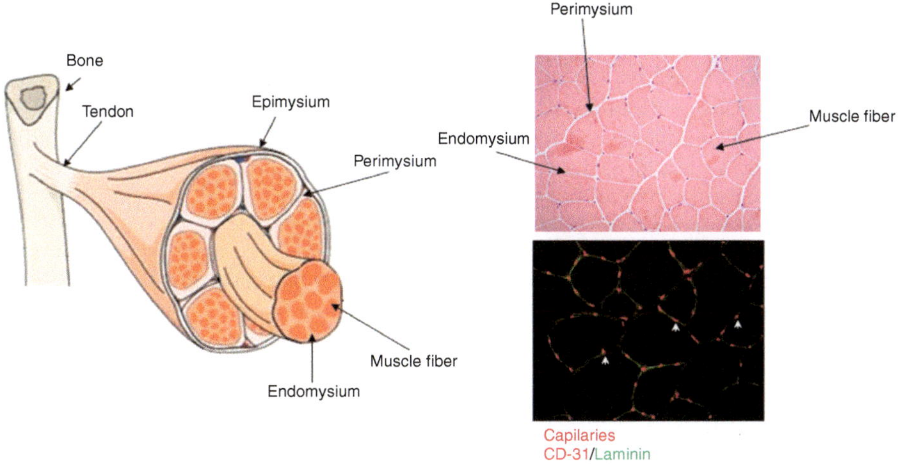

Fig. 1 *Left*: Schematic representation of skeletal muscle tissue, showing the basic components of SM. *Right*: Upper panel shows a transversal cross-section from gastrocnemius of wild-type mice stained with hematoxylin/eosin. Lower panel shows an immunofluorescence for CD31 that is present in capillaries and laminin that marks the basal lamina delineating the muscle fiber; note that there are several capillaries surrounding one fiber (photograph taken and donated by Dr. Daniela Rebolledo)

[11]. This brings more resistance to exercise performance because it can influence the blood flow and increase the oxidative metabolism in the SM [12].

The angiotensin type 1 (AT1) receptors are expressed in the vasculature of skeletal muscle of normal adult tissue [13] and its levels are increased in muscle byopsies obtained from DMD patients and AT1 can be found in the degenerating muscle fibers and fibroblasts as well [9]. AT1 receptors are also present in the SM satellite cells and its activation by angiotensin-II (Ang-II) reduces satellite cells proliferation and reduced regeneration after cardiotoxin (CTX) injury [14].

Angiotensin type 2 receptor (AT2) transcripts are detected in human fetal skeletal muscle tissues, but not in adult human skeletal muscle [13]; on the other hand, in mice, AT2 is detected in SM and shown to be increased by transforming growth factor beta (TGF-β1) injection in wild-type and in the *mdx* mice (the DMD murine model) [15]. AT2 receptor is increased in the regenerating muscle after CTX injury, and is necessary for proper satellite cell differentiation [16]. With regard to the opposite RAS axis, ACE2/angiotensin-(1-7)/Mas receptor (ACE2/Ang-(1-7)/Mas), there is evidence for ACE2 and Mas receptor presence in the SM [17–22].

ACE2 activity was present in mice SM, and it was localized in the sarcolemma and connective tissue of SM and was increased in the dystrophic muscle [18]. ACE2 is also playing an important role in the muscle metabolism, since its deletion leads to a reduced exercise performance [17]. Mas receptor is also found in mice SM and reported to increase in different muscle wasting conditions [20]. Interestingly, in *mdx* mice its expression was increased compared to wild-type mice (Fig. 1), which indicates that its levels are induced in this muscle pathology. At the cellular level, Mas receptor expression was found in myoblast and myotubes and also in SM-derived fibroblasts [23]. Mas receptor is downregulated by TGF-β1 in fibroblasts but not in myoblast or myotubes, suggesting differential regulation in the different SM cell types; interestingly, Mas expression was not modulated by Ang-II or connective tissue growth factor (CTGF/CCN2) [23]. Taken together, all these evidences support that the RAS is present and is an active player with critical and complex functions in the pathophysiology of SM.

RAS and SM Disease and Regeneration: Angiotensin-(1-7) Protective Role

Muscular Dystrophies

The classic arm of the RAS is related to skeletal muscle pathologies. In muscular dystrophies such as DMD, there are continuing cycles of progressive damage and regeneration of the fibers leading to a chronic inflammatory response that ultimately leads to the replacement of the fibers by fibrotic tissue. The main pro-fibrotic molecules in SM are (transforming growth factor type-beta) TGF-β [24–26], connective tissue growth factor (CTGF/CCN2) [26–29], and classic RAS components ACE/Ang-II/AT1 [25, 26].

Several studies have indicated that Ang-II plays an important role in the muscle fibrosis, since treatment with angiotensin type 1 receptor blockers (ARBs) and angiotensin-converting enzyme inhibitors (ACEi) general outcome is a reduction in the damage and fibrosis [30–33]. Losartan reduces TGF-β expression levels and smad-dependent signaling, an intracellular canonical branch activated upon TGF-β binding to its receptor [30, 31]. Both losartan and enalapril treatment reduce CTGF levels in *mdx* mice, and reduce the fibrotic response induced by CTGF overexpression [31, 32]. Ang-II through AT1 receptor leads to fibrosis by increasing TGF-β levels via activation of NAD(P)H oxidase-induced ROS and p38 phosphorylation that results in increased CTGF levels, increased extracellular matrix (ECM) proteins such as fibronectin, and collagen III [34, 35]; interestingly, the activation of NAD(P)H oxidase is PKC-dependent [35].

In muscular dystrophy, Ang-(1-7) systemic infusion by osmotic pumps or delivered orally by the formulation, cyclodextrin-conjugated Ang-(1-7) (CD-Ang-(1-7)), plays a protective role, reducing the damage, fibrosis, and restoring muscle strength in the *mdx* mice [21]; also in the model of sarcoglican-δ null muscular dystrophy, CD-Ang-(1-7) improves function and reduces the oxidative stress [36]. The mechanisms mediating the beneficial effects on dystrophy involve the reduction in TGF-β levels and smad-dependent signaling, observed by the reduction in phosphorylated smad3, decrease in the fibrotic miR-21, resulting in a reduction of TCF-4 positive fibroblasts. The Ang-(1-7) effects are mediated by the Mas receptor signaling since treatment with the Mas antagonist A-779 or the genetic deletion of Mas receptor in the *mdx* mice model results in a worse dystrophic phenotype, dramatically increasing damage, fibrosis, and smad3 signaling and reducing skeletal muscle strength [21], resembling the SM phenotype observed in DMD.

Another mechanism for the reduction of damage and fibrosis mediated by Ang-(1-7) in dystrophic muscle is by the decrease in CTGF levels (Fig. 3). Interestingly,

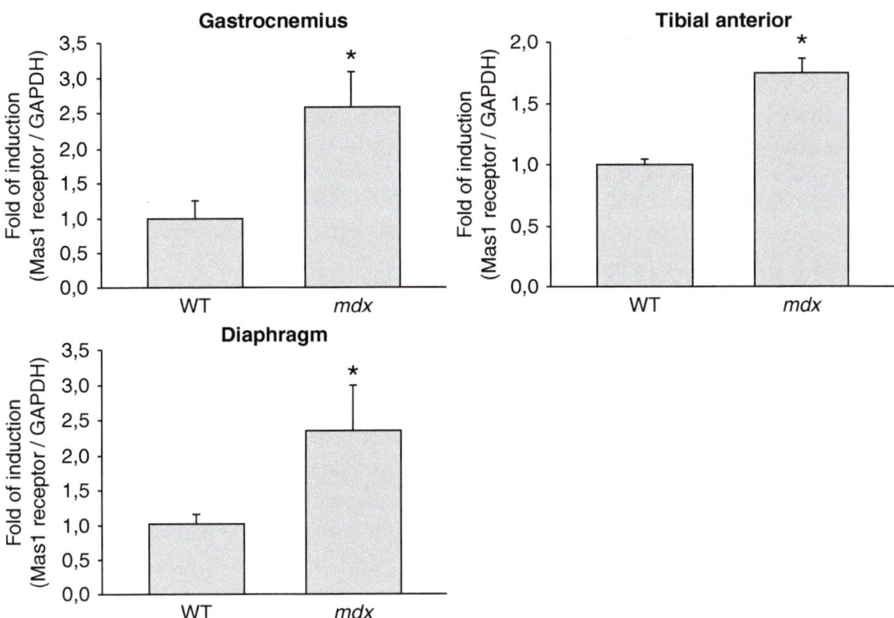

Fig. 2 Mas receptor expression is increased in skeletal muscle of *mdx* mice. The relative expression of Mas receptor was evaluated by qPCR using taqman probes for Mas1 and GAPDH as control, cDNA was obtained from total RNA extracts from wild-type (WT) and *mdx* mice Gastrocnemius, tibial anterior, and diaphragm muscles. T-test $*p < 0.05$ vs WT

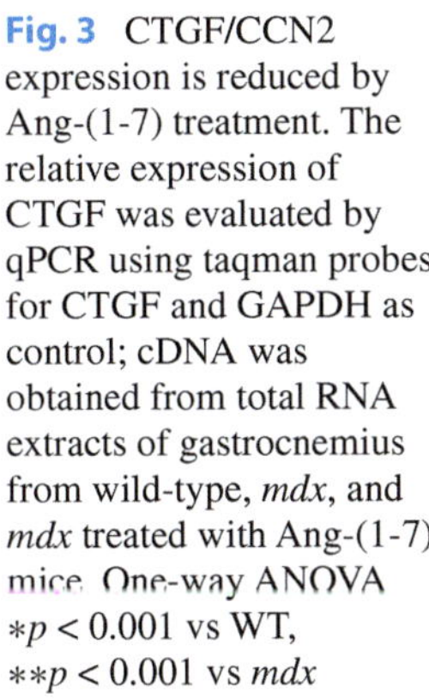

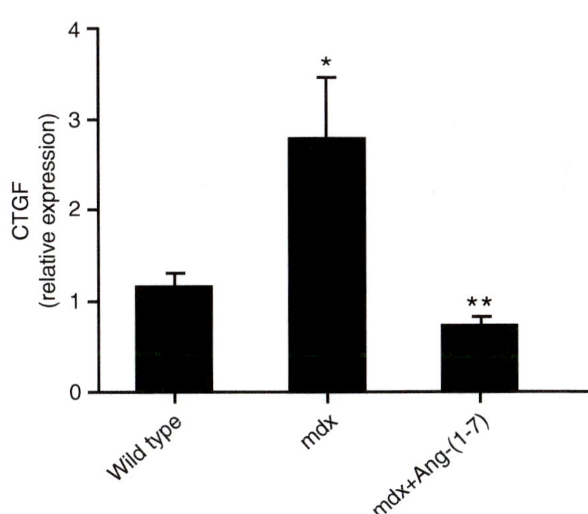

Fig. 3 CTGF/CCN2 expression is reduced by Ang-(1-7) treatment. The relative expression of CTGF was evaluated by qPCR using taqman probes for CTGF and GAPDH as control; cDNA was obtained from total RNA extracts of gastrocnemius from wild-type, *mdx*, and *mdx* treated with Ang-(1-7) mice. One-way ANOVA $*p < 0.001$ vs WT, $**p < 0.001$ vs *mdx*

Ang-(1-7) can also reduce muscle fibrosis in a model of radiation-induced fibrosis by reducing TGF-β and CTGF [37].

Overall, Ang-(1-7) protects muscle from damage and fibrosis, and since the *mdx* KO Mas mice had a worse phenotype [21], it seems that the endogenous Ang-(1-7)

protects from further damage, maybe by compensatory mechanisms, because ACE2 and MasR are increased in the *mdx* mice. It would be interesting to measure ACE2/Ang-(1-7)/MasR in DMD patient's samples. Interestingly, in the sarcoglican-δ null mice, there is a reduction of MasR levels compared with wild type that are restored by CD-Ang-(1-7) infusion [36], suggesting that the reduction in Ang-(1-7) receptor could lead to damage in this model.

Atrophy and Muscle Wasting Conditions

Muscle wasting, a condition observed in disuse, immobilization, aging, and some neuromuscular diseases, occurs by the increase of the catabolic pathways and reduction of the anabolic pathways. The muscle catabolic pathways upregulation is the result of increased activity of growth factors such as myostatin, which leads to protein degradation, and also of the induction of transcription factors such as NF-κB and FoxO3, which results in the expression of E3 ligases MuRF-1 and atrogin1 (also known as MAFbx) with consequent protein degradation by the ubiquitin proteasome pathway (UPS ubiquitin proteasome system) [38–41]. On the other hand, there is a reduction of anabolic pathways such as IGF-1 and its downstream signaling cascade PI3K/AKT/mTOR, leading to decreased protein synthesis and cell growth [39, 41, 43], the reduction of IGF-1 also activates the FoxO 1/3 transcription factors that induce atrogin1 [38, 42] tilting the balance toward the catabolic pathways, finally resulting in a decrease of SM mass.

Ang-II drives the muscle wasting by inducing the catabolism and inhibiting IGF-1 signaling [44]; it induces NF-κB, which in turns induces Atrogin1. Ang-II drives muscle wasting by increasing oxidative pathways [45, 46], leading to activation of catabolic pathways [47, 48]. Also by inducing protein phosphatase 2C alpha (PP2Cα), Ang-II reduces AMPK signaling also leading to muscle wasting [49]. Interestingly, the knockdown of PP2Cα reduces Ang-II-mediated atrophy by a mechanism involving mitochondrial recycling [50].

Cabello–Verrugio's group has studied Ang-(1-7) effects in several models of muscle atrophy. In a model of atrophy induced by Ang-II, the treatment with Ang-(1-7) prevented the reduction in fiber size, the reduction in strength, and maintained the levels of myosin heavy chain; these effects were dependent on Mas receptor since, in the presence of A-779 (the Mas receptor antagonist), the protective Ang-(1-7) effects were lost [51]. Ang-(1-7) prevented the increase of the expression of atrogin1 and MuRF1. Also, it was found that Ang-(1-7) induces the phosphorylation of AKT in a Mas-dependent fashion. In vitro, the inhibition of AKT inhibited the anti-atrophic effects of Ang-(1-7) [51]. Further experiments showed that Ang-(1-7) reduces the myonuclear apoptosis induced by Ang-II, as observed by the reduction in the apoptotic nuclei, reduction in the caspases 8 and 9, and reduction in the activity of caspase 3 and the Bax/Bcl-2 ratio. These effects were Mas-dependent because A-779 prevented Ang-(1-7) protection [52]. In a model of atrophy induced by limb immobilization, Ang-(1-7) also prevented the wasting effects, such as the reduction of fiber size decreased MHC levels, and reduced the E3 Ligases Atrogin-1 and MuRF-1 protein levels, by a mechanism that involved the phosphorylation of the IGF-1 receptor, AKT, and p70S6K resulting in the inhibition of FoxO3. The activation of IGF receptor (IGF-R) was necessary for the anti-atrophic effects of Ang-(1-7). Importantly, all these anti-atrophic effects of Ang-(1-7) were lost in the KO Mas mice [53]. Interestingly, a dendrimer carrying Ang-(1-7) (PAMAM-OH-Ang-(1-7)) that can encapsulate two molecules of Ang-(1-7) was administered intraperitoneally to mice, and also prevented the inmobilization-induced atrophy [54].

In a model of atrophy induced by the endotoxin lipopolysaccharide (LPS), Ang-(1-7) treatment prevented the atrophy induced by LPS, as seen by the maintenance of fiber size and MHC levels and the prevention in the reduction muscle force. Ang-(1-7) reduced the levels of atrogin-1 and MuRF-1 induced by LPS, and these effects were Mas receptor-dependent since A-779 inhibited the Ang-(1-7) effects. Interestingly, Ang-(1-7) reduced the p38 phosphorylation mediated by LPS and this was necessary for the anti-atrophic effects of Ang-(1-7) [55].

Ang-(1-7) through Mas receptor, also prevented the atrophy induced by TGF-β *in vivo* and *in vitro*, as observed by maintenance of fiber size, MHC levels, and a reduction in the MuRF-1 levels. Interestingly, Ang-(1-7) reduced ROS production induced by TGF-β preventing its atrophic effects [56].

Skeletal Muscle Regeneration

Muscle regeneration occurs after muscle injury or damage, and is a regulated process: different cell types participate in the regeneration – first, there is an acute inflammatory response that in turns activates the resident fibroblasts and FAPS that start producing ECM molecules to seal the injury site, then this ECM is degraded and fibroblasts die by apoptosis. During this process, the satellite cells begin to proliferate and ultimately differentiate to form new muscle fibers [2, 57]. Ang-II impairs muscle regeneration by reducing satellite cell proliferation and differentiation capacity [14]. Losartan treatment showed an improved regeneration capacity in CTX injury in a fibrilin-1 deficient mice, *mdx* mice, and in old mice with sarcopenia by a mechanism that involve TGF-β signaling reduction [30, 58].

The potential role of Ang-(1-7) participating in muscle regeneration has not been fully elucidated, since there are no reports to our knowledge. Nevertheless, this issue is being studied by us and so far we have found that systemic infusion with Ang-(1-7) accelerates the regeneration by modulating macrophages population and increasing satellite cells number, by a mechanism dependent of iNOS activation (Ramirez et al., ms in preparation)). Interestingly, in a model of chronic injury, by repeated cycles of BaCl₂ injection there are increased levels and activity of ACE2 [18], suggesting that Ang-(1-7) is present in the chronic damage mice model. What is the function played by Ang-(1-7) in skeletal muscle regeneration is still an open question.

Future Directions

There are still open questions about Ang-(1-7) role in skeletal muscle biology and disease. For example, we still don't know the mechanisms by which Ang-(1-7) has a protective role in SM disease downstream Mas receptor activation, we know that there are some signaling pathways involved like IGF-1 and AKT in the prevention of atrophy [51, 53], but it is still unclear if the same pathways are activated in the protection of damage and fibrosis in DMD. In regeneration, Ang-(1-7) induces iNOS but we do not know if it occurs directly or indirectly by activating another route. Recently, the Kallikrein Kinin system (KKS), a hypotensor and anti-fibrotic system [59], has been involved in DMD [60], so it would be interesting to elucidate if Ang-(1-7) could also be synergizing with Bradykinin, the main product of the KKS, to exert a beneficial effect.

Ang-(1-7) in SM has the opposite actions to Ang-II, and can reduce TGF-β and its signaling molecules, so it would be of interest to search what is the effect of Ang-(1-7) in other neuro- muscular diseases that have increased levels of TGF-β, such as ALS [61] or Marfan Syndrome [30]. Since Ang-(1-7) has anti-atrophic effects, there is also an interesting question if this peptide has any effect on sarcopenia or aged muscles.

There are efforts being made to test Ang-(1-7) analogs or Mas receptor agonists or different ways to deliver Ang-(1-7), that are less prone to degradation, since it could have a therapeutic potential; in this regard, cyclodextrin-delivered Ang-(1-7) has been proved and could be a good candidate [21, 36]; other way of delivering Ang-(1-7) was tested in a model of atrophy by the use of PAMAM-OH with promising results [54]. There are analogs, such as A-1317, being tested together with Robson Santo's group in DMD mice model to see if the same beneficial effects as for Ang-(1-7) can be achieved using this molecule. Other candidate to prove could be the cyclic Ang-(1-7) that has been tested in mice model of cardiac infarction [62]. AVE0991 a Mas agonist that has been tested in different pathologies such as heart failure, arthritis, gastric ulcers, and asthma [63–67] could also be tested in SM diseases.

Physical Exercise and ACE2/Angiotensin-(1-7)/Mas Axis

Physical exercise represents an important nonpharmacological tool for prevention and treatment of cardiovascular and metabolic disease. Skeletal muscle is the main organ affected by acute and chronic exercise and can act as an endocrine organ producing and secreting myokines [68]. The skeletal muscle contraction is a potent stimulus to induce exercise adaptations in many other organs such as brain, heart, lung, kidney, and liver. Here, we will describe:

- Angiotensin-(1-7) effects in trained animals.
- Modulation of ACE2/Ang-(1-7)/MAS axis in response to physical exercise.
- Angiotensin-(1-7) activation (treatment) or inactivation (Mas-deficiency) and exercise training.

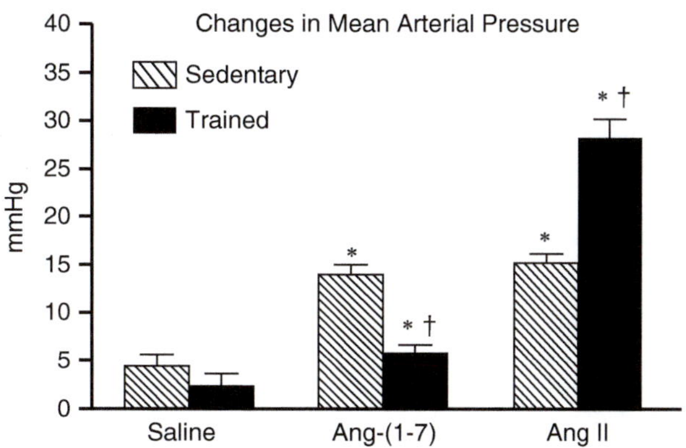

Fig. 4 Changes in mean arterial pressure (mm Hg) produced by microinjection of saline, Ang-II (50 ng) or Ang-(1-7) (50 ng) into the RVLM of sedentary and trained rats. $*p < 0.05$ as compared with saline

Angiotensin-(1-7) Effects in Trained Animals

The angiotensin-(1-7) effects in trained and sedentary brains and vessels have shown the role of this peptide in response to aerobic training exercise. These studies demonstrated that chronic exercise affects the acute actions of Ang-(1-7) (hemodynamic and vascular responses). In some pathological conditions as in hypertensive rats (SHR), these effects are more evident.

Brain

The first study showing the effects of angiotensin-(1-7) associated with exercise was published in 2005 by Becker et al. The authors demonstrated an involvement of Ang-(1-7) in physical exercise adaptations [69]. In this study, the rats performed 20 swimming exercise sessions, 1 hour per day, 5 days a week. After 4 weeks of training, cardiovascular effects produced by microinjections in rostroventrolateral medulla (RVLM) of the angiotensin peptides, Ang-II and Ang-(1-7) were evaluated in trained and sedentary animals. The exercise training enhanced Ang-II and attenuated Ang-(1-7) pressor effect (Fig. 4).

The spontaneous hypertensive rats (SHR) presented altered renin-angiotensin system in the RLVM and the effects of 12 weeks of running training (5 days per week; 60 min per day at 15–20 m/min) were evaluated by Ren et al. [70]. Aerobic training significantly reduced sarthan (antagonist of Ang-II) or increased A-779 (antagonist of Ang-(1-7)) cardiovascular responses to central application of each antagonist, respectively. The protein expression of MasR in the RVLM was significantly elevated in SHR following aerobic training. These results suggest that the central effect in the pressor response for Ang-(1-7) is modulated by physical exercise and future research needs to be made to understand the mechanisms involved in this modulation.

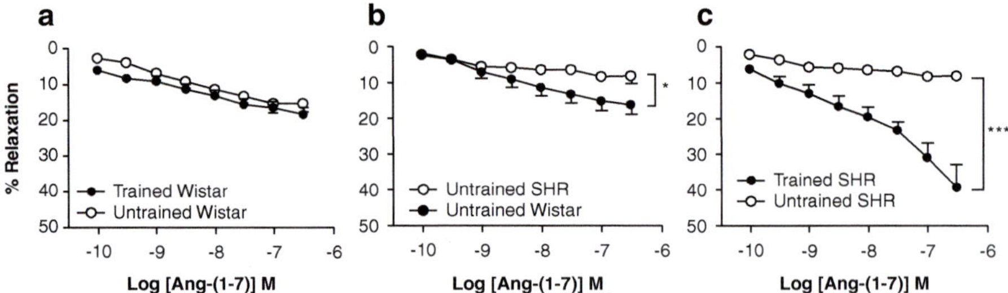

Fig. 5 Vasodilator effect of Ang-(1-7) in endothelium-intact aortic rings from (**a**) trained wistar and untrained wistar rats, (**b**) untrained SHR and untrained and (**c**) untrained and trained SHR rats, $*p < 0.05$; $***p < 0.001$

Fig. 6 Representative immunoblots of MasR in untrained and trained SHR aortas. Top shows representative Western blot signals. $*p < 0.05$ untrained vs. trained SHR

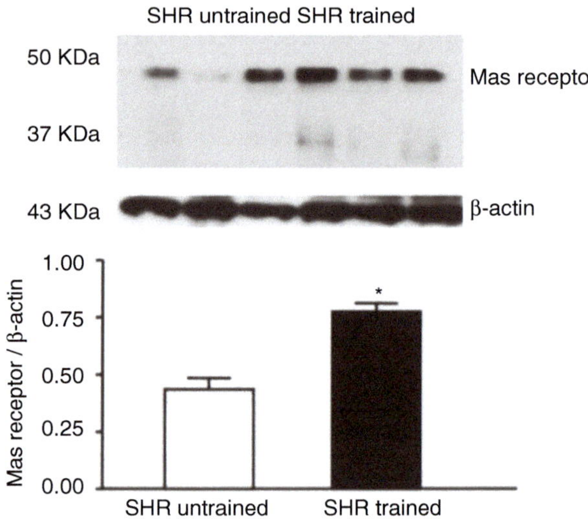

Vessels

The vascular effects of Ang-(1-7) in trained animals (8-weeks of overload 5% of the body weight swimming training) were investigated in the aorta of SHR-trained animals [71]. Untrained SHR had an impaired vasodilator response to Ang-(1-7) and exercise training could reverse this response (Fig. 5). The Ang-(1-7) vasodilator effect was abrogated by A-779 and d-Pro(7)-Ang-(1-7) (selective Ang-(1-7) receptor antagonists) and by removal of the endothelium.

This study also showed that only in SHR-trained animals, the aorta MasR protein expression was substantially increased and correlated with the Ang-(1-7) effect (Fig. 6).

The cardioprotective effects promoted by exercise training could implicate the activation of ACE2/Ang-(1-7)/Mas axis, especially in pathological conditions such as arterial hypertension.

Modulation of ACE2/ANG-(1-7)/MAS Axis in Response to Physical Exercise

The effects of physical exercise on ACE2/Ang-(1-7) /Mas activation in different pathologies/organs will be discussed in the following sections.

Heart, Hypertension, and Heart Failure

Aerobic exercise training induces several cardiac effects that culminate in an aerobic capacity improvement due in part to the increased ventricular stroke volume, cardiac output, and left ventricle hypertrophy [72]. The left-ventricle hypertrophy is dependent on volume training; therefore, Fernandes et al. investigated the modulation of ACE2 and Ang-(1-7) in rats following distinct swimming training protocols [73]:

- Low-intensity, moderate-volume exercise: 60-minute per session, 5 days a week, for 10 weeks
- Low-intensity, high-volume exercise: the same swimming training protocol described until the end of the eighth week and on the ninth week, rats were trained twice a day (60 min per session and an interval of 4 hours between sessions).

All exercise session was carried out with caudal dumbbells weighing 5% of animal body weight. ACE2, Ang-(1-7), and the ratio Ang-(1-7) in left ventricle increase significantly in both exercise protocols and involve regulatory MicroRNAs (miR-27a, miR-27b, and miR-143).

Physical training effect on MasR expression in hearts under different physiological and pathological conditions has been evaluated by Dias-Peixoto et al. [74]. The physiological stimulus was the swimming training performed 40–60 min per day, 5 days per week over 10 weeks. No changes in MasR expression in the trained-left ventricle Sprague–Dawley (SD) rats were observed. However, in some pathological conditions such as isoproterenol treatment and infarction, MasR downregulated responses have been evidenced. Since in hypertensive rats (SHR), the role of Ang-(1-7) mediating cardioprotective effects has been consistently demonstrated (please see Actions of Angiotensin-(1-7): Heart), the reduction of MasR expression showed by Dias-Peixoto et al. could explain in part the cardiac damage in these model.

The cardiac effects of 8 weeks period of 5% overload swimming training (1 hour per day, 5 days a week) and the role of Ang-(1-7) in normotensive (Wistar) and hypertensive (SHR) rats were evaluated by Filho et al. [75]. Interestingly, the plasma levels of Ang II reduced in both trained groups but only SHR-trained had a significant increase in left ventricle levels of Ang-(1-7). Cardiac (left ventricle) gene expression of MasR was significantly increased in trained SHR, but not in trained Wistar rats (Fig. 7).

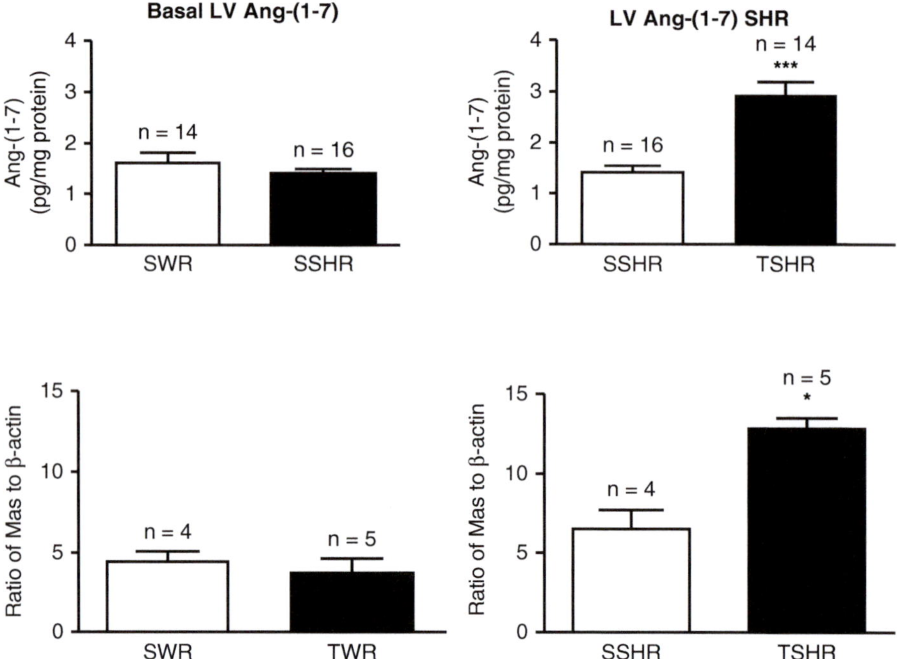

Fig. 7 Left ventricle Ang(1-7) levels in sedentary and trained Wistar rats and SHR (upper panel) and relative levels of MasR mRNA in left ventricle assessed by semi-quantitative RT-PCR (lower panel), $***p < 0.001$ compared with sedentary rats, $*p < 0.01$ compared with sedentary SHR. SWR sedentary Wistar, SSHR sedentary SHR, TSHR trained-SHR, TWR trained-Wistar

Additionally, it has been described that heart failure reduction of ACE and ACE2 levels in the brain can be normalized through chronic running exercise (30 min per day, 6 days a week for 3 weeks) [76].

Recently, Tyrankiewicz et al. analyzed systemic (plasma) and local (heart/aorta) changes in ACE/ACE-2 balance in Tgαq*44 mice in course of heart failure (HF) [77]. Tgαq*44 mice present cardiomyocyte 38 specific Gαq overexpression and develop late onset of HF. The HF development in this animal model is associated with systemic and local activation of ACE/Ang-II axis and this effect is counterbalanced by an important ACE2/Ang-(1-7) activation. In this study, they also evaluated voluntary wheels running performance in young and aged animals, but only in 12-month-old Tgαq*44 mice, the mean distance and time were significantly decreased. However, running wheels is not the best "model" to assess cardiovascular performance considering that neurotransmitter systems are involved in wheel-running behavior. In this line, running-wheel training in a ACE2-deficiency mice affects physical performance and impairs cardiac and skeletal muscle adaptations to exercise [17].

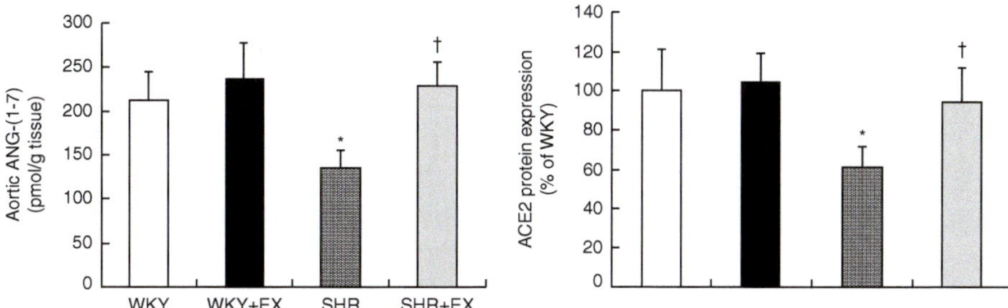

Fig. 8 Effects of exercise training on aortic Ang-(1-7) content and ACE2 expression in SHR. Values are represented as mean ± S.D. $*p < 0.05$ versus WKY; $\dagger p < 0.05$ versus SHR; $n = 10$–12 in each group

Another effect of ACE2 could involve tryptophan that was recently shown to stimulate the expression of myogenic genes [78]. The muscle strength in ACE2-knockout mice could be impaired by the reduction of tryptophan action, considering that ACE2 has an important effect on tryptophan uptake [79].

Moderate-intensity running training for 12 weeks modulates RAS axis by reducing AngII and increasing Ang-(1-7) in aorta of trained SHR [80]. Confirming the role of Ang-(1-7) promoting vasoprotective effect induced by physical exercise, Ang-(1-7) and ACE2 protein levels were normalized (Fig. 8).

The effects of running aerobic exercise training (60 min at 60% of peak VO_2, 5 days a week for 8 week) in a ischemic model of cardiac heart failure (CHF) confirm that exercise training causes a shift in the Ang-(1-7)/Mas axis in skeletal muscle of CHF rats [81]. CHF (left coronary artery ligation) reduced ACE2 serum activity; however, exercise training restored and increased the Ang-(1-7)/Ang-II ratio. Skeletal muscle ACE and ACE2 activity and protein did not change, but Ang-(1-7) in plantaris and MasR in the soleus of CHF mice significantly increased. It is important to note that the local RAS (skeletal muscle) is not directly affected by circulation levels of angiotensin peptides.

Another elegant study using SHR provides strong evidence that low-intensity aerobic training downregulates RAS not only in vessels but also in the kidney and plasma of normotensive and hypertensive rats [82].

Preeclampsia and Estrogen Deficiency

Exercise training can attenuate/prevent preeclampsia and these protective effects could be associated with RAS modulation [83]. In a previous study on preeclampsia mouse model (overexpressing human angiotensinogen and human renin), it was demonstrated that voluntary running is effective in attenuating blood pressure increase [84]. Preeclamptic mice presented lower MasR protein expression in aorta

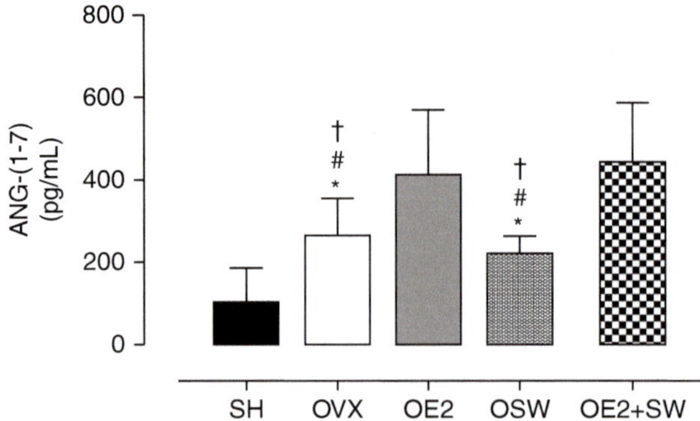

Fig. 9 ANG-(1-7) plasma levels in ovariectomized SHRs subjected to regular swimming or estrogen therapy. SH ($n = 7$), OVX ($n = 4$), OE2 ($n = 4$), OSW ($n = 6$), and OE2 + SW ($n = 5$) groups. Data are expressed as the means ± SEM. One-way ANOVA with Tukey's post-hoc test was used. $*p < 0.05$ vs. SH group, $\#p < 0.05$ vs. OE2 group, $\dagger p < 0.05$ vs. OE2 + SW group

and placenta compared to normotensive, and the voluntary running is able to "normalize" the MasR expression.

In estrogen deficiency and hypertensive conditions, some RAS components are altered. Endlich et al. demonstrated that ovariectomized spontaneously hypertensive trained rats (60 min per day, 5 days a week, lasted 8 weeks) increased constrictor responses to Ang-II and decreased dilatory responses to Ang-(1-7) independently of estrogen therapy [85]. A significant increase in Ang-(1-7) aorta was found in exercise trained-groups and in the estrogen therapy group. This study has demonstrated that in a model of menopause in rodents, the estrogen therapy and swimming training are able to decrease systolic blood pressure and increase Ang-(1-7) (Fig. 9).

Diabetes and Obesity

It is quite well established the protective metabolic effects of Ang-(1-7) in diabetes and obesity. However, little is known about the involvement of this heptapeptide in metabolic protective responses induced by chronic exercise. Somineni et al. investigated whether exercise training and/or metformin improve glucose homeostasis and downregulate renal ADAM17 and ACE2 shedding in db/db mice [86]. The training consisted 10 weeks of aerobic training and daily exercise (1 hour per day of forced exercise on walking wheel). Exercise training alone and in combination with metformin prevented shedding of renal ACE2 by decreasing ADAM17 protein, and this may partially contribute to renal protection.

The exercise training (8 weeks of training, 50–75% of maximal running speed, 60 min per day and 4 days a week) normalizes (prevents the increase) angiotensin-converting enzyme, ACE (activity and protein) and Ang-II in hepatic tissue in fructose overloaded rats [87]. In addition, ACE2, Ang(1-7), and Mas receptor increase in the liver leading to ACE/ACE2, Ang-II/Ang(1-7), and AT1R/Mas receptor ratios towards normal values. Interestingly, no changes in the systemic RAS components were detected.

The effects of High Intensity Interval Training (HIIT) mice fed high-fat or high-fructose in RAS components were recently evaluated [88]. The exercise training consisted of HIIT protocol: 2 min of high-intensity 45 m/min (90% VO2) running and 1 min of low intensity at 15 m/min (30% VO2) running 3 days a week on alternate days over 12 weeks. The authors showed that the exercise training enhanced the insulin sensitivity and these results could be related to reduced levels of the classic RAS components and increase in ACE2 and MasR in the HIIT mice compared with the nontrained group.

Asthma

Unpublished data suggested that the protective effects of chronic aerobic exercise in asthma involves the activation of MasR in a model of chronic asthma. In their study, Gregório et al. (2016), induced asthma through OVA albumin challenge and submitted the animals to running exercise (1 hour per day, 5 days a week, during 6 weeks). The IgG circulating levels in "asthmatic" animals were significantly elevated but running training was able to abolish this increase. If physical exercise is associated with MasR blockade (A-779 antagonist), the levels of IgG are significantly higher compared to asthmatic animals even when associated with exercise training. These results suggested that the beneficial effects of exercise could be missing when MasR actions were blocked.

Angiotensin-(1-7) Activation (Treatment) or Inactivation (Mas-Deficiency) and Exercise Training

Some studies investigated the additive effects of Ang-(1-7) treatment associated with physical training and the effects of exercise training in Mas-deficiency animals (Mas−/−) [89, 90]. Additionally, the protective effects induced by exercise training appear to be MasR-dependent. The Mas−/− mice did not decrease blood pressure and improve body composition (observed in Mas+/+ mice) when submitted to voluntary running through 6 weeks (Motta-Santos, unpublished data).

The Ang-(1-7) infusion effects associated or not to swimming training was evaluated in hypertensive rats (2K1C) [91]. Exercise training consisted of swimming training performed 1 hour per day, 5 days a week through 4 weeks. Intriguing, Ang-(1-7) treatment attenuated hypertension and cardiac hypertrophy only 2K1C-trained rats and MasR was upregulated only in the left-ventricles of trained 2K1C rats.

These results suggest that the beneficial effect of Ang-(1-7) is potentiated by physical performance.

The treatment with orally active Ang-(1-7) included in hydroxy-propyl-beta-cyclodextrin produces several cardioprotective effects (see Heart Chapter) and its effects combined or not to physical training in spontaneously hypertensive rats (SHR) were investigated by Bertagnolli et al. [90]. The SHR were divided in control (tap water) or treated with HPβ-CD/Ang-(1-7) with or without running exercise training (1 hour per day, 5 days per week, 10 weeks). Similar beneficial effects to the ones produced by exercise training were observed in HPβ-CD/Ang-(1-7) non-trained SHR. These effects include decreased arterial blood pressure (BP) and heart rate, improved left ventricular (LV) end-diastolic pressure, restored the maximum and minimum derivatives (dP/dT), and decreased cardiac hypertrophy index. Additionally, an improvement in autonomic control by attenuating sympathetic modulation on heart and vessels and the SAP variability, as well as increasing parasympathetic modulation and HR variability were observed in trained and Ang-(1-7)-treated SHR animals.

While activation of Ang-(1-7)/MAS promotes effects similar to those seen in trained animals, on the other hand, the MasR deficiency (Mas−/−) abolishes some benefits of the physical exercise.

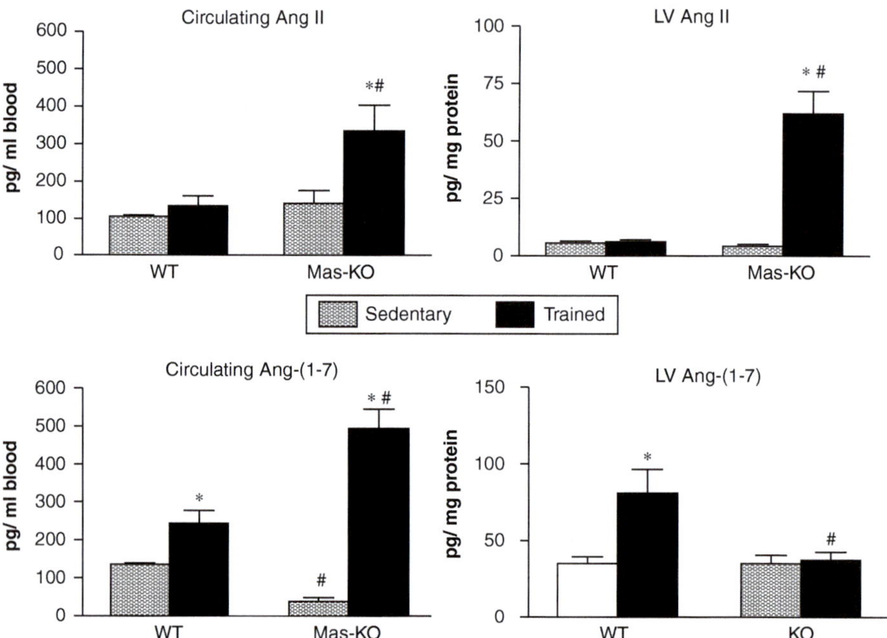

Fig. 10 Ang-II and Ang-(1-7) levels in the total blood and left ventricle (LV) of sedentary and trained Mas-KO and WT mice as determined by RIA. Statistically significant differences between the groups are indicated as *$p < 0.05$ in comparison to sedentary control and #$p < 0.05$ in comparison to WT

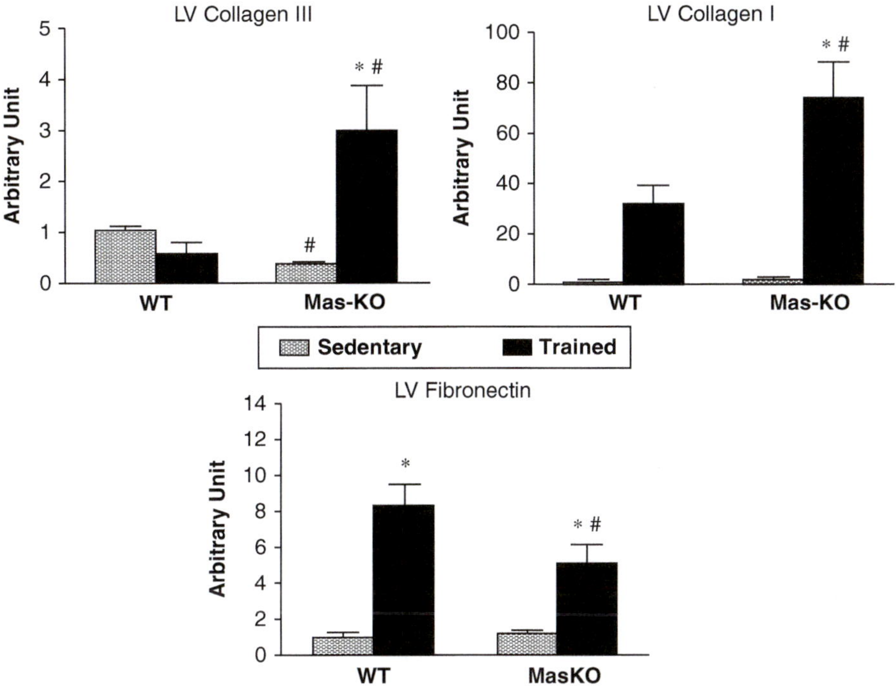

Fig. 11 Left ventricle (LV) mRNA expression of extracellular matrix proteins in sedentary and trained Mas-KO and WT mice. Data are presented as means ± SE. Statistically significant differences between the groups are indicated as $*p < 0.05$ in comparison to sedentary control and $\#p < 0.05$ in comparison to WT

Mas−/− mice presented heart and skeletal muscle remodeling alterations (please see Heart and Skeletal muscle chapter). Since Ang-(1-7) could be involved in the physiological cardiac remodeling induced by exercise training, Guimaraes et al. aimed to investigate the cardiac physical exercise effects in Mas−/− [89]. Six weeks of swimming training (5 days per week, once a day for 1 hour with an 80% of maximal load workload attached to the tail) induced similar increase (∼10%) in cardiomyocyte diameter in Mas−/− and Wild-Type (WT) animals. However, in sedentary groups, circulating levels of Ang-(1-7) were significantly lower in Mas−/− as compared to WT. Also, Ang-II levels in blood and LV increase only in the Mas-KO-trained (Fig. 10). Additionally, to a null increase in the cardiac Ang-(1-7) levels of Mas−/−trained group, they presented a higher collagen I and II gene expression compared to WT (Fig. 11). The authors concluded that exercise training was able to induce an increase in the Ang II/Ang-(1-7) blood ratio only in Mas-deficiency-trained, suggesting strong imbalance in circulating RAS with a predominance of Ang-II in Mas−/−.

The physical performance associated with asthma model in Mas−/− mice also suggests that the asthmatic Mas−/− animals presented a worse performance in a test of maximum physical exercise compared with WT asthmatic group [92].

Human Studies

A series of human studies have been conducted to verify the effects of acute and chronic exercise in Ang-(1-7) peptides levels. The preliminary and unpublished results suggest that Ang-(1-7) plasmatic levels after exercise (acute and chronic) is dependent of duration, intensity, and mode (resistance, aerobic, and combined exercise). The status of training is another point that can interfere in the systemic RAS modulation.

Acute eccentric physical exercise can promote some microlesions, inflammation, and pain (DOMS) in skeletal muscle but after a chronic period important adaptations can be reached. This include strength gain, power increase, and remodeling (hypertrophy). The effects of Ang-(1-7) oral compound associated with eccentric exercise induced muscle damage (squat exercise) was recently tested in young healthy subjects. Surprisingly, the Ang-(1-7)-treated group presented less pain and higher strength compared to placebo group [93]. The inflammatory marker responses suggest that the Ang-(1-7) can attenuate inflammation or maybe accelerate the recovery following eccentric exercise protocol. Further studies can elucidate the mechanism involved in this effect including local analysis directly to skeletal muscle.

Bibliography

1. Judson RN, Zhang RH, Rossi FM. Tissue-resident mesenchymal stem/progenitor cells in skeletal muscle: collaborators or saboteurs? FEBS J. 2013;280(17):4100–8.
2. Zammit PS. Function of the myogenic regulatory factors Myf5, MyoD, Myogenin and MRF4 in skeletal muscle, satellite cells and regenerative myogenesis. Semin Cell Dev Biol. 2017;72:19–32.
3. Pannerec A, Marazzi G, Sassoon D. Stem cells in the hood: the skeletal muscle niche. Trends Mol Med. 2012;18(10):599–606.
4. Latroche C, et al. Skeletal muscle microvasculature: a highly dynamic lifeline. Physiology (Bethesda). 2015;30(6):417–27.
5. Mounier R, Chretien F, Chazaud B. Blood vessels and the satellite cell niche. Curr Top Dev Biol. 2011;96:121–38.
6. Reneland R, Lithell H. Angiotensin-converting enzyme in human skeletal muscle. A simple in vitro assay of activity in needle biopsy specimens. Scand J Clin Lab Invest. 1994;54(2):105–11.
7. Ward PE, Russell JS, Vaghy PL. Angiotensin and bradykinin metabolism by peptidases identified in skeletal muscle. Peptides. 1995;16(6):1073–8.
8. Schaufelberger M, et al. Angiotensin-converting enzyme gene expression in skeletal muscle in patients with chronic heart failure. J Card Fail. 1998;4(3):185–91.
9. Sun G, et al. Intramuscular renin-angiotensin system is activated in human muscular dystrophy. J Neurol Sci. 2009;280(1–2):40–8.
10. Fairclough RJ, Wood MJ, Davies KE. Therapy for Duchenne muscular dystrophy: renewed optimism from genetic approaches. Nat Rev Genet. 2013;14(6):373–8.
11. Tiret L, et al. Evidence, from combined segregation and linkage analysis, that a variant of the angiotensin I-converting enzyme (ACE) gene controls plasma ACE levels. Am J Hum Genet. 1992;51(1):197–205.

12. Vaughan D, et al. The angiotensin converting enzyme insertion/deletion polymorphism modifies exercise-induced muscle metabolism. PLoS One. 2016;11(3):e0149046.
13. Malendowicz SL, et al. Angiotensin II receptor subtypes in the skeletal muscle vasculature of patients with severe congestive heart failure. Circulation. 2000;102(18):2210–3.
14. Yoshida T, et al. Angiotensin II inhibits satellite cell proliferation and prevents skeletal muscle regeneration. J Biol Chem. 2013;288(33):23823–32.
15. Painemal P, et al. Transforming growth factor type beta 1 increases the expression of angiotensin II receptor type 2 by a SMAD- and p38 MAPK-dependent mechanism in skeletal muscle. Biofactors. 2013;39(4):467–75.
16. Yoshida T, Huq TS, Delafontaine P. Angiotensin type 2 receptor signaling in satellite cells potentiates skeletal muscle regeneration. J Biol Chem. 2014;289(38):26239–48.
17. Motta-Santos D, et al. Effects of ACE2 deficiency on physical performance and physiological adaptations of cardiac and skeletal muscle to exercise. Hypertens Res. 2016;39(7):506–12.
18. Riquelme C, et al. ACE2 is augmented in dystrophic skeletal muscle and plays a role in decreasing associated fibrosis. PLoS One. 2014;9(4):e93449.
19. Echeverria-Rodriguez O, Del Valle-Mondragon L, Hong E. Angiotensin 1-7 improves insulin sensitivity by increasing skeletal muscle glucose uptake in vivo. Peptides. 2014;51:26–30.
20. Morales MG, et al. Expression of the Mas receptor is upregulated in skeletal muscle wasting. Histochem Cell Biol. 2015;143(2):131–41.
21. Acuna MJ, et al. Restoration of muscle strength in dystrophic muscle by angiotensin-1-7 through inhibition of TGF-beta signalling. Hum Mol Genet. 2014;23(5):1237–49.
22. Fernandes T, Hashimoto NY, Oliveira EM. Characterization of angiotensin-converting enzymes 1 and 2 in the soleus and plantaris muscles of rats. Braz J Med Biol Res. 2010;43(9): 837–42.
23. Cofre C, et al. Transforming growth factor type-beta inhibits Mas receptor expression in fibroblasts but not in myoblasts or differentiated myotubes; relevance to fibrosis associated to muscular dystrophies. Biofactors. 2015;41(2):111–20.
24. Pessina P, et al. Novel and optimized strategies for inducing fibrosis in vivo: focus on Duchenne muscular dystrophy. Skelet Muscle. 2014;4:7.
25. Serrano AL, Munoz-Canoves P. Fibrosis development in early-onset muscular dystrophies: mechanisms and translational implications. Semin Cell Dev Biol. 2017;64:181–90.
26. Smith LR, Barton ER. Regulation of fibrosis in muscular dystrophy. Matrix Biol. 2018;602:68–9.
27. Morales MG, et al. The pro-fibrotic connective tissue growth factor (CTGF/CCN2) correlates with the number of necrotic-regenerative foci in dystrophic muscle. J Cell Commun Signal. 2018;12(1):413–21.
28. Morales MG, et al. Reducing CTGF/CCN2 slows down mdx muscle dystrophy and improves cell therapy. Hum Mol Genet. 2013;22(24):4938–51.
29. Morales MG, et al. CTGF/CCN-2 over-expression can directly induce features of skeletal muscle dystrophy. J Pathol. 2011;225(4):490–501.
30. Cohn RD, et al. Angiotensin II type 1 receptor blockade attenuates TGF-beta-induced failure of muscle regeneration in multiple myopathic states. Nat Med. 2007;13(2):204–10.
31. Morales MG, et al. Inhibition of the angiotensin-converting enzyme decreases skeletal muscle fibrosis in dystrophic mice by a diminution in the expression and activity of connective tissue growth factor (CTGF/CCN-2). Cell Tissue Res. 2013;353(1):173–87.
32. Cabello-Verrugio C, et al. Angiotensin II receptor type 1 blockade decreases CTGF/CCN2-mediated damage and fibrosis in normal and dystrophic skeletal muscles. J Cell Mol Med. 2012;16(4):752–64.
33. Cozzoli A, et al. Enalapril treatment discloses an early role of angiotensin II in inflammation- and oxidative stress-related muscle damage in dystrophic mdx mice. Pharmacol Res. 2011;64(5):482–92.
34. Morales MG, et al. Angiotensin II-induced pro-fibrotic effects require p38MAPK activity and transforming growth factor beta 1 expression in skeletal muscle cells. Int J Biochem Cell Biol. 2012;44(11):1993–2002.

35. Cabello-Verrugio C, et al. Fibrotic response induced by angiotensin-II requires NAD(P)H oxidase-induced reactive oxygen species (ROS) in skeletal muscle cells. Biochem Biophys Res Commun. 2011;410(3):665–70.

36. Sabharwal R, Chapleau MW. Autonomic, locomotor and cardiac abnormalities in a mouse model of muscular dystrophy: targeting the renin-angiotensin system. Exp Physiol. 2014;99(4):627–31.

37. Willey JS, et al. Angiotensin-(1-7) attenuates skeletal muscle fibrosis and stiffening in a mouse model of extremity sarcoma radiation therapy. J Bone Joint Surg Am. 2016;98(1):48–55.

38. Sandri M, et al. Foxo transcription factors induce the atrophy-related ubiquitin ligase atrogin-1 and cause skeletal muscle atrophy. Cell. 2004;117(3):399–412.

39. von Haehling S, et al. Muscle wasting and cachexia in heart failure: mechanisms and therapies. Nat Rev Cardiol. 2017;14(6):323–41.

40. Ibebunjo C, et al. Genomic and proteomic profiling reveals reduced mitochondrial function and disruption of the neuromuscular junction driving rat sarcopenia. Mol Cell Biol. 2013;33(2):194–212.

41. Meng SJ, Yu LJ. Oxidative stress, molecular inflammation and sarcopenia. Int J Mol Sci. 2010;11(4):1509–26.

42. Brioche T, et al. Muscle wasting and aging: experimental models, fatty infiltrations, and prevention. Mol Asp Med. 2016;50:56–87.

43. Dennison EM, Sayer AA, Cooper C. Epidemiology of sarcopenia and insight into possible therapeutic targets. Nat Rev Rheumatol. 2017;13:340.

44. Brink M, et al. Angiotensin II induces skeletal muscle wasting through enhanced protein degradation and down-regulates autocrine insulin-like growth factor I. Endocrinology. 2001;142(4):1489–96.

45. Semprun-Prieto LC, et al. Angiotensin II induced catabolic effect and muscle atrophy are redox dependent. Biochem Biophys Res Commun. 2011;409(2):217–21.

46. Sukhanov S, et al. Angiotensin II, oxidative stress and skeletal muscle wasting. Am J Med Sci. 2011;342(2):143–7.

47. Delafontaine P, Yoshida T. The renin-angiotensin system and the biology of skeletal muscle: mechanisms of muscle wasting in chronic disease states. Trans Am Clin Climatol Assoc. 2016;127:245–58.

48. Yoshida T, et al. Molecular mechanisms and signaling pathways of angiotensin II-induced muscle wasting: potential therapeutic targets for cardiac cachexia. Int J Biochem Cell Biol. 2013;45(10):2322–32.

49. Tabony AM, et al. Angiotensin II upregulates protein phosphatase 2Calpha and inhibits AMP-activated protein kinase signaling and energy balance leading to skeletal muscle wasting. Hypertension. 2011;58(4):643–9.

50. Tabony AM, et al. Protein phosphatase 2C-alpha knockdown reduces angiotensin II-mediated skeletal muscle wasting via restoration of mitochondrial recycling and function. Skelet Muscle. 2014;4:20.

51. Cisternas F, et al. Angiotensin-(1-7) decreases skeletal muscle atrophy induced by angiotensin II through a Mas receptor-dependent mechanism. Clin Sci (Lond). 2015;128(5):307–19.

52. Meneses C, et al. The angiotensin-(1-7)/ Mas axis reduces myonuclear apoptosis during recovery from angiotensin II-induced skeletal muscle atrophy in mice. Pflugers Arch. 2015;467(9):1975–84.

53. Morales MG, et al. Angiotensin-(1-7) attenuates disuse skeletal muscle atrophy in mice via its receptor, Mas. Dis Model Mech. 2016;9(4):441–9.

54. Marquez-Miranda V, et al. The complex of PAMAM-OH dendrimer with angiotensin (1-7) prevented the disuse-induced skeletal muscle atrophy in mice. Int J Nanomedicine. 2017;12:1985–99.

55. Morales MG, et al. Endotoxin-induced skeletal muscle wasting is prevented by angiotensin-(1-7) through a p38 MAPK-dependent mechanism. Clin Sci (Lond). 2015;129(6):461–76.

56. Abrigo J, et al. Angiotensin-(1-7) prevents skeletal muscle atrophy induced by transforming growth factor type Beta (TGF-beta) via Mas receptor activation. Cell Physiol Biochem. 2016;40(1–2):27–38.

57. Bentzinger CF, et al. Cellular dynamics in the muscle satellite cell niche. EMBO Rep. 2013;14(12):1062–72.

58. Burks TN, et al. Losartan restores skeletal muscle remodeling and protects against disuse atrophy in sarcopenia. Sci Transl Med. 2011;3(82):82ra37.

59. Regoli D, Gobeil F. Kinins and peptide receptors. Biol Chem. 2016;397(4):297–304.

60. Acuna MJ, et al. Blockade of Bradykinin receptors worsens the dystrophic phenotype of mdx mice: differential effects for B1 and B2 receptors. J Cell Commun Signal. 2018;12(3): 589–601.

61. Gonzalez D, et al. ALS skeletal muscle shows enhanced TGF-beta signaling, fibrosis and induction of fibro/adipogenic progenitor markers. PLoS One. 2017;12(5):e0177649.

62. Seva Pessoa B, et al. Effect of a stable angiotensin-(1-7) analogue on progenitor cell recruitment and cardiovascular function post myocardial infarction. J Am Heart Assoc, 2015;4(2).

63. Cunha TM, et al. The nonpeptide ANG-(1-7) mimic AVE 0991 attenuates cardiac remodeling and improves baroreflex sensitivity in renovascular hypertensive rats. Life Sci. 2013;92(4–5):266–75.

64. Ferreira AJ, et al. The nonpeptide angiotensin-(1-7) receptor Mas agonist AVE-0991 attenuates heart failure induced by myocardial infarction. Am J Physiol Heart Circ Physiol. 2007;292(2):H1113–9.

65. da Silveira KD, et al. Anti-inflammatory effects of the activation of the angiotensin-(1-7) receptor, MAS, in experimental models of arthritis. J Immunol. 2010;185(9):5569–76.

66. Pawlik MW, et al. The renin-angiotensin system and its vasoactive metabolite angiotensin-(1-7) in the mechanism of the healing of preexisting gastric ulcers. The involvement of Mas receptors, nitric oxide, prostaglandins and proinflammatory cytokines. J Physiol Pharmacol. 2016;67(1):75–91.

67. Rodrigues-Machado MG, et al. AVE 0991, a non-peptide mimic of angiotensin-(1-7) effects, attenuates pulmonary remodelling in a model of chronic asthma. Br J Pharmacol. 2013;170(4):835–46.

68. Giudice J, Taylor JM. Muscle as a paracrine and endocrine organ. Curr Opin Pharmacol. 2017;34:49–55.

69. Becker LK, Santos RA, Campagnole-Santos MJ. Cardiovascular effects of angiotensin II and angiotensin-(1-7) at the RVLM of trained normotensive rats. Brain Res. 2005;1040(1–2):121–8.

70. Ren CZ, et al. Exercise training improves the altered renin-angiotensin system in the rostral ventrolateral medulla of hypertensive rats. Oxidative Med Cell Longev. 2016;2016:7413963.

71. Silva DM, et al. Swimming training improves the vasodilator effect of angiotensin-(1-7) in the aorta of spontaneously hypertensive rat. J Appl Physiol (1985). 2011;111(5):1272–7.

72. Bernardo BC, et al. Understanding key mechanisms of exercise-induced cardiac protection to mitigate disease: current knowledge and emerging concepts. Physiol Rev. 2018;98(1):419–75.

73. Fernandes T, et al. Aerobic exercise training-induced left ventricular hypertrophy involves regulatory MicroRNAs, decreased angiotensin-converting enzyme-angiotensin ii, and synergistic regulation of angiotensin-converting enzyme 2-angiotensin (1-7). Hypertension. 2011;58(2):182–9.

74. Dias-Peixoto MF, et al. The cardiac expression of Mas receptor is responsive to different physiological and pathological stimuli. Peptides. 2012;35(2):196–201.

75. Filho AG, et al. Selective increase of angiotensin(1-7) and its receptor in hearts of spontaneously hypertensive rats subjected to physical training. Exp Physiol. 2008;93(5):589–98.

76. Kar S, Gao L, Zucker IH. Exercise training normalizes ACE and ACE2 in the brain of rabbits with pacing-induced heart failure. J Appl Physiol (1985). 2010;108(4):923–32.

77. Tyrankiewicz U, et al. Activation pattern of ACE2/Ang-(1-7) and ACE/Ang II pathway in course of heart failure assessed by multiparametric MRI in vivo in Tgαq*44 mice. J Appl Physiol (1985). 2018;124(1):52–65.

78. Dukes A, et al. The aromatic amino acid tryptophan stimulates skeletal muscle IGF1/p70s6k/mTor signaling in vivo and the expression of myogenic genes in vitro. Nutrition. 2015;31(7–8):1018–24.
79. Hashimoto T, et al. ACE2 links amino acid malnutrition to microbial ecology and intestinal inflammation. Nature. 2012;487(7408):477–81.
80. Gu Q, et al. Contribution of renin-angiotensin system to exercise-induced attenuation of aortic remodeling and improvement of endothelial function in spontaneously hypertensive rats. Cardiovasc Pathol. 2014;23(5):298–305.
81. Gomes-Santos IL, et al. Effects of exercise training on circulating and skeletal muscle renin-angiotensin system in chronic heart failure rats. PLoS One. 2014;9(5):e98012.
82. Silva SD, et al. Downregulation of the vascular renin-angiotensin system by aerobic training - focus on the balance between vasoconstrictor and vasodilator axes. Circ J. 2015;79(6):1372–80.
83. Genest DS, et al. Impact of exercise training on preeclampsia: potential preventive mechanisms. Hypertension. 2012;60(5):1104–9.
84. Genest DS, et al. Novel role of the renin-angiotensin system in preeclampsia superimposed on chronic hypertension and the effects of exercise in a mouse model. Hypertension. 2013;62(6):1055–61.
85. Endlich PW, et al. Exercise modulates the aortic renin-angiotensin system independently of estrogen therapy in ovariectomized hypertensive rats. Peptides. 2017;87:41–9.
86. Somineni HK, Boivin GP, Elased KM. Daily exercise training protects against albuminuria and angiotensin converting enzyme 2 shedding in db/db diabetic mice. J Endocrinol. 2014;221(2):243–59.
87. Frantz EDC, et al. Exercise training modulates the hepatic renin-angiotensin system in fructose-fed rats. Exp Physiol. 2017;102(9):1208–20.
88. de Oliveira Sá G, et al. High-intensity interval training has beneficial effects on cardiac remodeling through local renin-angiotensin system modulation in mice fed high-fat or high-fructose diets. Life Sci. 2017;189:8–17.
89. Guimaraes GG, et al. Exercise induces renin-angiotensin system unbalance and high collagen expression in the heart of Mas-deficient mice. Peptides. 2012;38(1):54–61.
90. Bertagnolli M, et al. An orally active angiotensin-(1-7) inclusion compound and exercise training produce similar cardiovascular effects in spontaneously hypertensive rats. Peptides. 2013;51C:65–73.
91. Shah A, et al. Angiotensin-(1-7) attenuates hypertension in exercise-trained renal hypertensive rats. Am J Physiol Heart Circ Physiol. 2012;302(11):H2372–80.
92. Magalhaes GS, et al. Angiotensin-(1-7) attenuates airway remodelling and hyperresponsiveness in a model of chronic allergic lung inflammation. Br J Pharmacol. 2015;172(9):2330–42.
93. Becker LK, et al. Eccentric overload muscle damage is attenuated by a novel angiotensin-(1-7) treatment. Int J Sports Med. 2018;39(10):743–8.

Liver

Aline Silva de Miranda and Ana Cristina Simões e Silva

Introduction

Over the past decades, significant advances in the understanding of the renin–angiotensin system (RAS), especially regarding the local expression of RAS in several organs and tissues, including the kidney, brain, and liver, have hampered the classical view of RAS as a merely circulating hormonal system [1]. Particurlaly relevant for the reconceptualization of the RAS was the identification of the heptapeptide Ang-(1-7) [2], the ACE homolog enzyme responsible for the conversion of Ang II into Ang-(1-7), ACE2 [3, 4], and the Mas receptor, a G-protein coupled receptor, which mediates the main effects of Ang-(1-7) [5]. In this scenario, RAS is currently viewed as a system composed by two opposing axis: the classical one, including angiotensin converting enzyme (ACE)-Angiotensin (Ang) II-Ang type 1 (AT_1) receptor and the alternative one, comprising ACE2-Ang-(1-7)-Mas receptor [6].

It has also been often postulated that the classical arm mediates pro-inflammatory, pro-thrombotic, and pro-fibrotic processes, mainly through the activation of AT_1 receptors [6], whereas the alternative axis seems to play a protective role by frequently opposing Ang II actions via Mas receptors stimulation [7, 8]. An imbalance in the RAS classical and alternative axis components have

A. S. de Miranda
Interdisciplinary Laboratory of Medical Investigation, Faculty of Medicine, Federal University of Minas Gerais (UFMG), Belo Horizonte, MG, Brazil

Laboratory of Neurobiology, Department of Morphology, Institute of Biological Sciences, UFMG, Belo Horizonte, MG, Brazil

A. C. Simões e Silva (✉)
Interdisciplinary Laboratory of Medical Investigation, Faculty of Medicine, Federal University of Minas Gerais (UFMG), Belo Horizonte, Brazil

Department of Pediatrics, Interdisciplinary Laboratory of Medical Investigation, Faculty of Medicine, UFMG, Belo Horizonte, MG, Brazil
e-mail: ana@medicina.ufmg.br

© Springer Nature Switzerland AG 2019
R. A. S. Santos (ed.), *Angiotensin-(1-7)*,
https://doi.org/10.1007/978-3-030-22696-1_12

been implicated in the pathogenesis of a wide range of conditions including liver diseases [7–9]. Accordingly, therapeutic strategies have often been designed in order to inhibit ACE-Ang II-AT$_1$ receptor and to stimulate ACE2-Ang-(1-7)-Mas receptor activities [7, 9].

The involvement of both axes of the RAS in the pathogenesis of liver diseases has been supported by experimental and clinical studies [9, 10]. Herein, we will discuss current evidence regarding the role of RAS, mainly focusing on ACE2-Ang-(1-7)-Mas receptor arm, in liver physiological and clinical conditions as well as potential therapeutic role of RAS in liver diseases.

ACE2-ANG-(1-7)-MAS Receptor Axis Role in Liver Physiology

The local RAS concept opens the road for the hypothesis that RAS components activity might be tissue/organ specific and function-oriented [1]. Specifically, regarding the liver, the local hepatic RAS is not well defined, although studies about RAS involvement in hepatic diseases have supported a role for this system in liver function [11].

The liver, under physiological conditions, plays a pivotal role in metabolic homeostasis, by regulating glucose and lipid metabolisms [12, 13]. A great body of evidence has pointed out the RAS components as crucial regulators of hepatic-associated metabolic functions [12, 14–17]. For instance, the genetic deletion of the Mas receptor in FVB/N mice leads to a metabolic syndrome-like state characterized by dyslipidemia, increased abdominal fat mass, enhanced muscle triglycerides, glucose intolerance, and reduced insulin sensitivity, as well as a decrease in insulin-stimulated glucose uptake by adipocytes [16, 17]. In line with these findings, the absence of ACE2 expression in mice also increases liver insulin resistance and expression of hepatic lipogenic genes, and decreases the expression of fatty acid oxidation-related genes. These changes in hepatic metabolic activity were associated with enhanced liver oxidative stress and inflammation, all of which supporting the idea that ACE2 ameliorates hepatic insulin resistance, improves insulin signaling, and is involved in protection against oxidative stress in the liver [14, 15].

An in vitro approach also reinforced the protection of ACE2-Ang-(1-7)-Mas receptor arm in liver metabolic function. The exposure of HepG2 hepatocytes cells to the Ang-(1-7) increased liver glucose uptake and intracellular glycogen synthesis. The amelioration of insulin resistance in the liver was associated with the activation of the Akt/ PI3K/IRS-1/JNK insulin-signaling pathway. Importantly, the protective effect of Ang-(1-7) was partially blocked by the Mas receptor antagonist, A779, indicating that the beneficial effects of the alternative RAS arm are dependent of Mas receptor activation [15]. The same authors further demonstrated, by employing the same in vitro approach, that the activation of the ACE2-Ang-(1-7)-Mas receptor axis decreased liver oxidative stress, inflammation, and lipid accumulation partly by regulating lipid-metabolizing genes through ATP/P2 receptor/CaM signaling pathway [14]. A more recent study provided in vivo evidence of significant effects of Ang-(1-7) on metabolic pathways involved in lipid homeostasis. A transgenic rat overexpressing Ang-(1-7) presented a decrease in adiposity index along with a reduction in lipogenesis, suggesting a direct effect of Ang-(1-7) on adipose tissue lipid metabolism, independent of the stimulatory effect of insulin. Moreover, specifically in the liver, overexpression of Ang-(1-7) decreased the concentration of triacylglycerol and inhibited fatty acid synthase (FAS) and fatty acid transport protein (FATP) expression in the liver, suggesting a decrease in de novo fatty acid synthesis and fatty acid uptake [12]. Interestingly, exercise training prevented metabolic syndrome and nonalcoholic fatty liver disease (NAFLD) in fructose-fed rats by increasing hepatic ACE2-Ang-(1-7)-Mas receptor axis activity [18]. Of note, NAFLD is one of the most common chronic liver diseases worldwide and an important risk factor for nonalcoholic steatohepatitis, type 2 diabetes, and cardiovascular diseases [19, 20]. There is evidence that the activation of the counter-regulatory arm (ACE2-Ang-(1-7)-Mas) is beneficial in NAFLD and metabolic-associated syndromes [14, 15, 21–23].

Taken together, these studies provided strong evidence of the physiological role of RAS locally expressed in the liver in glycemic and lipid metabolisms, paving the road for the investigation of the ACE2-Ang-(1-7)-Mas receptor arm as a promising therapeutic strategy in liver diseases associated with metabolic dysfunctions.

ACE2-ANG-(1-7)-MAS Receptor Axis Role in the Pathophysiology of Liver Diseases

Liver diseases are major causes of morbidity and mortality worldwide. The leading causes of liver failure are hepatitis B and hepatitis C virus infections, alcohol use, and steatohepatitis related to obesity. Without proper treatment, all types of chronic hepatitis will progress to end-stage liver diseases, including cirrhosis, liver failure, and hepatocellular carcinoma, which ultimately lead to death [24, 25]. It is estimated that liver diseases account for a significant increase in the incidence of cirrhosis and for the death of at least 800,000 people worldwide annually [26].

Although the specific mechanisms underlying hepatic fibrosis pathophysiology remain to be fully revealed, some pathological characteristics of chronic liver diseases might include enhanced fibrosis, oxidative stress, and inflammatory markers [24, 27]. All together, these events lead to significant changes in hepatic perfusion, enhanced portal blood flow resistance as well as liver dysfunction. The end stage of progressive hepatic fibrosis, widely known as cirrhosis, culminates in liver architecture disruption due to fibrous scars and development of regenerating tissue, which, in turn, aggravates liver failure [28, 29].

Emerging evidence has supported the involvement of RAS components in hepatic fibrosis and cirrhosis. Particularly, an upregulation of classical RAS arm components, including angiotensinogen, renin, ACE, Ang II and AT1 receptors has been reported in experimental and clinical liver injury studies [30–34]. Accordingly, inhibition of RAS including the blockade of Ang II activity by lisinopril and captopril (ACE inhibitors) or losartan (AT1 receptor antagonist) prevented RAS profibrogenic effects and restored liver function [30, 35–39].

It is worth mentioning that liver fibrosis and hepatic cirrhosis seem to depend on the balance between the classical (ACE-Ang II-AT1 receptor) and the counter-regulatory (ACE2-Ang-(1-7)–Mas receptor) RAS axes [10, 40, 41]. Indeed, based on the concept that the RAS counter-regulatory axis opposes the classical arm actions, presenting anti-inflammatory, anti-oxidative, and anti-fibrotic effects, it is quite reasonable to expect a protective role for ACE2-Ang-(1-7)–Mas receptor axis in liver diseases [7, 8]. Moreover, considering that cirrhosis might be potentially reversible, particularly in a compensated stage [27], the ACE2-Ang-(1-7)–Mas receptor axis might represent a promise drug target in liver failure.

Insights from Preclinical Studies

Numerous studies have investigated the role of the counter-regulatory RAS arm as well as the mechanisms underlying its protective effects on liver function by employing different models of liver fibrosis, including bile duct ligation, carbon tetrachloride (CCl_4) treatment or continuous Ang-(1-7) infusion [42–49]. For instance, an increase in ACE2 expression in the liver parenchyma of rats submitted to bile duct ligation provided the first evidence of a potential role of the counter-regulatory RAS axis in chronic liver disease [47]. Similar findings were found with the progression of liver fibrosis induced by CCl_4 administration in rats. In this model, inhibition of ACE upregulated the mRNA expression of ACE2 and Mas receptor, contributing to liver protection [44]. It is worth noticing that ACE2 activity seems to be important as an endogenous negative regulator of RAS in chronic, but not acute, liver injury, primarily by promoting the conversion of Ang II into Ang (1-7). This statement is supported by the fact that ACE2 knockout mice only presented increased hepatic fibrosis 21 days after bile duct ligation or following chronic administration of CCl_4. On the other hand, no differences were found between ACE2 knockout mice and wild-type littermates when animals were subjected to acute liver injury. Moreover, genetic ablation of ACE2 in one-year-old mice resulted in spontaneous inflammatory cell infiltration and mild liver fibrosis [46].

The hepatic protection exerted by increased expression of ACE2 might rely on the fact that this enzyme catalyzes the pro-fibrotic peptide Ang II in the anti-fibrotic peptide Ang-(1-7). Thus, its catalytic action makes ACE2 a very interesting therapeutic target for liver fibrosis [44, 46]. In line with this hypothesis, an earlier study demonstrated, by employing a liver-specific adeno-associated viral genome 2 serotype 8 vector (rAAV2/8-ACE2) with a liver-specific promoter in bile duct ligation and CCL4 administration models, that the long-term therapeutic effect of recombinant ACE2 rapidly upregulated hepatic ACE2 and attenuated liver fibrosis. In parallel, the recombinant ACE therapy reduced hepatic Ang II levels concomitantly with an increase of Ang-(1-7) concentrations in liver tissue. This study also showed reductions in NADPH oxidase activity, oxidative stress, ERK1/2, and p38 phosphorylation, without unwanted systemic effects [45].

An even more attractive idea was to investigate the anti-fibrotic therapeutic capability of Ang-(1-7)–Mas receptor signaling in chronic liver disease models.

Accordingly, infusion of Ang-(1-7) markedly attenuated hepatic fibrosis in bile duct-ligated rats, decreased hydroxyproline content, and downregulated key genes involved in liver fibrosis and angiogenesis such as collagen 1A1, α-SMA (smooth muscle actin), VEGF (vascular endothelial growth factor), and CTGF (connective tissue growth factor) [43]. On the other hand, the pharmacological blockage of Mas receptors with the antagonist A-779, following a bile duct ligation in rats, induced an elevation in hepatic hydroxyproline and TGF-β1 concentrations, aggravating liver fibrosis [48]. There is evidence that Ang-(1-7) might exert its anti-fibrotic effects in liver tissue induced by bile duct ligation by means of the regulation of NLRP3 inflammasome/IL-1β/Smad pathway activation induced by Ang II-mediated reactive oxygen species (ROS) via redox balance modulation [42, 49]. A more recent study showed that the microRNA-21 (mir-21) mediates Ang-II-induced NLRP3 inflammasome activation via the Spry1/ERK/NF-κB, Smad7/Smad2/3/ NOX4 pathways contributing to liver fibrosis. The administration of Ang-(1-7) downregulated mir-21 expression, and protected against bile duct ligation and Ang-II infusion-induced hepatic fibrosis [50].

The protective role of Ang (1-7) in the liver has been also investigated in a murine model of hepatocellular carcinoma. The administration of Ang-(1-7) to H22 hepatoma-bearing mice prevented tumor growth by arresting tumor proliferation, promoting tumor apoptosis and inhibiting tumor angiogenesis. Interestingly, the treatment with Ang-(1-7) decreased AT1 receptor mRNA expression, upregulated mRNA levels of AT2 and Mas receptors, and suppressed H22 cell-endothelial cell communication. These findings suggest that benefits of Ang-(1-7) in hepatocellular carcinoma depend on the complex interaction between AT1, AT2, and Mas receptors [51]. A more recent study, by employing the same hepatocellular carcinoma model, investigated the long-term effects of adeno-associated virus (AAV) serotype-8-mediated Ang-(1-7) overexpression. The anti-tumoral activity of Ang-(1-7) was indicated by a persistent inhibition of the tumor growth and downregulation of angiogenesis along with a decrease in the levels of the proangiogenic factors phosphatidylinositol-glycan biosynthesis class F protein (PIGF) and VEGF [52]. Taken together, these studies reinforce the role of Ang-(1-7) as a promising drug target for liver diseases.

In vitro approaches also provided valuable evidence regarding the role of the counter-regulatory RAS axis in liver dysfunction. Culture hepatic stellate cells treated with Ang-(1-7) or the Mas receptor agonist, AVE 0991, expressed less α-SMA and hydroxyproline, while treatment with the Mas receptor antagonist, A779, induced opposite effects [53]. Accordingly, Ang-(1-7) through Mas receptor activation, in cultured hepatic stellate cells, inhibited Ang II-induced phosphorylation of extracellular signal-regulated kinase (ERK)1/2, a classical pathway of tissue fibrosis [46]. Ang-(1-7) also decreases Ang II-induced NLRP3 inflammasome/IL-1β/Smad pathway activation in hepatic stellate cells, thus preventing α-collagen I (Col1A1) accumulation. This finding suggests a novel potential mechanism by which Ang-(1-7) exerts its anti-fibrotic activity in liver tissue [42, 49]. Moreover, Ang-(1-7) seems to inhibit Ang II-induced NLRP3 inflammasome/IL-1β/Smad pathway activation in primary hepatic stellate cells also by suppressing mir-21 expression [50].

Insights from Clinical Studies

Data from experimental studies have broadened our knowledge regarding the involvement of ACE2-Ang-(1-7)–Mas receptor axis in liver physiology and pathophysiology. However, a translational approach to human studies seems to become a real challenge. Up to date, there is little evidence provided by clinical studies. A pivotal study first demonstrated widespread parenchymal expression of ACE2 in the liver of hepatitis C cirrhotic patients. The authors also reported increased levels of ACE2 in cultured human hepatocytes exposed to hypoxia [47]. In line with these findings, patients with cirrhosis induced by hepatitis C and mild-to-moderate liver disease presented enhanced circulating levels of Ang-(1-7), possibly as an attempt to counteract ACE-Ang II-AT1 receptor arm pro-fibrotic effects [40, 53]. A more recent study investigated the role of counter-regulatory RAS arm components in liver failure progression from fibrosis to cirrhosis to hepatocellular carcinoma. The concentrations of Ang II, Ang-(1-7), and VEGF were higher in the serum of patients compared with healthy subjects, and increased with the disease progression. Conversely, the liver mRNA expression of ACE2 gradually decreased with the increasing grade of disease severity. Importantly, higher liver expression of ACE2 was associated with patient's longer survival time, indicating that low expression of ACE2 may be a useful indicator of poor prognosis in hepatocellular carcinoma [54]. The evidence of ACE2-Ang-(1-7)–Mas receptor axis involvement in human liver diseases is scarce as well as its potential role as predictive biomarkers or drug targets. Further studies are necessary in order to better address this issue.

Concluding Remarks

The counter-regulatory RAS axis exerts anti-inflammatory, anti-oxidative, and anti-fibrotic effects in liver tissue. In general, ACE2-Ang-(1-7)-Mas axis opposes ACE-Ang II-AT1 receptor arm actions. The balance between both RAS axes may influence clinical and histopathological expression of liver diseases. Most data regarding ACE2-Ang-(1-7)-Mas axis role in hepatic pathophysiology as well as its therapeutic potential in liver diseases were generated from preclinical studies. To date, clinical research focused on the investigation of circulating and local concentrations of ACE2 and Ang-(1-7). Evidence regarding the interaction of AT1, AT2, and Mas receptor is still missing. Moreover, further studies that address the role of counter-regulatory RAS axis molecules as biomarkers of liver fibrosis and/or of disease prognosis as well as potential therapeutic targets are urgently necessary. The design of molecular or genetic methods to increase the expression of ACE2 and increased tissue levels of Ang-(1-7) and/or activation of the Mas receptor may, in turn, result in the development of new pharmacological approaches.

Acknowledgments The authors would like to thank FAPEMIG (Fundação de Amparo à Pesquisa do Estado de Minas Gerais, Brazil), CNPq (Conselho Nacional de Desenvolvimento Científico e Tecnológico, Brazil) and CAPES (Coordenação de Aperfeiçoamento de Pessoal de Nível Superior) for financial support. ACSS is a CNPq productivity fellowship recipient. ASM is a 2016 NARSAD Young Investigator Grant Awardee from the Brain and Behavior Research Foundation.

Conflict of interest None declared.

References

1. Paul M, Poyan Mehr A, Kreutz R. Physiology of local renin-angiotensin systems. Physiol Rev. 2006;86(3):747–803.
2. Santos RA, Brosnihan KB, Chappell MC, Pesquero J, Chernicky CL, Greene LJ, et al. Converting enzyme activity and angiotensin metabolism in the dog brainstem. Hypertension (Dallas, Tex: 1979). 1988;11(2 Pt 2):I153–7.
3. Donoghue M, Hsieh F, Baronas E, Godbout K, Gosselin M, Stagliano N, et al. A novel angiotensin-converting enzyme-related carboxypeptidase (ACE2) converts angiotensin I to angiotensin 1-9. Circ Res. 2000;87(5):E1–9.
4. Tipnis SR, Hooper NM, Hyde R, Karran E, Christie G, Turner AJ. A human homolog of angiotensin-converting enzyme. Cloning and functional expression as a captopril-insensitive carboxypeptidase. J Biol Chem. 2000;275(43):33238–43.
5. Santos RA, Simoes e Silva AC, Maric C, Silva DM, Machado RP, de Buhr I, et al. Angiotensin-(1-7) is an endogenous ligand for the G protein-coupled receptor Mas. Proc Natl Acad Sci U S A. 2003;100(14):8258–63.
6. Kamo T, Akazawa H, Komuro I. Pleiotropic effects of angiotensin II receptor signaling in cardiovascular homeostasis and aging. Int Heart J. 2015;56(3):249–54.
7. Rodrigues Prestes TR, Rocha NP, Miranda AS, Teixeira AL, Simoes ESAC. The anti-inflammatory potential of ACE2/Angiotensin-(1-7)/Mas Receptor axis: evidence from basic and clinical research. Curr Drug Targets. 2017;18(11):1301–13.
8. Simoes e Silva AC, Silveira KD, Ferreira AJ, Teixeira MM. ACE2, angiotensin-(1-7) and Mas receptor axis in inflammation and fibrosis. Br J Pharmacol. 2013;169(3):477–92.
9. Simoes ESAC, Miranda AS, Rocha NP, Teixeira AL. Renin angiotensin system in liver diseases: friend or foe? World J Gastroenterol. 2017;23(19):3396–406.
10. Grace JA, Herath CB, Mak KY, Burrell LM, Angus PW. Update on new aspects of the renin-angiotensin system in liver disease: clinical implications and new therapeutic options. Clin Sci (London, England: 1979). 2012;123(4):225–39.
11. Leung PS. The peptide hormone angiotensin II: its new functions in tissues and organs. Curr Protein Pept Sci. 2004;5(4):267–73.
12. Moreira CCL, Lourenco FC, Mario EG, Santos RAS, Botion LM, Chaves VE. Long-term effects of angiotensin-(1-7) on lipid metabolism in the adipose tissue and liver. Peptides. 2017;92:16–22.
13. Moreira de Macedo S, Guimaraes TA, Feltenberger JD, Sousa Santos SH. The role of renin-angiotensin system modulation on treatment and prevention of liver diseases. Peptides. 2014;62:189–96.
14. Cao X, Yang F, Shi T, Yuan M, Xin Z, Xie R, et al. Angiotensin-converting enzyme 2/angiotensin-(1-7)/Mas axis activates Akt signaling to ameliorate hepatic steatosis. Sci Rep. 2016;6:21592.
15. Cao X, Yang FY, Xin Z, Xie RR, Yang JK. The ACE2/Ang-(1-7)/Mas axis can inhibit hepatic insulin resistance. Mol Cell Endocrinol. 2014;393(1–2):30–8.

16. Clarke NE, Turner AJ. Angiotensin-converting enzyme 2: the first decade. Int J Hypertens. 2012;2012:12.
17. Santos SH, Fernandes LR, Mario EG, Ferreira AV, Porto LC, Alvarez-Leite JI, et al. Mas deficiency in FVB/N mice produces marked changes in lipid and glycemic metabolism. Diabetes. 2008;57(2):340–7.
18. Frantz EDC, Medeiros RF, Giori IG, Lima JBS, Bento-Bernardes T, Gaique TG, et al. Exercise training modulates the hepatic renin-angiotensin system in fructose-fed rats. Exp Physiol. 2017;102(9):1208–20.
19. Musso G, Cassader M, Cohney S, Pinach S, Saba F, Gambino R. Emerging liver-kidney interactions in nonalcoholic fatty liver disease. Trends Mol Med. 2015;21(10):645–62.
20. Musso G, Gambino R, Cassader M, Pagano G. Meta-analysis: natural history of non-alcoholic fatty liver disease (NAFLD) and diagnostic accuracy of non-invasive tests for liver disease severity. Ann Med. 2011;43(8):617–49.
21. de Macedo SM, Guimarares TA, Andrade JM, Guimaraes AL, Batista de Paula AM, Ferreira AJ, et al. Angiotensin converting enzyme 2 activator (DIZE) modulates metabolic profiles in mice, decreasing lipogenesis. Protein Pept Lett. 2015;22(4):332–40.
22. Feltenberger JD, Andrade JM, Paraiso A, Barros LO, Filho AB, Sinisterra RD, et al. Oral formulation of angiotensin-(1-7) improves lipid metabolism and prevents high-fat diet-induced hepatic steatosis and inflammation in mice. Hypertension (Dallas, Tex: 1979). 2013;62(2):324–30.
23. Santos SH, Andrade JM, Fernandes LR, Sinisterra RD, Sousa FB, Feltenberger JD, et al. Oral angiotensin-(1-7) prevented obesity and hepatic inflammation by inhibition of resistin/TLR4/MAPK/NF-kappaB in rats fed with high-fat diet. Peptides. 2013;46:47–52.
24. Ahmadian E, Pennefather PS, Eftekhari A, Heidari R, Eghbal MA. Role of renin-angiotensin system in liver diseases: an outline on the potential therapeutic points of intervention. Expert Rev Gastroenterol Hepatol. 2016;10(11):1279–88.
25. Wang FS, Fan JG, Zhang Z, Gao B, Wang HY. The global burden of liver disease: the major impact of China. Hepatology (Baltimore, MD). 2014;60(6):2099–108.
26. Kochanek KD, Murphy SL, Xu J, Tejada-Vera B. Deaths: Final Data for 2014. National vital statistics reports: from the Centers for Disease Control and Prevention, National Center for Health Statistics. Nat Vital Stat Syst. 2016;65(4):1–122.
27. Kim G, Baik SK. Overview and recent trends of systematic reviews and meta-analyses in hepatology. Clin Mol Hepatol. 2014;20(2):137–50.
28. Bataller R, Brenner DA. Liver fibrosis. J Clin Invest. 2005;115(2):209–18.
29. Friedman SL, Maher JJ, Bissell DM. Mechanisms and therapy of hepatic fibrosis: report of the AASLD Single Topic Basic Research Conference. Hepatology (Baltimore, MD). 2000;32(6):1403–8.
30. Bataller R, Sancho-Bru P, Gines P, Lora JM, Al-Garawi A, Sole M, et al. Activated human hepatic stellate cells express the renin-angiotensin system and synthesize angiotensin II. Gastroenterology. 2003;125(1):117–25.
31. Ikura Y, Ohsawa M, Shirai N, Sugama Y, Fukushima H, Suekane T, et al. Expression of angiotensin II type 1 receptor in human cirrhotic livers: its relation to fibrosis and portal hypertension. Hepatol Res. 2005;32(2):107–16.
32. Lubel JS, Herath CB, Burrell LM, Angus PW. Liver disease and the renin-angiotensin system: recent discoveries and clinical implications. J Gastroenterol Hepatol. 2008;23(9):1327–38.
33. Paizis G, Cooper ME, Schembri JM, Tikellis C, Burrell LM, Angus PW. Up-regulation of components of the renin-angiotensin system in the bile duct-ligated rat liver. Gastroenterology. 2002;123(5):1667–76.
34. Pereira RM, dos Santos RAS, da Costa Dias FL, Teixeira MM, ACSe S. Renin-angiotensin system in the pathogenesis of liver fibrosis. World J Gastroenterol. 2009;15(21):2579–86.
35. Bataller R, Gines P, Nicolas JM, Gorbig MN, Garcia-Ramallo E, Gasull X, et al. Angiotensin II induces contraction and proliferation of human hepatic stellate cells. Gastroenterology. 2000;118(6):1149–56.

36. Kim G, Kim J, Lim YL, Kim MY, Baik SK. Renin-angiotensin system inhibitors and fibrosis in chronic liver disease: a systematic review. Hepatol Int. 2016;10(5):819–28.
37. Koh SL, Ager E, Malcontenti-Wilson C, Muralidharan V, Christophi C. Blockade of the renin-angiotensin system improves the early stages of liver regeneration and liver function. J Surg Res. 2013;179(1):66–71.
38. Zhu Q, Li N, Li F, Zhou Z, Han Q, Lv Y, et al. Therapeutic effect of renin angiotensin system inhibitors on liver fibrosis. J Renin Angiotensin Aldosterone Syst. 2016;17(1):1470320316628717.
39. Bataller R, Schwabe RF, Choi YH, Yang L, Paik YH, Lindquist J, et al. NADPH oxidase signal transduces angiotensin II in hepatic stellate cells and is critical in hepatic fibrosis. J Clin Invest. 2003;112(9):1383–94.
40. Vilas-Boas WW, Ribeiro-Oliveira A Jr, Pereira RM, da Cunha Ribeiro R, Almeida J, Nadu AP, et al. Relationship between angiotensin-(1-7) and angiotensin II correlates with hemodynamic changes in human liver cirrhosis. World J Gastroenterol. 2009;15(20):2512–9.
41. Zhang W, Miao J, Li P, Wang Y, Zhang Y. Up-regulation of components of the renin–angiotensin system in liver fibrosis in the rat induced by CCL4. Res Vet Sci. 2013;95(1):54–8.
42. Cai SM, Yang RQ, Li Y, Ning ZW, Zhang LL, Zhou GS, et al. Angiotensin-(1-7) improves liver fibrosis by regulating the NLRP3 Inflammasome via redox balance modulation. Antioxid Redox Signal. 2016;24(14):795–812.
43. Herath CB, Warner FJ, Lubel JS, Dean RG, Jia Z, Lew RA, et al. Upregulation of hepatic angiotensin-converting enzyme 2 (ACE2) and angiotensin-(1-7) levels in experimental biliary fibrosis. J Hepatol. 2007;47(3):387–95.
44. Huang Q, Xie Q, Shi CC, Xiang XG, Lin LY, Gong BD, et al. Expression of angiotensin-converting enzyme 2 in CCL4-induced rat liver fibrosis. Int J Mol Med. 2009;23(6):717–23.
45. Mak KY, Chin R, Cunningham SC, Habib MR, Torresi J, Sharland AF, et al. ACE2 therapy using adeno-associated viral vector inhibits liver fibrosis in mice. Mol Ther. 2015;23(9):1434–43.
46. Osterreicher CH, Taura K, De Minicis S, Seki E, Penz-Osterreicher M, Kodama Y, et al. Angiotensin-converting-enzyme 2 inhibits liver fibrosis in mice. Hepatol (Baltimore, MD). 2009;50(3):929–38.
47. Paizis G, Tikellis C, Cooper ME, Schembri JM, Lew RA, Smith AI, et al. Chronic liver injury in rats and humans upregulates the novel enzyme angiotensin converting enzyme 2. Gut. 2005;54(12):1790–6.
48. Pereira RM, Dos Santos RA, Teixeira MM, Leite VH, Costa LP, da Costa Dias FL, et al. The renin-angiotensin system in a rat model of hepatic fibrosis: evidence for a protective role of angiotensin-(1-7). J Hepatol. 2007;46(4):674–81.
49. Zhang LL, Huang S, Ma XX, Zhang WY, Wang D, Jin SY, et al. Angiotensin(1-7) attenuated Angiotensin II-induced hepatocyte EMT by inhibiting NOX-derived H2O2-activated NLRP3 inflammasome/IL-1beta/Smad circuit. Free Radic Biol Med. 2016;97:531–43.
50. Ning ZW, Luo XY, Wang GZ, Li Y, Pan MX, Yang RQ, et al. MicroRNA-21 mediates angiotensin II-induced liver fibrosis by activating NLRP3 Inflammasome/IL-1beta axis via targeting Smad7 and Spry1. Antioxid Redox Signal. 2017;27(1):1–20.
51. Liu Y, Li B, Wang X, Li G, Shang R, Yang J, et al. Angiotensin-(1-7) suppresses hepatocellular carcinoma growth and angiogenesis via complex interactions of angiotensin II type 1 receptor, angiotensin II type 2 receptor and Mas receptor. Mol Med (Cambridge, Mass). 2015;21:626–36.
52. Mao Y, Pei N, Chen X, Chen H, Yan R, Bai N, et al. Angiotensin 1-7 overexpression mediated by a capsid-optimized AAV8 vector leads to significant growth inhibition of hepatocellular carcinoma in vivo. Int J Biol Sci. 2018;14(1):57–68.
53. Lubel JS, Herath CB, Tchongue J, Grace J, Jia Z, Spencer K, et al. Angiotensin-(1-7), an alternative metabolite of the renin-angiotensin system, is up-regulated in human liver disease and has antifibrotic activity in the bile-duct-ligated rat. Clin Sci (London, England: 1979). 2009;117(11):375–86.
54. Ye G, Qin Y, Lu X, Xu X, Xu S, Wu C, et al. The association of renin-angiotensin system genes with the progression of hepatocellular carcinoma. Biochem Biophys Res Commun. 2015;459(1):18–23.

Angiotensin 1-7 and Inflammation

Izabela Galvão, Flavia Rago, Isabella Zaidan Moreira,
and Mauro Martins Teixeira

Introduction

Inflammation is a physiological response of tissues to noxious stimulation of infectious nature or not. The immune system senses noxious stimuli and initiates an inflammatory process that ultimately aims to remove the threat and restore tissue structure and function homeostasis. Both processes, i.e., the initiation of the response and its resolution, are actively mediated by release of mediators of inflammation and recruitment and function of leukocytes and other cell types, including endothelial cells and tissue structural cells such as fibroblast [13, 32].

Acute inflammation has rapid onset and short duration. It is characterized initially by exudation of fluid and plasm proteins. During acute inflammation, cells from the innate immune system recognize damage or pathogen-associated molecular patterns. Recognition of these molecules by receptors on the surface of innate immune cells triggers several cellular responses, which result in the production of pro-inflammatory mediators (cytokines and chemokines) and consequent leukocyte recruitment to the site of injury, where they are able to deal with the inciting stimulus and initiate the repair of tissue damage. During acute inflammation, leukocyte accumulation is characterized predominantly by neutrophils [16].

Acute inflammation is induced and controlled by mediators produced by the host cells. These mediators act in the blood vessel to induce reversible changes, such as increased blood flow into the affected tissue, vessel dilatation, increased adhesiveness of leukocyte to the endothelial line, and increased permeability of capillaries and venules to plasm protein and fluid. All these changes are responsible for the

I. Galvão · F. Rago · M. M. Teixeira (✉)
Departamento de Bioquímica e Imunologia, Instituto de Ciências Biológicas, Universidade Federal de Minas Gerais, Belo Horizonte, MG, Brazil

I. Z. Moreira
Departamento de Análises Clínicas e Toxicológicas, Faculdade de Farmácia, Universidade Federal de Minas Gerais, Belo Horizonte, MG, Brazil

© Springer Nature Switzerland AG 2019
R. A. S. Santos (ed.), *Angiotensin-(1-7)*,
https://doi.org/10.1007/978-3-030-22696-1_13

cardinal signs of inflammation, as described by Celsus in the first century: Calor (heat), rubor (redness), tumor (swelling), dolor (pain). In the nineteenth century, Rudolf Virchow added the last cardinal sign: fuctio laesa (loss of function) [32].

In most cases, the acute inflammatory response is controlled in intensity, self-limited and accompanied by full restoration of tissue architecture and function. Once the stimulus that triggers inflammation is eliminated, activated cells and mediators are removed and degraded. This is normal resolution of inflammation. If the inciting stimulus is not properly eliminated or tissue injury is prolonged or excessive, the result can be the chronification of inflammation with serious consequences. Chronic inflammation is characterized by influx of lymphocytes and macrophages associated with vascular proliferation and fibrosis, and is seen in diseases such as rheumatoid arthritis, atherosclerosis and asthma [58]. During resolution of inflammation, biosynthesis of active mediators promotes the return to homeostasis by acting on specific targets to inhibit the neutrophil recruitment to the site of inflammation and promote the activation of apoptosis (programmed cell death in the recruited effector leukocytes), efferocytosis (clearance of apoptotic cell by macrophage), and reprogramming of macrophages from a pro-inflammatory to a resolutive phenotype. [46]. There has been much recent interest in the discovery of novel mediators of resolution of inflammation, as it is appreciated that there is much to be understood in the biology of resolution.

One of the endogenous mediators that has been studied in the context of inflammation is Angiotensin 1-7 (Ang-(1-7)). Ang-(1-7) is a biologically active peptide synthetized from the action of ACE2 on Angiotensin I (Angio I) and Angiotensin II (Angio II). It binds to a 7-transmembrane G-protein-coupled receptor, MAS, and exerts many beneficial actions in the context of acute inflammation. In this chapter, we will summarize the relevance of Ang-(1-7) and its MAS receptor in the context of inflammation, highlighting the advances and potential clinical use of this system for the treatment of inflammatory diseases.

The Evidence for the Anti-Inflammatory Actions of Angio-(1-7)

The renin-angiotensin system (RAS) is a very complex and dynamic system, composed of a cascade of enzymes and peptides that are believed to play an important role in many physiological processes such as blood pressure regulation and water balance [54]. The RAS has been described to participate in two opposite axes: one acting on AT1 receptors mediating pro-inflammatory effects and another one acting on Mas receptors mediating anti-inflammatory effects [35]. Angiotensin II (AngII) has pro-inflammatory and pro-fibrotic effects caused by the activation of its AT_1receptor. On the other hand, AngI and AngII can be cleaved by Neprilysin and angiotensin converting enzyme 2 (ACE2), respectively, generating Angiotensin-(1-7) [Ang-(1-7)]. Ang-(1-7) binds to its Mas receptor activating anti-inflammatory and anti-fibrotic processes [11, 33]. AngI can also undergo the action of the ACE 2 to produce Ang-(1-9) which binds on AT2 receptor and exerts anti-inflammatory actions [40]. Ang-(1-9) may form Angi-(1-7) through ACE in a less efficient way (([12, 61]

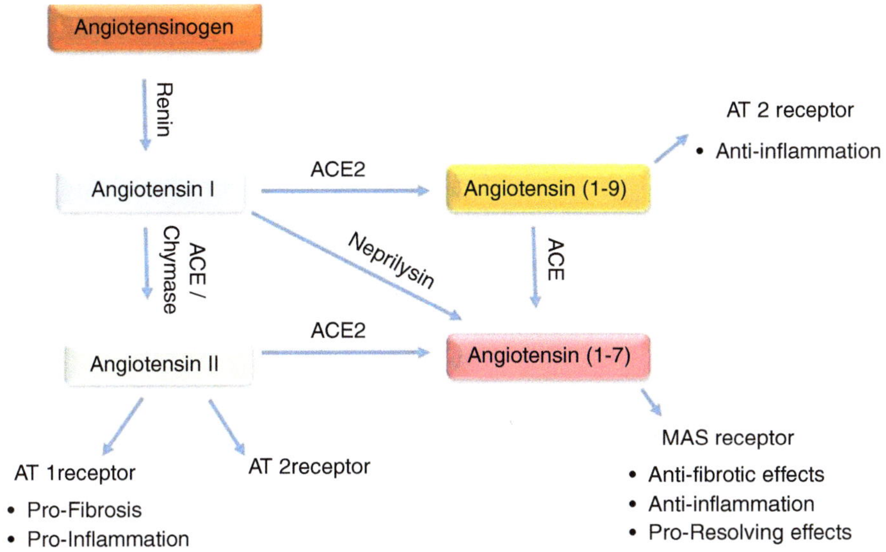

Fig. 1 Ang-(1-7) formation and the renin-angiotensin system in the context of inflammation. ACE angiotensin converting enzyme, ACE2 angiotensin converting enzyme 2, AT1 receptor angiotensin receptor type 1, AT2 receptor angiotensin receptor type 2, MAS Mas receptor

(Fig. 1). Other recently found end-products of RAS system besides Ang-(1-7) and Ang-(1-9) include Ang-(1-5), Ang A, Ang III, Ang IV and alamandine [3].

There is much evidence to link the function of the renin-angiotensin system (RAS) and that of the immune system. Much initial work was focused on the ACE/AngII/AT_1 axis until we suggested that the ACE2/Ang-(1-7)/Mas receptor axis could also play a role in the context of inflammation [51, 52]. While the action of the ACE/AngII/AT_1 axis is mostly pro-inflammatory and pro-fibrotic, the action of the ACE2/Ang-(1-7)/Mas receptor axis is mostly anti-inflammatory and anti-fibrogenic [54].

The effects of Ang-(1-7) have been studied in several disease models in which there is an inflammatory component [7, 14, 27, 28, 43, 44, 56]. These data are summarized in Table 1.

In an Arthritis model, the treatment with Ang-(1-7) decreased neutrophils recruitment and increased efferocytosis of apoptotic human neutrophils by macrophages [4]. In a DSS-induced colitis model, Ang-(1-7) was able to reduce inflammation by modulating plasma levels of cytokines and chemokines such as IL-1α, TNF-α, IL-5, IL-6, IL-13, GM-CSF, M-CSF, C5/C5a, MCP-5, and MIP-1. Daily treatment with Ang-(1-7) previously improved the severity of colitis, showed a significant reduction in the circulating levels of several cytokines and chemokines, and recruitment of neutrophils into the colon tissue [20].

In acute lung injury (ALI) model, Ang-(1-7) reduced lung edema, myeloperoxidase activity, histological lung injury score, and pulmonary vascular resistance [21]. Besides that, Ang-(1-7) reduced the release of pro-inflammatory cytokine and

Table 1 Studies evaluating the relevance of Ang-(1-7) and its MAS receptor in the context of inflammation

Organ/model	Compounds or strategy used	Effects	References
Hepatic fibrosis	A779	↑ Levels of OH-proline and TGF-β	[44]
Ischemia/reperfusion injury	Expression of ACE2/Ang-(1-7)/Mas receptor axis	↓ Levels of mRNA of Ang-(1-7) and ACE2 ↑↑ Expression of Mas receptor	[53]
Lung fibrosis and pulmonary hypertension	Lentivirus (Lenti-ACE2 or Lenti-Ang-(1-7)	↓ Lung fibrosis ↓ Levels of TGF-β ↓ Levels of TNFα, IL-6 and IL-1β	[49]
Arthritis	Ang-(1-7), AVE0991 and Mas receptor null mice	↓ Neutrophils accumulation ↓ Levels of TNFα, CXCL1, and IL-1β ↓ Histological parameters ↑ Levels of TNF-α, CXCL1, and neutrophils accumulation	[53]
Asthma	Ang-(1-7) and A779	↓ Recruitment of neutrophils, lymphocytes, eosinophils, and macrophage ↓ Fibrosis Prevented the improvement caused by Ang-(1-7) treatment	[12]
Renal ischemia/reperfusion (I/R)	AVE0991 and Mas receptor null mice	↓ Levels of CXCL1 ↓ MP0 ↓ Tissue damage No effect in I/R model	[9]
Atherosclerosis	AVE0991	↓ CD86, CD80, CD40 in macrophage and dendritic cells ↓ CD69 expression in CD4+ T cells	[18]
Hyperalgesia model	Ang-(1-7)	↓ Nociception	[10]
High-fat diet (HFD)-induced hepatic steatosis	Ang-(1-7)	↓ Levels of IL-6 and TNF-α	[14]
Acute lung injury model	Ang-(1-7)	↓ Lung edema ↓ MPO ↓ Histological score ↓ Pulmonary vascular resistance	[21]
Asthma	AVE0991	↓ Levels of IL-5 ↑ Levels of IL-10	[48]
Intracranial aneurysms	Ang-(1-7)	↓ Levels of MMP-9 and TNF-α ↑ Levels of HGF and COX-2	[43]

Table 1 (continued)

Organ/model	Compounds or strategy used	Effects	References
Ehrlich's ascites carcinoma (EAC)	Ang-(1-7)	↓ Tumor weight and levels of IGF-I and VEGF	[1]
Lung Fibrosis	Ang-(1-7)	↓ Lung fibrosis ↓ Levels of CTGF and collagen I ↓ Levels of TNF-α and IL-6	[34]
Type 2 diabetes (db/db)	Ang-(1-7)	↓ Renal fibrosis ↓ Reactive oxygen species ↓ Macrophage infiltration	[38]
Lung fibrosis	Ang-(1-7) and lentivirus (Lenti-ACE-2)	↓ Collagen deposition ↓ Production of NOX4 protein and H2O2 ↓ Production of NOX4 and H2O2 ↓ Levels of ACE and AT1R	[33]
Intracranial aneurysms	Ang-(1-7)	↓ Levels of IL-1β and TNF-α	[50]
Hyperalgesia model	Ang-(1-7)	↓ Peripheral nociception	[8]
Chronic allergic lung inflammation	Ang-(1-7)	↓ Levels of IgE ↓ Total number of cells ↓ Levels of IL-4, IL-5, GM-CSF, CCL5, and CCL2 ↓ Histology score of inflammation ↓ Collagen deposition ↓ Pulmonary vascular hyperplasia/hypertrophy	[31]
Acute respiratory distress syndrome	Ang-(1-7)	↓ Recruitment of leukocytes ↓ Collagen deposition	[68]
Autoimmune encephalomyelitis (EAE)	Mas receptor deficiency and AVE0991	↑ M(LPS/INFγ) profile genes ↓ M(IL-4/IL-13) gene expression ↑ T cells proliferation ↓ M(LPS/INFγ) expression ↑ M(IL-4/IL-13) expression	[17]
Chronic allergic pulmonary inflammation	Mas receptor deficiency	↑ Total cells ↑ Levels of eosinophils and mononuclear cells ↑ Histology score of inflammatory cells infiltrate ↑ IL-13TNF-α, CCL2/MCP-1, and CCL5/RANTES ↑ Deposition of airway extracellular matrix	[29]

(continued)

Table 1 (continued)

Organ/model	Compounds or strategy used	Effects	References
Type 2 diabetes (db/db)	Ang-(l-7)	↓ Fibrosis ↓ IL-1β ↓ Macrophage infiltration in the lungs ↓ Oxidative stress	[42]
Atherosclerosis	AVE0991	↓ Macrophage, T cells and NK cells recruitment ↓ CCL2, CCL5, CXCL10, TNF-α and IL-6 ↓ M1 population	[55]
High salt (HS) diet	Expression of ACE2/Ang-(l-7)/Mas receptor axis and on the expression of ACE/Angll/ATIR axis	Expression of ACE2 and Mas receptor ↑ Expression of Angll and AT1R	[7]
Antenatal corticosteroid (ANCS) treatment	Expression of Angll and Ang-(l-7)	↑ Urinary Angll ↑ Urinary Angll/Ang-(l-7) ↑ Plasma Ang-(l-7)	[56]
Renal injury	Ang-(l-7)	↓Oxidative stress ↓ Extracellular matrix proteins ↓ Levels of CTGF and TGF-β ↓ Fibrosis ↓ IL-6 and TNF-α	[27]
Cardiomyopathy	Adenovirus carrying the murine ACE2 (Ad-ACE2)	↑ Survival rate ↑ Expression of ACE2 ↓ Recruitment of inflammatory cells, loss of myofibrils, and disorganization ↓ Cell death ↑ pAMPK(Thrl72)/AMPK, pP13K/P13K, and pAKT/AKT radio ↓ c-caspase3/caspase3 ratio ↓ VCAM, TNF-α, and ICAM-1 ↓ p-ERK, NOX2, P47, and iNOX ↓ Collagen deposition and TGF-β ↑ MMP-9	[28]
Diabetic cardiomyopathy	AT1 inhibition (Azilsartan)	↑ Expression of ACE2 and Mas receptor ↓ Oxidative stress cardioprotection	[59]

Table 1 (continued)

Organ/model	Compounds or strategy used	Effects	References
Arthritis	Ang-(l-7) and A779	↑ Apoptosis of neutrophils ↓ NFκB activation ↑ Efferocytosis of apoptotic neutrophils blocked the effects of Ang-(l-7) treatment	[4]
Arthritis	Expression of ACE2 and Ang-(l-7)	↓ Levels of cardiac and kidney ACE2 and Ang-(l-7) NSAIDs restored the levels of cardiac Ang-(l-7)	[2]
Asthma	Ang-(l-7)	↓ Levels of eosinophils ↓ EPO activity ↓ Apoptosis eosinophils ↑ Efferocytosis is PMN cells ↓ Levels of NFκB ↓ ERK1/2, IκB-α, and GATA3 ↓ Extracellular matrix deposition ↓ Collagen 1 and collagen III mRNA	[30]

Legend: *A779* antagonist of Mas receptor, *Ang I* angiotensin I, *Ang II* angiotensin II, *Ang(1-7)* angiotensin 1-7, *ACE* angiotensin converting enzyme, *ACE2* angiotensin converting enzyme 2, *AT1* angiotensin receptor type 1, *AVE0991* nonpeptide agonist of Mas receptor, *MAS* Mas receptor, *MMP9* matrix metallopeptidase, *EPO* eosinophilic peroxidase, *TGF-β* transforming growth factor beta, *TNF-α* tumor necrosis factor alfa, *IL* interleukin, *CXCL* chemokine, *MMP-9* matrix metallopeptidase 9, *HGF* hepatocyte growth factor, *COX-2* cyclooxygenase 2, *IGF-1* insulin growth factor 1, *VEGF* vascular endothelial growth factor, *CTGF* connective tissue growth factor, *AT1R* angiotensin II receptor type 1, *GM-CSF* granulocyte-macrophage colony-stimulating factor, *LPS* lipopolysaccharides, *INF-γ* interferon gamma, *c-caspase 3* cleaved caspase 3, *VCAM* vascular cell adhesion protein 1, *ICAM1* intercellular adhesion molecule 1, *MMP-9* matrix metallopeptidase 9, *NFκB* factor nuclear kappa B, *NSAIDs* nonsteroidal anti-inflammatory drugs

suppressed the expression of Nox4 and its subunits in the lungs in a model of hypoxia [26]. In a model of pulmonary fibrosis, treatment with Ang-(1-7) decreased lung fibrosis, the production of type I collagen, and the production of connective tissue growth factor (CTGF). In addition, it decreased the levels of TNF-α and IL-6 [34].

In asthma models, Ang-(1-7) reduces the production of pro-inflammatory cytokines and the activation of downstream pathways. For example, the treatment of Ang-(1-7) decreased immune cells recruitment and fibrosis [12]; decreased the levels of erythropoietin (EPO) activity in the lung; increased apoptotic eosinophils and its efferocytosis; induced resolution of inflammation by the down-expression of ERK1/2, IκB-α and GATA3 [30]. The treatment with Ang-(1-7) prevented the increase of plasma IgE and pro-inflammatory cytokines such as IL-4, IL-5, GM-CSF,

CCL5 and CCL2 in lungs of OVA-challenged mice. These effects were associated with reduced levels of p-ERK1/2 and p-JNK in lungs [31].

In a mouse model of type 2 diabetes (db/db), treatment with Ang-(1-7) reduced the levels of circulating pro-inflammatory cytokines, decreased lung fibrosis, oxidative stress and macrophage infiltration in the lungs [42]. Mori et al., [38] suggested that Ang-(1-7) represents a promising therapy for diabetic nephropathy by exerting renoprotective effects associated with reduction of oxidative stress, inflammation and fibrosis. Similar results were observed by Lu et al., [27], where the treatment with Ang-(1-7) reduced oxidative stress, extracellular matrix proteins, pro-inflammatory cytokines and fibrosis in a renal injury model. Ang-(1-7) also induced neuroprotection by reduction of TNF-α and IL-1β levels, attenuation of oxidative stress and reduction of phosphorylation of IκB and NFκB p65 subunit [19].

Further studies are needed to elucidate the mechanisms by which Ang-1-7 cooperates with other mediators to modulate inflammation. Nevertheless, current knowledge do support the possibility that drugs which mimic or enhance the function of the ACE2/Ang-(1-7)/Mas axis may be beneficial for the treatment of inflammatory diseases [54].

Further evidence for the relevance of the ACE2/Ang-(1-7)/Mas receptor axis in the control of inflammatory response derives from studies using the compound AVE 0991, a nonpeptide Mas receptor agonist [47]. It has been demonstrated that AVE0991 treatment decreased neutrophil accumulation and pro-inflammatory cytokines production in a model of arthritis [53]. During renal ischemia and reperfusion, the administration of AVE 0991 promoted renoprotective effects, such as decrease of tissue injury, leukocyte infiltration and release of CXCL1 [9].

Treatment with AVE 0991 significantly reduced disease incidence and slightly ameliorated the clinical course of experimental autoimmune encephalomyelitis (EAE) [17]. In a model of spontaneous atherosclerosis, the use of a nonpeptide Ang-(1-7), AVE 0991, inhibited perivascular inflammation by reducing chemokine expression and monocyte/macrophage activation [55] and also reduced the expression of co-stimulatory molecules in macrophage and dendritic cells, consequently reducing T-cell activation [18]. The use of AVE0991 in a murine model of asthma reversed the increased airway wall and pulmonary vasculature thickness, reduced IL-5 and increased IL-10 levels [48].

Moreover, there is evidence to suggest that that Ang-(1-7) has anti-nociceptive effects by Mas receptor activation. Besides the blockage of hypernociception in arthritis [53], Ang-(1-7) attenuated cancer-induced bone pain [15]. Inhibition of Mas receptor improved neuropathic pain [70] and the absence of Mas receptor reduced hyperalgesia induced by carrageenan and prostaglandin E2 [10]. Other studies also demonstrated that Ang-(1-7) has an anti-nociceptive role via Mas receptor activation [8, 10].

In contrast to the overall anti-inflammatory effects of the ACE2/Ang-(1-7)/Mas receptor axis, the ACE/AngII/AT$_1$ axis is, in general, associated with pro-inflammatory and pro-fibrotic responses [54]. In this regard, blockade of the ACE/AngII/AT$_1$ axis is expected to have anti-inflammatory effects. The activation of AT1 can be inhibited through specific and competitive angiotensin II receptor blockers (ARBs) called sartans. Sartans, together with inhibitors of ACE, are efficient inhibitors of the ACE/AngII/AT$_1$ axis [60]. Interestingly, some of the anti-inflammatory effects observed after treatment with losartan were dependent on Ang-(1-7)/Mas receptor. The long-term administration of losartan exerts an antithrombotic effect mediated by Ang-(1-7) [22] and treatment with telmisartan and losartan effectively increased the plasma levels of Ang-(1-7) [71]. In a model of Adriamycin-induced renal injury, the protective effects of losartan were ablated in Mas receptor-deficient mice ([51] PLOS one), whereas the effects of this compound were Mas receptor-independent in a model of antigen-induced arthritis [51, 52]. These studies clearly show that blockade of the ACE/AngII/AT$_1$ axis may decrease inflammation by facilitating the release of An-(1-7) and activation of Mas receptors.

In addition to the biological effects of Ang-(1-7) in the context of different animal models of human diseases, several studies have now shown that Ang-(1-7) through its MAS receptor may modify the function of cells associated with the inflammatory response (Table 2). Akin to the in vivo data, most studies in cell types

Table 2 Studies of Ang-(1-7) effects on MAS receptor in different cell types associated with inflammation

Cell type	Compounds or strategy used	Effects	References
Human lung adenocarcinoma cells (AS49)	Ang-(1-7)	↓ Cell migration ↓ MMP-2 mRNA expression. ↓ P13K/Akt, JNK1/2 and p38MAPK phosphorylation	[39]
Peritoneal macrophage	Ang-(1-7)	↓ IL-6 and TNF-α mRNA levels ↓ Src kinases activity	[57]
Astrocytes	Ang-(1-7)	↓ IL-1β and IL-6mRNA. ↓ COX-2 and GFAP protein expression ↑ DUSP1	[36]
Skeletal muscle cells	Ang-(1-7)	↓ AngII-induced TGF-β1 ↓ AngII-induced ROS and NOX subunit p47/phox protein levels ↓ AngII-induced p38, and smad-2 phosphorylation, and smad-4 nuclear translocation	[37]
Rat pancreatic acinar AR42J cells	Ang-(1-7) and *A779*	↑ Levels of IL-10 ↓ IL-6 and IL-8 *↑ Levels of IL-6 and IL-8 P13K/ AKT pathway and eNOS*	[63]
Human brain vascular smooth muscle cells (HBVSMC)	Ang-(1-7) and *A779*	↓ NFκB ↑ IκBα ↓ TNF-α, MCP1, IL-8 *Blocked Ang-(1-7) effects*	[5]
Umbilical vein endothelial cells (HUVECs)	Ang-(1-7) and *A779*	↓ ICAM-1, VCAM-1, and MCP-1 expression and secretion ↓ NFκB and p38 activation *Blocked Ang-(1-7) effects*	[24]

(continued)

Table 2 (continued)

Cell type	Compounds or strategy used	Effects	References
Human vascular smooth muscle cells	Ang-(l-7) and *A779*	↓ iNOS elicited by IL-1β, and this effect was blocked by A779 ↓ NF-κB activation - blocked Ang-(l-7) effects	[62]
Vascular smooth muscle cells (VSMCs)	Ang-(l-7) and *A779*	↓ IVSMC proliferation induced by Ang II ↓ AngII-induced Akt and ERK1/2 phosphorylation *Blocked Ang-(l-7) effects*	[69]
Bone marrow-derived macrophage (BMDM)	AVE 0991	↑ Alternative activated macrophage ↓ T cell activation mediated by macrophage	[17]
Microglia	Ang-(l-7)	↓ IL-1β and TNF-α mRNA levels ↑ IL-10 mRNA levels ↓ Expression of NF-κB subunits	[25]
Human peripheral blood isolated neutrophils	Ang-(l-7)	↑ Apoptosis ↓ NFκB activation	[4]
Bone marrow-derived neutrophil and spleen-derived mononuclear cells	Ang-(l-7) and *A779*	↑ Neutrophil and mononuclear cells apoptosis ↓ Neutrophil chemotactic migration ↓ Superoxide release by neutrophils *Blocked Ang-(l-7) effects*	[20]
THP-1 Monocyte/ macrophage	AVE 0991	↓ TNF-α, IL-1β, CCL2, and CXCL-l0mRNA ↓ Monocyte/macrophage activation and migration ↓ Differentiation in M1 phenotype	[55]
Human aortic endothelial cells (HAECs)	Ang-(l-7)	↓ Monocyte/macrophage adhesion and migration ↓ Reactive oxygen species (ROS)	[41]
Mouse pancreatic acinar cancer (MPC-83)	Ang-(l-7) and *A779*	↓ TNF-α, IL-6, and IL-8 ↑ Levels of IL-10 ↓ p38 MAPK and NFκB signaling pathway *Blocked Ang-(l-7) effects*	[67]
Rat pancreatic acinar AR42J cells	Ang-(1-7)	↓ TLR4/NF-κB signaling pathway ↓ TNF-α, IL-6, and IL-8 mRNA levels ↓ IL-10 mRNA levels	[64]

Legend: *A779* antagonist of Mas receptor, *Ang II* angiotensin II, *Ang (1-7)* Angiotensin 1-7, *AVE0991* nonpeptide agonist of Mas receptor, *MAS* Mas receptor, *MMP-2* matrix metallopeptidase 2, *IL* interleukin, *TNF-α* tumor necrosis factor alfa, *TGF-β* transforming growth factor beta, *ROS* reactive oxygen species, *eNOS* endothelial nitric oxide synthase, *NFκB* factor nuclear kappa B, *MCP-1* monocyte chemoattractant protein-1, *ICAM-1* intercellular adhesion molecule 1, *VCAM-1* vascular cell adhesion protein 1, *iNOS* inducible nitric oxide synthase, *TLR-4* Toll-like receptor 4, *WKYMV* fMLP-like peptide suggest that the overarching effects of Ang-(1-7) is to decrease cell functions associated with active pro-inflammatory responses. For example, in neutrophils, Ang-(1-7) reduced survival and induced apoptosis, and reduced recruitment and NFκB activation [4, 20].

The treatment of Ang-(1-7) or AVE0991 in macrophage attenuated the expression of TNF-α, IL-6, IL-1β, CCL2 and CXCL10 pro-inflammatory cytokines, and reduced Src kinase activity [55, 57], M1 polarization, and the number of proliferating T cells [17, 55]. Ang-(1-7) has also shown anti-proliferative effects in human peripheral blood mononuclear cells (HPBMC) [12].

In cultured hypothalamic microglia, treatment with Ang-(1-7) decreased the basal levels of mRNA for the pro-inflammatory cytokines such as IL-1β and TNF-α and increased in basal levels the anti-inflammatory cytokine IL-10 [25].

In human lung adenocarcinoma epithelial cells (A549), Ang-(1-7) reduced migration and phosphorylation of PI3K/AKT, JNK1/2, and p38 MAPK signaling pathways [39]. Ang-(1-7) decreases the activation of oxidative stress in epithelial cells [33]. It was also demonstrated in pancreatic cells that the treatment with Ang-(1-7) reduced pro-inflammatory cytokine release, increased IL-10 levels and reduced pro-survival signaling pathways including PI3K/AKT [63], TLR4/NFκB [65] and p38 MAPK [67].

It was demonstrated that pre-treatment with Ang-(1-7) in human aortic endothelial cells (HAECs) prevented monocyte adhesion and migration impairment induced by thrombin via downregulation of reactive oxygen species (ROS) production [41]. Similar effects were observed in Umbilical vein endothelial cells (HUVECs). The treatment with Ang-(1-7) reduced cell adhesion molecule expression and NFκB and p38 activation [24]. Ang-(1-7) could counteract the pro-inflammatory effects of Ang II in skeletal muscle cells [23, 37]. Ang-(1-7) attenuated the induction of iNOS through its binding to Mas receptor [62].

Resolution of Inflammation and Ang-(1-7)

During inflammation, leukocytes interact with the endothelial cells, scanning the tissue for molecular cues for migration, a process called rolling and this is dependent on a group of adhesion molecules named selectins. Once these leukocytes find a chemoattractant molecule to which they bind, integrins on their surface switch into an active conformation state. The activation of integrins allows firm adhesion to endothelial cells and consequent migration into tissues [45]. There is now much interest in understanding not only the mechanisms by which cells migrate into tissues, but also the understanding of the mechanisms responsible for keeping cells there or clearing them from tissue. There is a feeling that novel therapies may be derived from the knowledge of the mechanisms that resolve inflammation. The resolution of acute inflammation is an active process, which is characterized by active biosynthesis pro-resolving mediators that limit the duration of inflammatory response and induce the return to homeostasis [6]. Therefore, different from anti-inflammatory therapy, pro-resolving strategies should balance the inflammatory response to reach homeostasis.

The key steps for the induction of resolution of inflammation include reduction or blockade of neutrophil recruitment to the site of inflammation, induction of neutrophil apoptosis, increase expression of find-me and eat-me signals, induction of phagocytosis of apoptotic neutrophil by macrophages (efferocytosis), a non-phlogistic recruitment of monocytes, reprogramming of macrophages from classically activated to alternative activated, instructs the adaptive immune system and induction of repair and regeneration [46, 58].

Pro-resolving mediators initiate resolution programs by acting on specific cell surface G-protein-coupled receptor to drive cellular response to restore the homeostasis [46]. Of note, the Mas receptor is a G-protein-coupled receptor [66]. As discussed above, there is much evidence to suggest that Ang-(1-7) and its Mas receptor have potent anti-inflammatory effects. More recently, we have shown that this molecule also has relevant pro-resolving activity. Indeed, Ang-(1-7) induced apoptosis of neutrophils and increased their clearance by macrophage, therefore enhancing efferocytosis. The resolution of neutrophilic inflammation was associated with a decrease of NFκB phosphorylation [4]. The same was observed for in a model of eosinophilic inflammation. Ang-(1-7) increased the number of apoptotic eosinophil, which was associated with decreased NFκB, ERK1/2 and GATA3 expression [30]. The treatment with Ang-(1-7) also significantly improved apoptosis of immune cells, and reduced neutrophil chemotaxis and superoxide release in vitro [20].

Altogether, these studies suggest that Ang-(1-7) has a role in the resolution of inflammation by fulfilling some fundamental criteria: limitation of neutrophil recruitment, counter regulation of chemokine and cytokines, induction of apoptosis of neutrophils and their subsequently efferocytosis and reprogramming of macrophages. Molecules that fulfil these criteria are qualified as pro-resolving mediators [58] (Fig. 2).

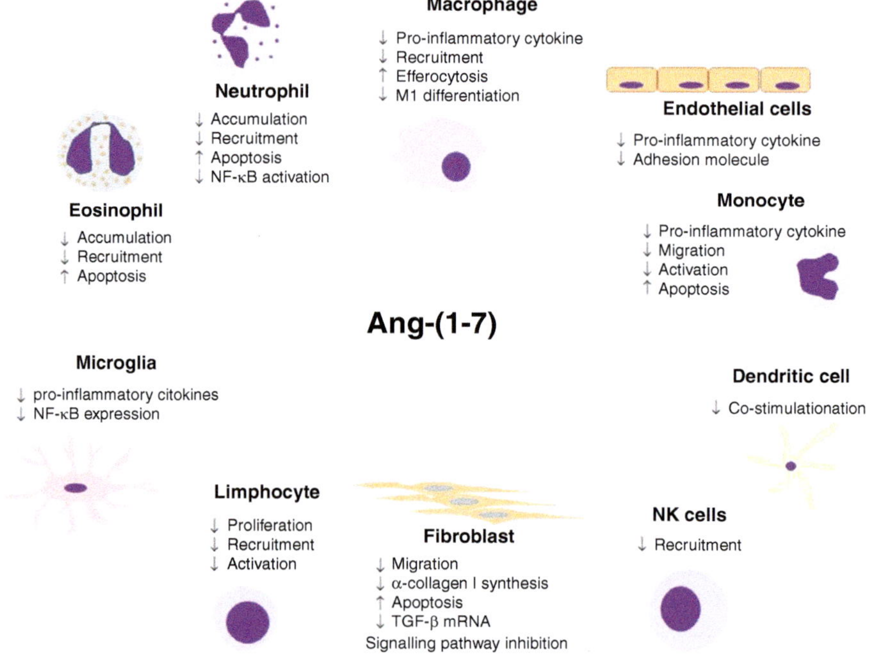

Fig. 2 Ang-(1-7) effects in different type of cells

Concluding Remarks

In the last decade, many insights of renin-angiotensin system (RAS) axis have been revealed. An important advance in the understanding of RAS was the recognition of Ang-(1-7) as a biologically active peptide produced from the cleavage of Ang-II by the angiotensin converting enzyme type 2 (ECA2) and acts through a receptor coupled to the specific G-protein, the Mas receptor. Identification of the Mas and MrgD protein as the Ang-(1-7) receptor provided an important molecular basis for the biological significance of this peptide, although there is no evidence of the downstream signaling.

Ang-(1-7) has been demonstrated to have many beneficial actions in the context of inflammatory response. In most studies, Ang-(1-7) reduced leukocyte recruitment, and the production and expression of chemokines, cytokines and adhesion molecules. Moreover, there was downregulation of signaling pathways, such as PI3K/Akt, p38 MAPK and NFκB, usually associated with an active inflammatory response. In addition to preventing crucial aspects of the productive phase of the inflammatory response, the binding of Ang-(1-7) to Mas receptor increases neutrophil apoptosis, efferocytosis and macrophage reprogramming from classically active to alternatively activated. All these functions appear to contribute to the capacity of this molecule to induce the resolution of inflammation in various animal models of inflammation.

It is unclear and further studies are needed to elucidate the downstream events triggered by Ang-(1-7)/Mas receptor and that modulate inflammation. In addition, further studies are needed to evaluate the role and relevance of Ang-(1-7) in the context of adaptive immunity and T cell function.

Understanding the ACE2/Ang-(1-7)/Mas receptor pathway may represent a valuable pharmacological opportunity to reveal new strategies to attenuate the pro-inflammatory environment that promotes and sustains the development of various chronic diseases. Activation of this pathway may not only attenuate pro-inflammatory responses but may also improve the resolution of inflammation and minimize inflammatory tissue damage and disease.

References

1. Abd-Alhaseeb MM, Zaitone SA, Abou-El-Ela SH, Moustafa YM. Olmesartan potentiates the anti-angiogenic effect of sorafenib in mice bearing Ehrlich's ascites carcinoma: role of angiotensin (1-7). PLoS One. 2014;9:e85891.
2. Asghar W, Aghazadeh-Habashi A, Jamali F. Cardiovascular effect of inflammation and non-steroidal anti-inflammatory drugs on renin–angiotensin system in experimental arthritis. Inflammopharmacology. 2017;25:543–53.
3. Bader M. Tissue renin-angiotensin-aldosterone systems: targets for pharmacological therapy. Annu Rev Pharmacol Toxicol. 2010;50(1):439–65. https://doi.org/10.1146/annurev.pharmtox.010909.105610.
4. Barroso LC, Magalhaes GS, Galvão I, Reis AC, Souza DG, Sousa LP, Santos RAS, Campagnole-Santos MJ, Pinho V, Teixeira MM. Angiotensin-(1-7) promotes resolution of neutrophilic inflammation in a model of antigen-induced arthritis in mice. Front Immunol. 2017;8:1596. https://doi.org/10.3389/fimmu.2017.01596.

 5. Bihl JC, et al. Angiotensin-(1-7) counteracts the effects of Ang II on vascular smooth muscle cells, vascular remodeling and hemorrhagic stroke: Role of the NFκB inflammatory pathway. Vasc Pharmacol. 2015;73:115–23.
 6. Buckley CD, Gilroy DW, Serhan CN. Proresolving lipid mediators and mechanisms in the resolution of acute inflammation. Immunity. 2014;40(3):315–27. https://doi.org/10.1016/j. immuni.2014.02.009.
 7. Cao G, Della Penna SL, Kouyoumdzian NM, Choi MR, Gorzalczany S, Fernández BE, Toblli JE, Rosón MI. Immunohistochemical expression of intrarenal renin angiotensin system components in response to tempol in rats fed a high salt diet. World J Nephrol. 2017;6(1):29–40. https://doi.org/10.5527/wjn.v6.i1.29.
 8. Castor MGM, Santos RAS, Duarte IDG, Romero TRL. Angiotensin-(1-7) through Mas receptor activation induces peripheral antinociception by interaction with adrenoreceptors. Peptides. 2015;69:80–5. https://doi.org/10.1016/j.peptides.2015.04.011.
 9. Corrêa Barroso L, Daniela Silveira K, Xavier Lima C, Borges V, Bader M, Rachid M, Augusto Souza Santos R, Gloria Souza D, Cristina Simões Silva A, Martins Teixeira M. Renoprotective effects of AVE0991, a nonpeptide Mas receptor agonist, in experimental acute renal injury. Int J Hypertension. Hindawi Publishing Corporation. 2012;8. https://doi.org/10.1155/2012/808726.
10. Costa ACO, Becker LK, Moraes ER, Romero TRL, Guzzo L, Santos RAS, Duarte IDG. Angiotensin-(1-7) induces peripheral antinociception through mas receptor activation in an opioid-independent pathway. Pharmacology. 2012;89(3–4):137–44. https://doi.org/10.1159/000336340.
11. Domenig O, Manzel A, Grobe N, Königshausen E, Kaltenecker CC, Kovarik JJ, Stegbauer J, Gurley SB, van Oyen D, Antlanger M, Bader M, Motta-Santos D, Santos RA, Elased KM, Säemann MD, Linker RA, Poglitsch M. Neprilysin is a mediator of alternative renin-angiotensin-system activation in the murine and human kidney. Sci Rep. 2016;6(1):33678. https://doi.org/10.1038/srep33678.
12. El-Hashim AZ, Renno WM, Raghupathy R, Abduo HT, Akhtar S, Benter IF. Angiotensin-(1-7) inhibits allergic inflammation, via the MAS1 receptor, through suppression of ERK1/2- and NF-κB-dependent pathways. Br J Pharmacol. 2012;166(6):1964–76. https://doi.org/10.1111/j.1476-5381.2012.01905.x.
13. Fagundes CT, Amaral FA, Teixeira AL, Souza DG, Teixeira MM. Adapting to environmental stresses: the role of the microbiota in controlling innate immunity and behavioral responses. Immunol Rev. 2012;245(1):250–64. https://doi.org/10.1111/j.1600-065X.2011.01077.x.
14. Feltenberger JD, Andrade JMO, Paraíso A, Barros LO, Filho ABM, Sinisterra RDM, Sousa FB, Guimarães ALS, de Paula AMB, Campagnole-Santos MJ, Qureshi M, dos Santos RAS, Santos SHS. Oral formulation of angiotensin-(1-7) improves lipid metabolism and prevents high-fat diet-induced hepatic steatosis and inflammation in mice. Hypertension (Dallas, Tex: 1979). 2013;62(2):324–30. https://doi.org/10.1161/HYPERTENSIONAHA.111.00919.
15. Forte BL, Slosky LM, Zhang H, Arnold MR, Staatz WD, Hay M, Largent-Milnes TM, Vanderah TW. Angiotensin-(1-7)/Mas receptor as an antinociceptive agent in cancer-induced bone pain. Pain. 2016;157(12):2709–21. https://doi.org/10.1097/j.pain.0000000000000690.
16. Fullerton JN, Gilroy DW. Resolution of inflammation: a new therapeutic frontier. Nat Rev Drug Discov. 2016;15(8):551–67. https://doi.org/10.1038/nrd.2016.39.
17. Hammer A, Yang G, Friedrich J, Kovacs A, Lee D-H, Grave K, Jörg S, Alenina N, Grosch J, Winkler J, Gold R, Bader M, Manzel A, Rump LC, Müller DN, Linker RA, Stegbauer J. Role of the receptor Mas in macrophage-mediated inflammation in vivo. Proc Natl Acad Sci. 2016;113(49):14109–14. https://doi.org/10.1073/pnas.1612668113.
18. Jawien J, Toton-Zuranska J, Gajda M, Niepsuj A, Gebska A, Kus K, Suski M, Pyka-Fosciak G, Nowak B, Guzik TJ, Marcinkiewicz J, Olszanecki R, Korbut R. Angiotensin-(1-7) receptor Mas agonist ameliorates progress of atherosclerosis in apoE-knockout mice. J Physiol Pharmacol. 2012;63(1):77–85. Available at: http://www.ncbi.nlm.nih.gov/pubmed/22460464. Accessed: 23 April 2018.
19. Jiang T, Yu J-T, Zhu X-C, Zhang Q-Q, Tan M-S, Cao L, Wang H-F, Lu J, Gao Q, Zhang Y-D, Tan L. Angiotensin-(1-7) induces cerebral ischaemic tolerance by promoting brain

angiogenesis in a Mas/eNOS-dependent pathway. Br J Pharmacol. 2014;171(18):4222–32. https://doi.org/10.1111/bph.12770.

20. Khajah MA, Fateel MM, Ananthalakshmi KV, Luqmani YA. Anti-inflammatory action of angiotensin 1-7 in experimental colitis may be mediated through modulation of serum cytokines/chemokines and immune cell functions. Dev Comp Immunol. 2017;74:200–8. https://doi.org/10.1016/j.dci.2017.05.005.

21. Klein N, Gembardt F, Supé S, Kaestle SM, Nickles H, Erfinanda L, Lei X, Yin J, Wang L, Mertens M, Szaszi K, Walther T, Kuebler WM. Angiotensin-(1-7) protects from experimental acute lung injury. Crit Care Med. 2013;41(11):e334–43. https://doi.org/10.1097/CCM.0b013e31828a6688.

22. Kucharewicz I, Pawlak R, Matys T, Pawlak D, Buczko W. Antithrombotic effect of captopril and losartan is mediated by angiotensin-(1-7). Hypertension (Dallas, Tex: 1979). 2002;40(5):774–9. Available at: http://www.ncbi.nlm.nih.gov/pubmed/12411476. Accessed: 21 April 2018.

23. Della Latta V, Cecchettini A, Del Ry S, Morales MA. Bleomycin in the setting of lung fibrosis induction: from biological mechanisms to counteractions. Pharmacol Res. Elsevier Ltd. 2015;97:122–30. https://doi.org/10.1016/j.phrs.2015.04.012.

24. Liang B, Wang X, Zhang N, Yang H, Bai R, Liu M, Bian Y, Xiao C, Yang Z. Angiotensin-(1-7) attenuates angiotensin II-induced ICAM-1, VCAM-1, and MCP-1 expression via the MAS receptor through suppression of P38 and NF-κB pathways in HUVECs. Cell Physiol Biochem. 2015;35(6):2472–82. https://doi.org/10.1159/000374047.

25. Liu M, Shi P, Sumners C. Direct anti-inflammatory effects of angiotensin-(1-7) on microglia. J Neurochem. 2016;136(1):163–71. https://doi.org/10.1111/jnc.13386.

26. Lu W, Kang J, Hu K, Tang S, Zhou X, Yu S, Li Y, Xu L. Angiotensin-(1-7) inhibits inflammation and oxidative stress to relieve lung injury induced by chronic intermittent hypoxia in rats. Braz J Med Biol Res. 2016;49(10):e5431. https://doi.org/10.1590/1414-431X20165431.

27. Lu W, Kang J, Hu K, Tang S, Zhou X, Yu S, Xu L. Angiotensin-(1-7) relieved renal injury induced by chronic intermittent hypoxia in rats by reducing inflammation, oxidative stress and fibrosis. Braz J Med Biol Res. 2017;50(1). https://doi.org/10.1590/1414-431x20165594.

28. Ma H, Kong J, Wang Y-L, Li J-L, Hei N-H, Cao X-R, Yang J-J, Yan W-J, Liang W-J, Dai H-Y, Dong B. Angiotensin-converting enzyme 2 overexpression protects against doxorubicin-induced cardiomyopathy by multiple mechanisms in rats. Oncotarget. 2017;8(15):24548–63. https://doi.org/10.18632/oncotarget.15595.

29. Magalhães GS, et al. Chronic allergic pulmonary inflammation is aggravated in angiotensin-(1–7) Mas receptor knockout mice. Am J Physiol Cell Mol Physiol. 2016;311:L1141–8.

30. Magalhaes GS, Barroso LC, Reis AC, Rodrigues-Machado MG, Gregório JF, Motta-Santos D, Oliveira AC, Perez DA, Barcelos LS, Teixeira MM, Santos RAS, Pinho V, Campagnole-Santos MJ. Angiotensin-(1-7) promotes resolution of eosinophilic inflammation in an experimental model of asthma. Front Immunol. 2018;9:58. https://doi.org/10.3389/fimmu.2018.00058.

31. Magalhães GS, Rodrigues-Machado MG, Motta-Santos D, Silva AR, Caliari MV, Prata LO, Abreu SC, Rocco PRM, Barcelos LS, Santos RAS, Campagnole-Santos MJ. Angiotensin-(1-7) attenuates airway remodelling and hyperresponsiveness in a model of chronic allergic lung inflammation. Br J Pharmacol. 2015;172(9):2330–42. https://doi.org/10.1111/bph.13057.

32. Medzhitov R. Inflammation 2010: new adventures of an old flame. Cell. 2010;140(6):771–6. https://doi.org/10.1016/j.cell.2010.03.006.

33. Meng Y, Li T, Zhou GS, Chen Y, Yu CH, Pang MX, Li W, Li Y, Zhang WY, Li X. The angiotensin-converting enzyme 2/angiotensin (1-7)/Mas axis protects against lung fibroblast migration and lung fibrosis by inhibiting the NOX4-derived ROS-mediated RhoA/Rho kinase pathway. Antioxid Redox Signal. 2015;22(3):241–58. https://doi.org/10.1089/ars.2013.5818.

34. Meng Y, Yu CH, Li W, Li T, Luo W, Huang S, Wu PS, Cai SX, Li X. Angiotensin-converting enzyme 2/angiotensin-(1-7)/mas axis protects against lung fibrosis by inhibiting the MAPK/NF-κB pathway. Am J Respir Cell Mol Biol. 2014;50(4):723–36. https://doi.org/10.1165/rcmb.2012-0451OC.

35. Miranda AS, Simões e Silva AC. Serum levels of angiotensin converting enzyme as a biomarker of liver fibrosis. World J Gastroenterol. 2017;23(48):8439–42. https://doi.org/10.3748/wjg.v23.i48.8439.

36. Moore ED, Kooshki M, Metheny-Barlow LJ, Gallagher PE, Robbins ME. Angiotensin-(1-7) prevents radiation-induced inflammation in rat primary astrocytes through regulation of MAP kinase signaling. Free Radic Biol Med. 2013;65:1060–8.

37. Morales MG, Abrigo J, Meneses C, Simon F, Cisternas F, Rivera JC, Vazquez Y, Cabello-Verrugio C. The Ang-(1-7)/Mas-1 axis attenuates the expression and signalling of TGF-β1 induced by AngII in mouse skeletal muscle. Clin Sci (Lond). 2014;127(4):251–64. https://doi.org/10.1042/CS20130585.

38. Mori J, Patel VB, Ramprasath T, Alrob OA, DesAulniers J, Scholey JW, Lopaschuk GD, Oudit GY. Angiotensin 1-7 mediates renoprotection against diabetic nephropathy by reducing oxidative stress, inflammation, and lipotoxicity. Am J Physiol Renal Physiol. 2014;306(8):F812–21. https://doi.org/10.1152/ajprenal.00655.2013.

39. Ni L, Feng Y, Wan H, Ma Q, Fan L, Qian Y, Li Q, Xiang Y, Gao B. Angiotensin-(1-7) inhibits the migration and invasion of A549 human lung adenocarcinoma cells through inactivation of the PI3K/Akt and MAPK signaling pathways. Oncol Rep. 2012;27(3):783–90. https://doi.org/10.3892/or.2011.1554.

40. Ocaranza MP, Moya J, Barrientos V, Alzamora R, Hevia D, Morales C, Pinto M, Escudero N, García L, Novoa U, Ayala P, Díaz-Araya G, Godoy I, Chiong M, Lavandero S, Jalil JE, Michea L. Angiotensin-(1-9) reverses experimental hypertension and cardiovascular damage by inhibition of the angiotensin converting enzyme/Ang II axis. J Hypertens. 2014;32(4):771–83. https://doi.org/10.1097/HJH.0000000000000094.

41. Pai W-Y, Lo W-Y, Hsu T, Peng C-T, Wang H-J. Angiotensin-(1-7) inhibits thrombin-induced endothelial phenotypic changes and reactive oxygen species production via NADPH oxidase 5 downregulation. Front Physiol. 2017;8:994. https://doi.org/10.3389/fphys.2017.00994.

42. Papinska AM, Soto M, Meeks CJ, Rodgers KE. Long-term administration of angiotensin (1-7) prevents heart and lung dysfunction in a mouse model of type 2 diabetes (db/db) by reducing oxidative stress, inflammation and pathological remodeling. Pharmacol Res. 2016;107:372–80. https://doi.org/10.1016/j.phrs.2016.02.026.

43. Pena Silva RA, Kung DK, Mitchell IJ, Alenina N, Bader M, Santos RAS, Faraci FM, Heistad DD, Hasan DM. Angiotensin 1-7 reduces mortality and rupture of intracranial aneurysms in mice. Hypertension. 2014;64(2):362–8. https://doi.org/10.1161/HYPERTENSIONAHA.114.03415.

44. Pereira RM, dos Santos RAS, Teixeira MM, Leite VHR, Costa LP, da Costa Dias FL, Barcelos LS, Collares GB, Simões e Silva AC. The renin–angiotensin system in a rat model of hepatic fibrosis: evidence for a protective role of angiotensin-(1-7). J Hepatol. 2007;46(4):674–81. https://doi.org/10.1016/j.jhep.2006.10.018.

45. Peres RS, Menezes GB, Teixeira MM, Cunha FQ. Pharmacological opportunities to control inflammatory diseases through inhibition of the leukocyte recruitment. Pharmacol Res. 2016;112:37–48. https://doi.org/10.1016/j.phrs.2016.01.015.

46. Perretti M, Cooper D, Dalli J, Norling LV. Immune resolution mechanisms in inflammatory arthritis. Nat Rev Rheumatol. 2017;13(2):87–99. https://doi.org/10.1038/nrrheum.2016.193.

47. Pinheiro SVB, Simoes e Silva AC, Sampaio WO, de Paula RD, Mendes EP, Bontempo ED, Pesquero JB, Walther T, Alenina N, Bader M, Bleich M, RAS S. Nonpeptide AVE 0991 is an angiotensin-(1-7) receptor Mas agonist in the mouse kidney. Hypertension. 2004;44(4):490–6. https://doi.org/10.1161/01.HYP.0000141438.64887.42.

48. Rodrigues-Machado MG, Magalhães GS, Cardoso JA, Kangussu LM, Murari A, Caliari MV, Oliveira ML, Cara DC, Noviello MLM, Marques FD, Pereira JM, Lautner RQ, Santos RAS, Campagnole-Santos MJ. AVE 0991, a non-peptide mimic of angiotensin-(1-7) effects, attenuates pulmonary remodelling in a model of chronic asthma. Br J Pharmacol. 2013;170(4):835–46. https://doi.org/10.1111/bph.12318.

49. Shenoy V, et al. The angiotensin-converting enzyme 2/angiogenesis-(1-7)/Mas axis confers cardiopulmonary protection against lung fibrosis and pulmonary hypertension. Am J Respir Crit Care Med. 2010;182:1065–72.

50. Shimada K, et al. Angiotensin-(1-7) protects against the development of aneurysmal subarachnoid hemorrhage in mice. J Cereb Blood Flow Metab. 2015;35:1163–8.

51. Silveira KD, Barroso LC, Vieira AT, Cisalpino D, Lima CX, Bader M, Arantes RME, dos Santos RAS, Simões-e-Silva AC, Teixeira MM. Beneficial effects of the activation of the angiotensin-(1-7) Mas receptor in a murine model of adriamycin-induced nephropathy. PLoS One. Edited by J-C Dussaule. 2013a;8(6):e66082. https://doi.org/10.1371/journal.pone.0066082.

52. Silveira KD, Coelho FM, Vieira AT, Barroso LC, Queiroz-Junior CM, Costa VV, Sousa LFC, Oliveira ML, Bader M, Silva TA, Santos RAS, Silva ACSE, Teixeira MM. Mechanisms of the anti-inflammatory actions of the angiotensin type 1 receptor antagonist losartan in experimental models of arthritis. Peptides. 2013b;46:53–63. https://doi.org/10.1016/j.peptides.2013.05.012.

53. da Silveira KD, Coelho FM, Vieira AT, Sachs D, Barroso LC, Costa VV, Bretas TLB, Bader M, de Sousa LP, da Silva TA, dos Santos RAS, Simoes e Silva AC, Teixeira MM. Anti-inflammatory effects of the activation of the angiotensin-(1-7) receptor, Mas, in experimental models of arthritis. J Immunol. 2010;185(9):5569–76. https://doi.org/10.4049/jimmunol.1000314.

54. Simões e Silva A, Silveira K, Ferreira A, Teixeira M. ACE2, angiotensin-(1-7) and Mas receptor axis in inflammation and fibrosis. Br J Pharmacol. 2013;169(3):477–92. https://doi.org/10.1111/bph.12159.

55. Skiba DS, Nosalski R, Mikolajczyk TP, Siedlinski M, Rios FJ, Montezano AC, Jawien J, Olszanecki R, Korbut R, Czesnikiewicz-Guzik M, Touyz RM, Guzik TJ. Anti-atherosclerotic effect of the angiotensin 1-7 mimetic AVE0991 is mediated by inhibition of perivascular and plaque inflammation in early atherosclerosis. Br J Pharmacol. 2017;174(22):4055–69. https://doi.org/10.1111/bph.13685.

56. South AM, Nixon PA, Chappell MC, Diz DI, Russell GB, Snively BM, Shaltout HA, Rose JC, O'Shea TM, Washburn LK. Antenatal corticosteroids and the renin-angiotensin-aldosterone system in adolescents born preterm. Pediatr Res. 2017;81(1–1):88–93. https://doi.org/10.1038/pr.2016.179.

57. Souza LL, Costa-Neto CM. Angiotensin-(1-7) decreases LPS-induced inflammatory response in macrophages. J Cell Physiol. 2012;227(5):2117–22. https://doi.org/10.1002/jcp.22940.

58. Sugimoto MA, Sousa LP, Pinho V, Perretti M, Teixeira MM. Resolution of inflammation: what controls its onset? Front Immunol. 2016;7:160. https://doi.org/10.3389/fimmu.2016.00160.

59. Sukumaran V, Tsuchimochi H, Tatsumi E, Shirai M, Pearson JT. Azilsartan ameliorates diabetic cardiomyopathy in young db/db mice through the modulation of ACE-2/ANG 1–7/Mas receptor cascade. Biochem Pharmacol. 2017;144:90–9.

60. Uemura H, Ishiguro H, Kubota Y. Pharmacology and new perspectives of angiotensin II receptor blocker in prostate cancer treatment. Int J Urol. 2008;15(1):19–26. https://doi.org/10.1111/j.1442-2042.2007.01937.x.

61. Vickers C, Hales P, Kaushik V, Dick L, Gavin J, Tang J, Godbout K, Parsons T, Baronas E, Hsieh F, Acton S, Patane M, Nichols A, Tummino P. Hydrolysis of biological peptides by human angiotensin-converting enzyme-related carboxypeptidase. J Biol Chem. 2002;277(17):14838–43. https://doi.org/10.1074/jbc.M200581200.

62. Villalobos LA, San Hipólito-Luengo Á, Ramos-González M, Cercas E, Vallejo S, Romero A, Romacho T, Carraro R, Sánchez-Ferrer CF, Peiró C. The angiotensin-(1-7)/mas Axis counteracts angiotensin II-dependent and -independent pro-inflammatory Signaling in human vascular smooth muscle cells. Front Pharmacol. 2016;7:482. https://doi.org/10.3389/fphar.2016.00482.

63. Wang J, Liu R, Qi H, Wang Y, Cui L, Wen Y, Li H, Yin C. The ACE2-angiotensin-(1-7)-Mas axis protects against pancreatic cell damage in cell culture. Pancreas. 2015;44(2):266–72. https://doi.org/10.1097/MPA.0000000000000247.

64. Wang Y, et al. Angiotensin 1-7 ameliorates caerulein-induced inflammation in pancreatic acinar cells by downregulating Toll-like receptor 4/nuclear factor-κB expression. Mol Med Rep. 2018;17:3511–8.

65. Yang L-W, Qin D-Z, James E, McKallip RJ, Wang N-P, Zhang W-W, Zheng R-H, Han Q-H, Zhao Z-Q. CD44 deficiency in mice protects the heart against angiotensin II-induced cardiac fibrosis. Shock. 2018:1. https://doi.org/10.1097/SHK.0000000000001132.

66. Young D, Waitches G, Birchmeier C, Fasano O, Wigler M. Isolation and characterization of a new cellular oncogene encoding a protein with multiple potential transmembrane domains. Cell. 1986;45(5):711–9. Available at: http://www.ncbi.nlm.nih.gov/pubmed/3708691. Accessed: 23 April 2018.
67. Yu X, Cui L, Hou F, Liu X, Wang Y, Wen Y, Chi C, Li C, Liu R, Yin C. Angiotensin-converting enzyme 2-angiotensin (1-7)-mas axis prevents pancreatic acinar cell inflammatory response via inhibition of the p38 mitogen-activated protein kinase/nuclear factor-κB pathway. Int J Mol Med. 2018;41(1):409–20. https://doi.org/10.3892/ijmm.2017.3252.
68. Zambelli V, et al. Angiotensin-(1-7) improves oxygenation, while reducing cellular infiltrate and fibrosis in experimental Acute Respiratory Distress Syndrome. Intensive Care Med Exp. 2015;3(8)
69. Zhang F, Ren X, Zhao M, Zhou B, Han Y. Angiotensin-(1-7) abrogates angiotensin II-induced proliferation, migration and inflammation in VSMCs through inactivation of ROS-mediated PI3K/Akt and MAPK/ERK signaling pathways. Sci Rep. 2016;6:34621.
70. Zhao Y, Qin Y, Liu T, Hao D. Chronic nerve injury-induced Mas receptor expression in dorsal root ganglion neurons alleviates neuropathic pain. Exp Ther Med. 2015;10(6):2384–8. https://doi.org/10.3892/etm.2015.2801.
71. Zong W, Yang X, Chen X, Huang H, Zheng H, Qin X, Yong Y, Cao K, Huang J, Lu X. Regulation of angiotensin-(1-7) and angiotensin II type 1 receptor by telmisartan and losartan in adriamycin-induced rat heart failure. Acta Pharmacol Sin. 2011;32(11):1345–50. https://doi.org/10.1038/aps.2011.96.

The Role of Angiotensin–(1-7) in Cancer

Ana Cristina Simões e Silva and Walkyria O. Sampaio

Introduction

One of the first indirect evidence for the role of Angiotensin-(1-7) [Ang-(1-7)] in cancer was that hypertensive patients treated with angiotensin-converting enzyme (ACE) inhibitors have a decreased risk of cancer [1] and ACE inhibitors cause a significant elevation in both tissue and circulating Ang-(1-7) [2–4]. Considering that Ang-(1-7) inhibits the growth of several cell lines [5–9], it has been suggested that the heptapeptide may also reduce the proliferation of cancer cells and tumors.

However, the direct role of Ang-(1-7) in tumor process was first described by Rodgers and coworkers [10, 11]. These authors showed that treatment with Ang-(1-7) accelerates hematopoietic recovery by increasing both the number of white blood cells and myeloid progenitors in the peripheral blood and bone marrow after chemotherapy [10, 11]. Soon after, Gallagher and Tallant [12] reported that Ang-(1-7) inhibits lung cancer cell growth through the activation of Mas receptor.

In this chapter, we summarize studies on the role of Ang-(1-7) in different types of cancer.

A. C. Simões e Silva (✉)
Department of Pediatrics, Interdisciplinary Laboratory of Medical Investigation, Faculty of Medicine, UFMG, Belo Horizonte, MG, Brazil

Interdisciplinary Laboratory of Medical Investigation, Faculty of Medicine, Federal University of Minas Gerais (UFMG), Belo Horizonte, MG, Brazil
e-mail: ana@medicina.ufmg.br

W. O. Sampaio
National Institute of Science and Technology in Nanobiopharmaceutics, Department of Physiology and Biophysics, Institute of Biological Sciences, Federal University of Minas Gerais, Belo Horizonte, Brazil

© Springer Nature Switzerland AG 2019
R. A. S. Santos (ed.), *Angiotensin-(1-7)*,
https://doi.org/10.1007/978-3-030-22696-1_14

Angiotensin-(1-7) in Lung Cancer

Lung cancer is one of the most frequent types of cancer in humans and a leading cause of death [13]. The high mortality rates have been associated with late diagnosis, which results in elevated frequency of metastasis [13]. Therefore, despite all developments in therapeutic approach, the investigation of novel treatments that control neoplastic cell migration, proliferation and metastasis is urgently needed.

In this regard, Ang-(1-7) has emerged as a potential therapeutic target [14]. The first evidence was provided by an in vitro study. Three lung cancer cell lines including human adenocarcinoma SK-LU-1 and A549 cells and non-small cell lung cancer SK-MES-1 cells were incubated with Ang-(1-7) to determine the effect of the heptapeptide on cell growth [12]. Ang-(1-7) potently inhibited the growth of all cell lines of lung cancer at subnanomolar concentrations [12]. The antiproliferative effect of Ang-(1-7) was associated with reduction in serum-stimulated phosphorylation of the MAP kinases ERK1/2 [12]. In addition, the inhibitory effect of Ang-(1-7) was mediated by Mas receptor, since it was blocked by the Mas receptor antagonist D-Ala7-Ang-(1-7) (A-779) and not affected by AT_1 or AT_2 [12]. In vivo evidence was obtained by the intravenous administration of Ang-(1-7) in mice with A549 human lung tumors [15]. Ang-(1-7) infusion was very well tolerated by the mice and resulted in a reduction of tumor volume by 30% compared to the size prior to treatment [15]. These findings were associated with a decrease in the proliferation marker Ki67 [15].

Two other mechanisms that may contribute to antitumor effect of Ang-(1-7) in lung cancer are cyclooxygenase 2 (COX-2) inhibition and antiangiogenic activity. COX-2 is overexpressed in 70–90% of adenocarcinomas [16]. Clinical trials with nonselective inhibitors of COX-2 decreased the risk for lung cancer, suggesting that a reduction in COX-2 is associated with inhibition of lung tumor growth [17, 18]. Ang-(1-7) significantly reduced COX-2 mRNA and protein in both A549 tumor xenografts and A549 cells in culture [15]. It should be pointed that Ang-(1-7) has significant advantages over the administration of nonselective and selective COX-2 inhibitors for lung cancer since the heptapeptide-mediated reduction in COX-2 is associated with antithrombotic and antiinflammatory activities without the side effects related to COX-2 inhibitors [19]. In regard to antiangiogenic activity, athymic mice with A549 lung tumors were injected daily with 1000 µg/kg of Ang-(1-7) for 5 days, followed by a 2-day drug-free, and sacrificed after 42 days [20]. Ang-(1-7) markedly decreased vascular endothelial growth factor (VEGF) protein and mRNA, vessel density and A549 lung tumor growth [20]. The antiangiogenic effect of Ang-(1-7) was also mediated by Mas receptor [20].

More recently, the expression pattern of microRNAs (miRNAs) in lung tumor cells has been investigated to elucidate the mechanisms by which Ang-(1-7) controls tumor migratory processes [21, 22]. It was found that miRNA-149-3p plays a role in cellular migration processes [21] and miRNA-513a-3p controls the protein level and molecular function of integrin-β8, thus reducing cell migration and inflammation [22]. Another recent line of investigation is the use viral vectors to deliver Ang-(1-7). Chen and coworkers [23] constructed a mutant adeno-associated viral vector AAV8 (Y733F) that produced stable and highly efficient Ang-(1-7) expression in a xenograft tumor model. AAV8-mediated Ang-(1-7) overexpression inhibited tumor growth in vivo by downregulating Cdc6 and anti-angiogenesis. These findings provide useful information for future investigations on drug development.

Angiotensin-(1-7) in Breast Cancer

Among all the malignant diseases, breast cancer is considered as one of the most important causes of death in postmenopausal women, accounting for 23% of all cancer deaths [24]. There are three major types of breast cancer: estrogen receptor-positive (ER-positive) breast cancer, which can be treated with selective estrogen receptor modifiers (SERMs) or aromatase inhibitors; human epidermal growth factor receptor 2 (HER2)-overexpressing breast cancer, which can be treated with an antibody to the HER2 receptor; and triple-negative breast cancers, which lack both estrogen and progesterone receptors and also do not overexpress HER2, very frequently being refractory to conventional treatments [24]. In spite of targeted treatments for ER-positive and HER2-overexpressing breast cancers, there is still need for novel therapies for both primary and metastatic diseases.

The effect of Ang-(1-7) was first investigated in human ZR-75-1 ER-positive and BT474 ER-positive/HER2-overexpressing breast tumors [25]. Ang-(1-7) significantly reduced tumor volume and weight in both ZR-75-1 and BT474 breast tumors [25]. In addition, treatment with Ang-(1-7) markedly decreased fibroblast growth, interstitial fibrosis within the tumors and perivascular fibrosis surrounding tumor vessels, in association with a decrease in immunoreactive collagen I deposition [25]. The antifibrotic effect of Ang-(1-7) was associated with increase in MAP kinase phosphatase and reductions in phosphorylated ERK1/2 and in the production of transforming growth factor-beta (TGF-β) [25]. These findings indicate that Ang-(1-7) inhibits cancer-associated fibroblasts growth and tumor fibrosis in breast cancer.

The role of ACE2/Ang-(1-7)/MAS receptor axis was also investigated in the metastasis of breast cancer [26]. Yu and coworkers [26] detected that ACE2 protein level is negatively associated with the metastatic ability of breast cancer cells and breast tumor grade. Furthermore, stimulation of the ACE2/Ang-(1-7)/Mas receptor axis reduced breast cancer cell migration and invasion in vivo and in vitro by means of the inhibition of store-operated calcium entry and PAK1/NF-κB/Snail1 pathways, and the induction of E-cadherin expression [26].

The counterregulatory role of Ang-(1-7) against deleterious actions of Ang II was also observed in breast cancer. In this regard, Cambados and coworkers [27] have investigated the effect of Ang-(1-7) on Ang II-induced protumorigenic features on normal murine mammary epithelial cells NMuMG and on breast cancer cells MDA-MB-231. Ang II stimulated PI3K/AKT pathway leading to epithelial-mesenchymal transition in NMuMG cells via AT$_1$ receptor activation [27].

Simultaneous administration of Ang II and Ang-(1-7) abolished the effects of Ang II in NMuMG cells. In addition, Ang-(1-7) annulled Ang II-induced migration and invasion of the MDA-MB-231 cells and inhibited proangiogenic process by reducing VEGF expression [27].

Angiotensin-(1-7) in Prostate Cancer

Prostate cancer is the second most important cause of cancer deaths in men [28]. Treatment options for localized prostate cancer include surgery, radiation therapy, and hormone ablation therapy. Although treatment is encouraging for primary prostate cancer, metastatic prostate cancer, predominantly to the bone, is often fatal [28].

Ang-(1-7) was administered for 54 days to athymic mice with human LNCaP prostate cancer cells injected into the flank [29]. Ang-(1-7) treatment significantly reduced the volume and weight of LNCaP xenograft tumors [29]. Histological analysis of the tumor showed that Ang-(1-7) decreased Ki67 immunoreactivity, ERK1/2 activities and vessel density. The reduced angiogenesis was associated with less concentration of VEGF and of PlGF and increased levels of the soluble fraction of VEGF receptor 1 (sFlt-1). sFlt-1 acts as a decoy receptor that catches VEGF and PlGF, making the ligands unavailable to membrane-bound VEGF receptors and preventing activation of proangiogenic signaling [29].

In order to investigate the effect of Ang-(1-7) on metastatic prostate cancer to the bone, human PC3 prostate cancer cells were injected into the aortic arch of mice pretreated with Ang-(1-7) or into the tibia of athymic mice, subsequently administered with Ang-(1-7) for 5 weeks beginning 2 weeks after cancer cells injection [30]. When PC3 cells were injected, the mice developed tumors in the submandibular bone, the spinal column, or the long bone of the leg. In sharp contrast, pretreatment with Ang-(1-7) prevented metastatic tumor formation. Ang-(1-7) administered 2 weeks after human PC3 prostate cancer cells also attenuated intratibial tumor growth. In addition, bone marrow cells were extruded from the long bone of untreated mice, differentiated to induce osteoclastogenesis and treated with Ang-(1-7). The heptapeptide reduced by 50% osteoclastogenesis in bone marrow cells, suggesting that Ang-(1-7) treatment impedes the formation of osteolytic lesions to reduce tumor survival in the bone microenvironment [30].

Angiotensin-(1-7) in Hepatocellular Cancer

The effects of Ang-(1-7) were also investigated in hepatocellular carcinoma [31, 32]. For this purpose, H22 hepatoma-bearing mice were randomly divided into five groups: vehicle-treated group, mice receiving low-dose of Ang-(1-7), high-dose of Ang-(1-7), high-dose of Ang-(1-7) plus A779, and high-dose of Ang-(1-7) plus PD123319. Ang-(1-7) inhibited tumor growth in a time- and dose-dependentl manner [31]. The antitumoral mechanisms elicited by Ang-(1-7) include reduction of cell proliferation and of angiogenesis as well as induction of tumor cell apoptosis. The effects of Ang-(1-7) on tumor cell proliferation and apoptosis were reversed by coadministration with A779 or PD123319, whereas the effects on tumor angiogenesis were completely blocked by A779 but not by PD123319. Moreover, Ang-(1-7) downregulated mRNA for AT_1 receptor, upregulated mRNA for AT_2 and Mas receptors [31].

As previously described for lung cancer, the use viral vectors to deliver Ang-(1-7) was also employed in hepatocellular carcinoma also. Thus, the effects of Adeno-associated virus serotype-8 (AAV8)-mediated Ang-(1-7) overexpression were investigated in hepatocellular carcinoma [32]. The authors first generated three different mutants of AAV8 (Y447F, Y703F, Y708F) and evaluated in vivo transduction efficiencies. The antitumor effects of Ang-(1-7) delivered by Y703F, the most efficient vector, was evaluated in H22 hepatoma-bearing mice. AAV-Ang-(1-7) persistently inhibited angiogenesis and the growth of hepatocellular carcinoma [32].

Angiotensin-(1-7) in Glioblastoma

Glioblastoma multiforme (GBM) is the most primary brain tumor, specially characterized with the damage of blood–brain barrier [33]. Ang-(1-7) inhibited the growth and invasiveness of GBM [34]. To investigate the mechanisms beyond antitumor effect of Ang-(1-7) in GBM, Liu and coworkers evaluated the crosstalk between Podocalyxin (PODX) and Ang-(1-7)/Mas receptor signaling in GBM cells, and examined its effect on GBM cell invasion and proliferation [35]. It has been previously reported that PODX enhances invasion in many human cancers including GBM. The authors found a significant negative correlation between the expression of PODX and Mas in GBM tumor tissues from 10 patients ($r = -0.768$, $p < 0.01$) [35]. The stable overexpression of PODX in LN-229 and U-118 MG human GBM cells decreased the mRNA and protein expression of Mas receptor, resulting in low density of Ang-(1-7)-binding Mas on the cell membrane. Overexpression and knockdown of PODX respectively reversed and enhanced the inhibitory effects of Ang-(1-7) on the expression/activity of matrix metalloproteinase-9 and cell invasion and proliferation in GBM cells. PODX inhibited Ang-(1-7)/Mas signaling by downregulating the expression of Mas through a PI3K-dependent mechanism in GBM cells. This effect increased GBM cell invasion and proliferation [35].

Besides the inhibitory effect of Ang-(1-7) on GBM growth, the heptapeptide significantly relieved the damage of blood–brain barrier in rats with intracranial U87 gliomas [36]. Furthermore, treatment with Ang-(1-7) restored the function of blood–brain barrier and avoided edema formation. Similarly, Ang-(1-7) also decreased U87 glioma cells barrier permeability in vitro. The protective effect of Ang-(1-7) on blood–brain barrier was associated with inhibition of JNK pathway and a consequent return of tight junction protein (claudin-5 and ZO-1) expression to normal levels both in rats and in U87 glioma cells [36].

Angiotensin-(1-7) in Other Cancers

Basal and interleukin (IL)-1β-stimulated expression of components of ACE2/Ang(1-7)/Mas receptor axis were evaluated in U-2 OS and MNNG-HOS osteosarcoma cells analyzed as well as the effect of Mas receptor on proliferation and/or migration of these cells [37]. The two cell lines expressed Ang-(1-7)-generating peptidases, including ACE2, neutral endopeptidase 24.11 and prolyl-endopeptidase together with Mas receptor. IL-1β stimulated mRNA and protein expression for Mas receptor, which was associated with a reduction of proliferation and migration. On the other hand, RNA-mediated knockdown of Mas expression led to increased cell proliferation, supporting a beneficial role of ACE2/Ang(1-7)/Mas receptor axis in the treatment of osteosarcoma [37].

The treatment of nasopharyngeal carcinoma (NPC) has been associated with several side effects [38]. Therefore, the investigation on novel treatment modalities for NPC is of utmost importance. It was found that Mas receptor in significantly upregulated in NPC specimens and NPC cell lines [39]. Viral vector-mediated expression of Ang-(1-7) markedly inhibited NPC cell proliferation and migration in vitro. These effects were mediated by Mas receptor since they were completely blocked by A-779 [39]. In addition, Ang-(1-7) significantly reduced the growth and the vessel density of human nasopharyngeal xenografts [39]. Mechanistic investigations revealed that, also in this tumor, Ang-(1-7) inhibited the expression of the proangiogenic factors VEGF and PlGF. The effects and signaling pathways involved in the Ang-(1-7)/Mas receptor axis in NPC were further investigated both in vitro and in vivo [40]. Ang-(1-7) inhibited cell proliferation, migration, and invasion in NPC-TW01 cells. Ang-(1-7) induced autophagy by increasing the levels of the autophagy marker LC3-II and by enhancing p62 degradation via activation of the Beclin-1/Bcl-2 signaling pathway with participation of the PI3K/Akt/mTOR and p38 pathways [40]. Pretreatment with Ang-(1-7) also inhibited tumor growth in NPC xenografts by stimulating autophagy, while no autophagy was observed following Ang-(1-7) posttreatment [40]. To sum up, Ang-(1-7) plays a role in autophagy downstream signaling pathways in NPC, supporting its therapeutic potential for reducing the incidence of primary NPC tumors and for preventing recurrent disease [40].

A beneficial role of Ang-(1-7) was also reported in head and neck squamous cell carcinoma (HNSCC) [41]. Hinsley and coworkers showed that Ang II promotes the invasion and migration of HNSCC cells both in an autocrine manner. The effects of Ang II were mediated by AT_1 receptors and inhibited by Ang-(1-7) [41].

Angiotensin-(1-7) for Cancer Pain and Side Effects of the Treatment

Besides antitumor actions, Ang-(1-7) may also be useful to control cancer pain [42] and side effects secondary to radiation therapy [43].

Several solid tumors metastasize to the bone and induce intense pain. Cancer-induced bone pain is often severe due to accentuated inflammation, rapid bone degradation, and disease progression [44]. Opioids are recommended to manage this pain, but these medications may enhance bone loss and increase tumor proliferation [44]. The antinociceptive effect of Ang-(1-7) was investigated in a murine model of breast cancer-induced bone pain. Breast cancer cells were implanted into the femur of BALB/c mice [42]. Spontaneous and evoked pain behaviors were examined before and after acute and chronic administration of Ang-(1-7). Cancer inoculation increased spontaneous pain behaviors by day 7. Both single injection and sustained administration of Ang-(1-7) significantly reduced pain. Preadministration of A-779 impeded this reduction, while pretreatment with an AT_2 receptor antagonist had no effect. However, the use of an AT_1 antagonist potentiated the antinociceptive effect of Ang-(1-7). Ang-(1-7) via Mas receptor activation significantly attenuated pain without the side effects seen with opioids [42].

Radiation-induced fibrosis of musculoskeletal tissue is a common complication of radiation therapy for extremity soft-tissue sarcoma, without a strategy for prevention and treatment [45]. In this regard, Ang-(1-7) was tested as a radioprotectant agent for radiation-induced fibrosis and stiffening of irradiated muscles [43]. Male CD-1 mice were randomized to three treatment groups: control, simulated sarcoma radiation therapy to the gastrocnemius and soleus muscles, or radiation therapy along with continuous Ang-(1-7) infusion initiated 3 days before radiation therapy. Ang-(1-7) attenuated radiation-induced fibrosis, stiffening, and reduced the production of profibrotic cytokines that were elevated in mouse skeletal muscles after radiation therapy [43].

Clinical Trials with Ang-(1-7) or TXA127

A total of nine clinical trials with Ang-(1-7) or TXA127 are registered in NIH (3 completed, 3 withdrawn, 2 active and 1 terminated). TXA127 is a pharmaceutical grade formulation of the naturally occurring peptide Ang-(1-7), which Tarix Pharmaceuticals is developing for the treatment of a number of diseases.

The first trial registered was a phase I study that treated patients with metastatic or unresectable solid tumors with Ang-(1-7) [46]. Eighteen patients were enrolled in this trial. Dose-limiting toxicities found at the 700 microg/kg included stroke (grade 4) and reversible cranial neuropathy (grade 3). Other side effects were generally mild. One patient developed a 19% reduction in tumor measurements. Three additional patients showed clinical benefit with stabilization of disease lasting more than 3 months. Ang-(1-7) administration reduced circulating levels of plasma placental growth factor (PlGF) levels in patients with clinical improvement, but not in patients without clinical benefit [46]. Further results of this trial were not reported or published (ClinicalTrials.gov identifier: NCT00471562).

The second completed clinical trial is a phase II study on the role of Ang-(1-7) as second-line therapy or third-line therapy in treating patients with metastatic sarcoma that cannot be removed by surgery (ClinicalTrials.gov Identifier: NCT01553539). This study enrolled 20 adult patients with different types of sarcoma. Patients received Ang-(1-7) subcutaneous once daily in the absence of disease progression or unacceptable toxicity. Results revealed low rate of significant adverse effects and slight reduction in the concentrations of VEGF and PIGF at day 22 after the beginning of the treatment with Ang-(1-7).

The third completed clinical trial is a phase IIb, multicenter, randomized, double-blind, placebo-controlled study comparing safety and efficacy of two dose levels of TXA127 when administered during 6 cycles of combination gemcitabine and platinum-based chemotherapy (ClinicalTrials.gov Identifier: NCT00771810). This study intends to investigate the effectiveness of TXA127 for the mitigation of severity and/or incidence of thrombocytopenia, as well as safety when administered as a self-injected, subcutaneous solution. TXA127 was administered to patients with breast cancer in the adjuvant setting to determine the effect of Ang-(1-7) on cytopenia [47]. No dose-limiting toxicities were reported, and Ang-(1-7) reduced thrombocytopenia and lymphopenia [47]. Patients with ovarian, Fallopian tube, or peritoneal carcinoma receiving gemcitabine and carboplatin or cisplatin were also treated with TXA127. Once more, Ang-(1-7) reduced thrombocytopenia following gemcitabine and platinum chemotherapy [48]. These data suggest that Ang-(1-7) may be beneficial in attenuating multilineage cytopenias following bone marrow-toxic chemotherapy.

The three withdrawn clinical trials were phase II studies and had the objective to investigate the role of Ang-(1-7) or TXA127 in hematologic malignancies. The first registered study aimed to examine the safety and the efficacy of TXA127 to enhance engraftment in pediatric patients undergoing single or double umbilical cord blood transplantation (ClinicalTrials.gov identifier: NCT01554254). The second study aimed to evaluate the efficacy of TXA127 to reduce the incidence (Grade II-IV) of acute Graft-versus-Host Disease (aGVHD) in adult subjects undergoing double umbilical cord blood transplantation (ClinicalTrials.gov identifier: NCT01882374). The third study had the purpose to evaluate the efficacy of TXA127 to reduce the incidence (Grade II-IV) of aGVHD in adult subjects undergoing allogeneic peripheral blood stem cell transplantation (ClinicalTrials.gov identifier: NCT01882387). In these three studies, no results were reported, and the studies were withdrawn before participants were enrolled.

The two active clinical trials are investigating the role of TXA127 in hematologic malignancies. The first registered is a randomized, double-blind, placebo-controlled study phase II with the purpose to determine the safety and effectiveness of TXA127 in accelerating the time it takes for patients to recover their platelet counts following an Autologous Peripheral Blood Stem Cell transplant (ClinicalTrials.gov identifier: NCT01121120). The second is a phase I study with the aim to examine the safety and efficiency of TXA127 (two injected doses: 300 or 1000 mcg/kg/day for 28 days) in enhancement of engraftment in adult double cord blood transplantation for the treatment of a variety of hematologic malignancies for whom there is no available therapy with substantive antidisease effect (ClinicalTrials.gov identifier: NCT01300611). No results are posted for both studies.

The single terminated clinical trial was a phase 1 study that aimed to determine safety and tolerability of TXA127 (300, 600, or 900 µg/kg daily by subcutaneous injection) in thrombocytopenic subjects with low or Intermediate-1 risk myelodysplastic syndrome (ClinicalTrials.gov identifier: NCT01362036). No results were reported.

Concluding Remarks

Studies in vitro and in vivo experimental models showed that Ang-(1-7) reduced proliferation of human cancer cells and xenograft tumors. The antitumor effect of Ang-(1-7) was due to reduction of angiogenesis, cancer-associated fibrosis, osteoclastogenesis, tumor-induced inflammation, and metastasis, as well as inhibition of cancer cell growth and proliferation. In clinical trials, Ang-(1-7) was well tolerated with limited toxic or quality-of-life side effects and showed clinical benefit in cancer patients with solid tumors. In conclusion, these findings so far suggest that Ang-(1-7) may act as chemotherapeutic agent, reducing cancer growth and metastases by pleiotropic mechanisms as well as targeting the tumor microenvironment. Further clinical trials are needed to confirm safety, and to determine doses and clinical indications.

References

1. Lever AF, Hole DJ, Gillis CR, McCallum IRMGT, MacKinnon PL, Meredith PA, et al. Do inhibitors of angiotensin-I-converting enzyme protect against risk of cancer? Lancet. 1998;352:179–84.
2. Luque M, Martin P, Martell N, Fernandez C, Brosnihan KB, Ferrario CM. Effects of captopril related to increased levels of prostacyclin and angiotensin-(1-7) in essential hypertension. J Hypertens. 1996;14:799–805.
3. Iyer SN, Ferrario CM, Chappell MC. Angiotensin-(1-7) contributes to the antihypertensive effects of blockade of the renin-angiotensin system. Hypertension. 1998;31:356–61.
4. Simões e Silva AC, Diniz JS, Pereira RM, Pinheiro SV, Santos RAS. Circulating renin angiotensin system in childhood chronic renal failure: marked increase of angiotensin-(1-7) in end-stage renal disease. Pediatr Res. 2006;60:734–9.
5. Freeman EJ, Chisolm GM, Ferrario CM, Tallant EA. Angiotensin-(1-7) inhibits vascular smooth muscle cell growth. Hypertension. 1996;28:104–8.
6. McCollum LT, Gallagher PE, Tallant EA. Angiotensin-(1-7) abrogates mitogen-stimulated proliferation of cardiac fibroblasts. Peptides. 2012;34:380–8.
7. Strawn WB, Ferrario CM, Tallant EA. Angiotensin-(1-7) reduces smooth muscle growth after vascular injury. Hypertension. 1999;33(part II):207–11.
8. Langeveld B, Van Gilst WH, Gio RA, Zijlstra F, Roks AJ. Angiotensin-(1-7) attenuates neointimal formation after stent implantation in the rat. Hypertension. 2005;45:138–41.
9. Tallant EA, Ferrario CM, Gallagher PE. Angiotensin-(1-7) inhibits growth of cardiac myocytes through activation of the *mas* receptor. Am J Physiol Heart Circ Physiol. 2005;289:1560–6.
10. Rodgers KE, Xiong S, diZerega GS. Accelerated recovery from irradiation injury by angiotensin peptides. Cancer Chemother Pharmacol. 2002;49:403–11.
11. Rodgers K, Xiong S, DiZerega GS. Effect of angiotensin II and angiotensin(1-7)on hematopoietic recovery after intravenous chemotherapy. Cancer Chemother Pharmacol. 2003;51:97–106.
12. Gallagher PE, Tallant EA. Inhibition of human lung cancer cell growth by angiotensin-(1-7). Carcinogenesis. 2004;25:2045–52.
13. Hirsch FR, Scagliotti GV, Mulshine JL, Kwon R, Curran WJ Jr, Wu YL, Paz-Ares L. Lung cancer: current therapies and new targeted treatments. Lancet. 2017;389:299–311.
14. Gallagher PE, Arter AL, Deng G, Tallant EA. Angiotensin-(1-7): a peptide hormone with anti-cancer activity. Curr Med Chem. 2014;21:2417–23.
15. Menon J, Soto-Pantoja DR, Callahan MF, Cline JM, Ferrario CM, Tallant EA, Gallagher PE. Angiotensin-(1-7) inhibits growth of human lung adenocarcinoma xenografts in nude mice through a reduction in cyclooxygenase-2. Cancer Res. 2007;67:2809–15.

16. Hida T, Yatabe Y, Achiwa H, Muramatsu H, Kozaki K, Nakamura S, et al. Increased expression of cyclooxygenase 2 occurs frequently in human lung cancers, especially in adenocarcinomas. Cancer Res. 1998;58:3761–4.
17. Harris RE, Beebe-Donk J, Schuller HM. Chemoprevention of lung cancer by non-steroidal anti-inflammatory drugs among cigarette smokers. Oncol Rep. 2002;9:693–5.
18. Lee EO, Lee HJ, Hwang HS, Ahn KS, Chae C, Kang KS, et al. Potent inhibition of Lewis lung cancer growth by heyneanol A from the roots of *Vitis amurensis* through apoptotic and anti-angiogenic activities. Carcinogenesis. 2006;27:2059–69.
19. Mukherjee D, Topol EJ. Cox-2: where are we in 2003?—Cardiovascular risk and Cox-2 inhibitors. Arthritis Res Ther. 2003;5:8–11.
20. Soto-Pantoja DR, Menon J, Gallagher PE, Tallant EA. Angiotensin-(1-7) inhibits tumor angiogenesis in human lung cancer xenografts with a reduc- tion in vascular endothelial growth factor. Mol Cancer Ther. 2009;8:1676–83.
21. de Oliveira da Silva B, Lima KF, Gonçalves LR, Silveira MB, Moraes KC. MicroRNA profiling of the effect of the heptapeptide Angiotensin-(1-7) in A549 lung tumor cells reveals a role for miRNA149-3p in cellular migration processes. PLoS One. 2016;11:e0162094. https://doi.org/10.1371/journal.pone.0162094.
22. Silveira MB, Lima KF, Silva AR, Santos RAS, Moraes KC. Mir-513a-3p contributes to the controlling of cellular migration processes in the A549 lung tumor cells by modulating integrin β-8 expression. Mol Cell Biochem. 2018;444:43–52.
23. Chen X, Chen S, Pei N, Mao Y, Wang S, Yan R, et al. AAV-Mediated angiotensin 1-7 overexpression inhibits tumor growth of lung cancer in vitro and in vivo. Oncotarget. 2017;8:354–63.
24. Akram M, Iqbal M, Danlyal M, Khan AU. Awareness and current knowledge of breast cancer. Biol Res. 2017;50:33. https://doi.org/10.1186/s40659-017-0140-9.
25. Cook KL, Metheny-Barlow LJ, Tallant EA, Gallagher PE. Angiotensin-(1-7) reduces fibrosis in orthotopic breast tumors. Cancer Res. 2010;70:8319–28.
26. Yu C, Tang W, Wang Y, Shen Q, Wang B, Cai C, et al. Downregulation of ACE2/Ang-(1-7)/ Mas axis promotes breast cancer metastasis by enhancing store-operated calcium entry. Cancer Lett. 2016;376:268–77.
27. Cambados N, Walther T, Nahmod K, Tocci JM, Ribinstein N, Böhme I, et al. Angiotensin-(1-7) counteracts the transforming effects triggered by angiotensin II in breast cancer cells. Oncotarget. 2017;8:88475–87.
28. Dong L, Zieren RC, Xue W, de Reijke TM, Pienta KJ. Metastatic prostate cancer remains incurable, why? Asian J Urol. 2019;6:26–41.
29. Krishnan B, Torti FM, Gallagher PE, Tallant EA. Angiotensin-(1-7) reduces proliferation and angiogenesis of human prostate cancer xenografts with a decrease in angiogenic factors and an increase in sFlt-1. Prostate. 2013;73:60–70.
30. Krishnan B, Smith TL, Dubey P, Zapadka ME, Torti FM, Willingham MC, et al. Angiotensin-(1-7) attenuates metastatic prostate cancer and reduces osteoclastogenesis. Prostate. 2013;73:71–82.
31. Liu Y, Li B, Wang X, Li G, Shang R, Yang J, et al. Angiotensin-(1-7) Suppresses Hepatocellular Carcinoma Growth and Angiogenesis via Complex Interactions of Angiotensin II Type 1 Receptor, Angiotensin II Type 2 Receptor and Mas Receptor. Mol Med. 2015;21:626–36.
32. Mao Y, Pei N, Chen X, Chen H, Yan R, Bai N, et al. Angiotensin 1-7 overexpression mediated by a capsid-optimized AAV8 vector leads to significant growth inhibition of hepatocellular carcinoma *in vivo*. Int I Biol Sci. 2018;14:57–68.
33. Raucher D. Tumor targeting peptides: novel therapeutic strategies in glioblastoma. Curr Opin Pharmacol. 2019;47:14–9.
34. Garcia-Espinosa MA, Lesser GJ, Debinski W, Tallant EA, Gallagher PE. Angiotensin-(1-7), a peptide hormone with therapeutic potential for the treatment of glioblastomas (abstract). In: Proceedings of the 103rd annual meeting of the American Association for Cancer Research; 31 Mar–Apr 4, 2012. Chicago/Philadelphia: AACR. Cancer Res. 1938;72(Suppl 8):Abs.
35. Liu B, Liu Y, Jiang Y. Podocalyxin promotes glioblastoma multiforme cell invasion and proliferation by inhibiting angiotensin-(1-7)/Mas signaling. Oncol Rep. 2015;33:2583–91.

36. Li X, Wang X, Xie J, Liang B, Wu J. Suppression of angiotensin-(1-7) on the disruption of blood-brain barrier in rat of brain glioma. Pathol Oncol Res. 2019;25:429–35.
37. Ender SA, Dallmer A, Lässig F, Lendeckel U, Wolke C. Expression and function of the ACE2/angiotensin(1-7)/Mas axis in osteosarcoma cell lines U-2 OS and MNNG-HOS. Mol Med Rep. 2014;10:804–10.
38. Chua MLK, Wes JTS, Hui EP, Chan ATC. Nasopharyngeal carcinoma. Lancet. 2016;387:1012–24.
39. Pei N, Wan R, Chen X, Li A, Zhang Y, Li J, et al. Angiotensin-(1-7) decreases cell growth and angiogenesis of human nasopharyngeal carcinoma xenografts. Mol Cancer Ther. 2016;15:37–47.
40. Lin YT, Wang HC, Chuang HC, Hsu TC, Yang MY, Chien CY. Pre-treatment with angiotensin-(1-7) inhibits tumor growth via autophagy by downregulating PI3K/Akt/mTOR signaling in human nasopharyngeal carcinoma xenografts. J Mol Med (Berl). 2018;96:1407–18.
41. Hinsley EE, de Oliveira CE, Hunt S, Coletta RD, Lambert DW. Angiotensin 1-7 inhibits angiotensin II-stimulated head and neck cancer progression. Eur J Oral Sci. 2017;125:247–57.
42. Forte BL, Slosky LM, Zhang H, Arnold MR, Staatz WD, Hay M, et al. Angiotensin-(1-7)/Mas receptor as an antinociceptive agent in cancer-induced bone pain. Pain. 2016;157:2709–21.
43. Willey JS, Barcey DN, Gallagher PE, Tallant EA, Wiggins WF, Callahan MF, et al. Angiotensin-(1-7) attenuates skeletal muscle fibrosis and stiffening in a mouse model of extremity sarcoma radiation therapy. J Bone Joint Surg Am. 2016;98:48–55.
44. Ahmad I, Ahmed MM, Ahsraf MF, Naeem A, Tasleem A, Ahmed M, Farooqi MS. Pain management in metastatic bone disease: a literature review. Cureus. 2018;10:e3286. https://doi.org/10.7759/cureus.3286.
45. Straub JM, New J, Hamilton CD, Lominska C, Shnayder Y, Thomas SM. Radiation-induced fibrosis: mechanisms and implications for therapy. J Cancer Res Clin Oncol. 2015;141:1985–94.
46. Petty WJ, Miller AA, McCoy TP, Gallagher PE, Tallant EA, Torti FM. Phase I and pharmacokinetic study of angiotensin-(1-7), an endogenous antiangiogenic hormone. Clin Cancer Res. 2009;15:7398–404.
47. Rodgers KE, Oliver J, diZerega GS. Phase I/II dose escalation study of angiotensin 1-7 [A(1-7)] administered before and after chemotherapy in patients with newly diagnosed breast cancer. Cancer Chemother Pharmacol. 2006;57:559–68.
48. Pham H, Schwartz BM, Delmore JE, Reed E, Cruickshank S, Drummond L, Rodgers KE, Peterson KJ, diZerega GS. Pharmacodynamic stimulation of thrombogenesis by angiotensin (1-7) in recurrent ovarian cancer patients receiving gemcitabine and platinum-based chemotherapy. Cancer Chemother Pharmacol. 2013;71:965–72.

Concluding Remarks

Robson Augusto Souza Santos

As illustrated in many chapters of this book, angiotensin-(1-7) is now well established and considered an important player of the renin–angiotensin system (RAS). Due to its pleiotropic aspect and consequent beneficial effects in the body, this peptide is a target for developing new medications aimed to treat cardiovascular, renal, and metabolic diseases. The use of Ang-(1-7) by itself as a drug has been hampered by the naive concept that since it has a short half-life in plasma, it cannot be used as a drug. However, the action of a peptide in the body is not proportional to its half-life in the blood as the traditional allopathic drugs. For example, the signaling cascade activated by a peptide acting on its receptor may have a kinetics completely different from the half-life of the free peptide in the blood. Likewise, the receptor-bound peptide does not follow the classical linear relationship expected from evaluations using conventional pharmacokinetics models. Actually, as illustrated in different chapters, there are many publications showing that a single daily oral administration of the inclusion compound Ang-(1-7)/HPB-cyclodextrin is capable of producing beneficial effects in rodents [1–13, 15–18], including improved heart function after myocardial infarction and reduction in blood pressure in SHR) More recently the same formulation was tested in humans and a significant improvement of recovery from a supramaximal physical exercise was observed [4]. These data did not fit with a half-life of seconds for the peptide. The potential role of clinical use of the stimulation of the Ang-(1-7) receptor Mas was recently emphasized by an outstanding publication of a Mayo Clinics' group. In their manuscript, a single

R. A. S. Santos (✉)
National Institute of Science and Technology in Nano Biopharmaceutics –
Department of Physiology and Biophysics, Biological Sciences Institute,
Federal University of Minas Gerais, Belo Horizonte, MG, Brazil

Department of Physiology and Pharmacology, Biological Sciences Institute,
Federal University of Minas Gerais, Belo Horizonte, MG, Brazil
e-mail: santos@icb.ufmg.br

© Springer Nature Switzerland AG 2019
R. A. S. Santos (ed.), *Angiotensin-(1-7)*,
https://doi.org/10.1007/978-3-030-22696-1_15

polypeptide comprising a fragment of BNP associated with the angiotensin-(1-7) sequence was demonstrated to behave as a Mas agonist and more important, to have a dramatic gain in the efficacy on the blood pressure and other parameters. The effect of the polypeptide was blocked by the Mas antagonist A-779 [14]. This manuscript opens new venues to explore the potential of Mas agonists as beneficial drugs in the cardiovascular field. In addition to these exciting new findings, the rather recent discovery of the angiotensin derivative alamandine by our research group [11] adds new possibilities to explore the physiology and pharmacological potential of the protecting arm of the RAS.

References

1. Acuña M, Pessina P, Olguin H, Cabrera D, Vio CP, Bader M, Muñoz-Canoves P, Santos RA, Cabello-Verrugio C, Brandan E. Restoration of muscle strength in dystrophic muscle by angiotensin-1-7 through inhibition of TGF-β signalling. Hum Mol Genet. 2014;23:1237–49. https://doi.org/10.1093/hmg/ddt514.
2. Andrade J, de Lemos F, da Pires S, Millán R, de Sousa F, Guimarães A, Qureshi M, Feltenberger J, de Paula A, Neto J, Lopes M, de Andrade H, Santos R, Santos S. Proteomic white adipose tissue analysis of obese mice fed with a high-fat diet and treated with oral angiotensin-(1-7). Peptides. 2014;60:56–62. https://doi.org/10.1016/j.peptides.2014.07.023.
3. Andrade J, Paraíso A, Garcia Z, Ferreira A, Sinisterra R, Sousa FB, Guimarães A, de Paula A, Campagnole-Santos M, dos Santos R, Santos S. Cross talk between angiotensin-(1-7)/Mas axis and sirtuins in adipose tissue and metabolism of high-fat feed mice. Peptides. 2014;55:158–65. https://doi.org/10.1016/j.peptides.2014.03.006.
4. Becker L, Totou N, Moura S, Kangussu L, Millán R, Campagnole-Santos M, Coelho D, Motta-Santos D, Santos R. Eccentric overload muscle damage is attenuated by a Novel Angiotensin-(1-7) Treatment. Int J Sports Med. 2018;39:743–8. https://doi.org/10.1055/a-0633-8892.
5. Bennion DM, Jones CH, Donnangelo LL, Graham JT, Isenberg JD, Dang AN, Rodriguez V, Sinisterra RD, Sousa FB, Santos RA, Sumners C. Neuroprotection by post-stroke administration of an oral formulation of angiotensin-(1-7) in ischaemic stroke. Exp Physiol. 2018;103:916–23. https://doi.org/10.1113/ep086957.
6. Bertagnolli M, Casali KR, Sousa FB, Rigatto K, Becker L, Santos S, Dias LD, Pinto G, Dartora DR, Schaan BD, Milan R, Irigoyen M, Santos R. An orally active angiotensin-(1-7) inclusion compound and exercise training produce similar cardiovascular effects in spontaneously hypertensive rats. Peptides. 2014;51:65–73. https://doi.org/10.1016/j.peptides.2013.11.006.
7. Feltenberger J, Andrade J, Paraíso A, Barros L, Filho A, Sinisterra R, Sousa FB, Guimarães A, de Paula A, Campagnole-Santos M, Qureshi M, dos Santos R, Santos S. Oral formulation of angiotensin-(1-7) improves lipid metabolism and prevents high-fat diet–induced hepatic steatosis and inflammation in Mice. Hypertension. 2013;62:324–30. https://doi.org/10.1161/hypertensionaha.111.00919.
8. Fraga-Silva R, Costa-Fraga FP, Sousa FB, Alenina N, Bader M, Sinisterra RD, Santos RA. An orally active formulation of angiotensin-(1-7) produces an antithrombotic effect. Clinics. 2011;66:837–41. https://doi.org/10.1590/s1807-59322011000500021.
9. Fraga-Silva R, Savergnini S, Montecucco F, Nencioni A, Caffa I, Soncini D, Costa-Fraga F, Sousa F, Sinisterra R, Capettini L, Lenglet S, Galan K, Pelli G, Bertolotto M, Pende A, Spinella G, Pane B, Dallegri F, Palombo D, Mach F, Stergiopulos N, Santos R, da Silva R. Treatment with Angiotensin-(1-7) reduces inflammation in carotid atherosclerotic plaques. Thromb Haemost. 2014;111:736–47. https://doi.org/10.1160/th13-06-0448.
10. Fraga-Silva RA, Costa-Fraga FP, Savergnini SQ, Sousa FB, Montecucco F, Silva D, Sinisterra RD, Mach F, Stergiopulos N, Silva RF, Santos R. An oral formulation of angiotensin-(1-7)

reverses Corpus Cavernosum damages induced by hypercholesterolemia. J Sex Med. 2013;10:2430–42. https://doi.org/10.1111/jsm.12262.

11. Lautner R, Villela DC, Fraga-Silva RA, Silva N, Verano-Braga T, Costa-Fraga F, Jankowski J, Jankowski V, Sousa F, Alzamora A, Soares E, Barbosa C, Kjeldsen F, Oliveira A, Braga J, Savergnini S, Maia G, Peluso A, Passos-Silva D, Ferreira A, Alves F, Martins A, Raizada M, Paula R, Motta-Santos D, Klempin F, Kemplin F, Pimenta A, Alenina N, Sinisterra R, Bader M, Campagnole-Santos M, Santos RA. Discovery and characterization of Alamandine. Circ Res. 2013;112:1104–11. https://doi.org/10.1161/circresaha.113.301077.

12. Marques FD, Ferreira AJ, Sinisterra R, Jacoby BA, Sousa FB, Caliari MV, Silva G, Melo MB, Nadu AP, Souza LE, Irigoyen M, Almeida AP, Santos R. An oral formulation of angiotensin-(1-7) produces cardioprotective effects in infarcted and isoproterenol-treated rats. Hypertension. 2011;57:477–83. https://doi.org/10.1161/hypertensionaha.110.167346.

13. Marques FD, Melo MB, Souza LE, Irigoyen MC, Sinisterra RD, de Sousa FB, Savergnini SQ, Braga VB, Ferreira AJ, Santos RA. Beneficial effects of long-term administration of an oral formulation of angiotensin-(1-7) in infarcted rats. Int J Hypertens. 2012;2012:795452. https://doi.org/10.1155/2012/795452.

14. Meems LM, Andersen IA, Pan S, Harty G, Chen Y, Zheng Y, Harders GE, Ichiki T, Heublein DM, Iyer SR, Sangaralingham JS, McCormick DJ, Burnett JC. Design, synthesis, and actions of an innovative bispecific designer peptide. Hypertension. 2019;73(4):900–9. https://doi.org/10.1161/hypertensionaha.118.12012.

15. Sabharwal R, Cicha MZ, nisterra RD, Sousa FB, Santos RA, Chapleau MW. Chronic oral administration of Ang-(1-7) improves skeletal muscle, autonomic and locomotor phenotypes in muscular dystrophy. Clin Sci. 2014;127:101–9. https://doi.org/10.1042/cs20130602.

16. Santos C, Santos S, Ferreira A, Botion LM, Santos R, Campagnole-Santos M. Association of an oral formulation of angiotensin-(1-7) with atenolol improves lipid metabolism in hypertensive rats. Peptides. 2013;43:155–9. https://doi.org/10.1016/j.peptides.2013.03.009.

17. Santos S, Andrade J, Fernandes L, Sinisterra R, Sousa FB, Feltenberger J, Alvarez-Leite J, Santos R. Oral angiotensin-(1-7) prevented obesity and hepatic inflammation by inhibition of resistin/TLR4/MAPK/NF-κB in rats fed with high-fat diet. Peptides. 2013;46:47–52. https://doi.org/10.1016/j.peptides.2013.05.010.

18. Santos SH, Giani JF, Burghi V, Miquet JG, Qadri F, Braga JF, Todiras M, Kotnik K, Alenina N, Dominici FP, Santos RA, Bader M. Oral administration of angiotensin-(1-7) ameliorates type 2 diabetes in rats. J Mol Med. 2014;92:255–65. https://doi.org/10.1007/s00109-013-1087-0.

Index